Genetic variability

Genetic variability

CHRISTOPHER WILLS

CLARENDON PRESS • OXFORD
1981

Oxford University Press, Walton Street, Oxford OX2 6DP

OXFORD LONDON GLASGOW
NEW YORK TORONTO MELBOURNE WELLINGTON
KUALA LUMPUR SINGAPORE JAKARTA HONG KONG TOKYO
DELHI BOMBAY CALCUTTA MADRAS KARACHI
NAIROBI DAR ES SALAAM CAPE TOWN

Published in the United States by Oxford University Press, New York

British Library Cataloguing in Publication Data

Wills, Christopher
Genetic variability.
1. Human population genetics
2. Human evolution
I. Title
575.1 GN289 80-40761

ISBN 0-19-857570-X

Printed in the United States of America

Preface

This book addresses questions of central concern to biologists who are interested in the evolutionary process. What is the nature of genetic variability in populations of organisms? How much is there, how is it stored, and how much actually contributes at any given time to adaptive differences between organisms? How does this pool of variation contribute to long-term evolutionary change?

After more than a hundred years of intensive study, these questions have been answered in many different, and in some cases starkly conflicting, ways. A concensus has yet to emerge, and seems unlikely to do so in the near future. Indeed, it is possible to give diametrically opposite interpretations to many results in the field of experimental population genetics. Some data can be viewed both as providing evidence for strong and widespread selection, and as providing evidence against it! It is not suprising that, as a result, a pall of gloom and uncertainty has spread over the field. The time is therefore ripe for an investigation of less-orthodox theoretical and experimental approaches. Though these approaches may not be fully explored as yet, details of several of them have been available in the literature for some time. Others are, I think, new, and I hope that they will be challenging.

The approaches set forth in this book are colored by my experience as an experimentalist. The reader will almost certainly disagree with some, though not, I hope, with all of them. In view of the wide spectrum of opinions in the field, it is safe to predict that different readers will disagree with different parts of the book.

The picture of the nature of genetic variability presented here is as self-consistent as I can make it, and it is also consistent with a wide variety of experimental observations. Further, it has predictive value, leading to suggestions for experiments. Some of these possible experimental directions are set forth in Chapters 7 and 8, and I hope that others will occur to the reader. If some of the predictions made here are tested, then regardless of the outcome the book will have served its purpose.

There are two central themes running through this book.

(a) Genetic variability can be subdivided into many different classes, and the properties of these classes may be strikingly different

Genetic variants affecting viability can easily be subdivided into dominant and recessive lethals, dominant and recessive severe-subvitals, genes with varying degrees of dominance with much smaller individual effect, and so on. The same spectrum of effects can be seen for genes affecting fertility and fecundity. It is, of course, straightforward to assign selection coefficients, measuring the degree of effect, to lethal and severe-subvital genes. But it has long been realized that genes of smaller effect are far more likely to be subject to a variety of forces, depending on circumstance.

For example, the frequencies of the alleles in the population may have a bearing on the direction and amount of selection. Such alleles are subject to frequency-dependent selection. If the most common allele is selected against, a stable and nontrivial internal point of equilibrium may be reached. Whether it is reached depends on the strength of the selection and the size of the population. The alleles at the locus being measured may also interact *epistatically* with alleles at other loci, so that the rate and direction of change may depend on many other loci. The influence of the *environment* may be profound, and may alter both the amount and the direction of selection—the case of sickle-cell anemia is a well-studied example. These and many other effects of circumstance on selection can be studied theoretically, with varying degrees of difficulty.

What is not generally realized, however, is that certain classes of selected genetic variation may accumulate in a population to very high levels. If selection is such as to lead to an internal equilibrium and the selective differences are entirely dependent on the severity of the environment, the background genotype, and the genotypes of the other organisms in the population, then polymorphic gene loci may be maintained in a population in immense numbers. The numbers may be many orders of magnitude more than can be maintained by the more commonly considered type of selection, in which selection coefficients are presumed to be invariant. Similarly, a great many more loci can be swept nearly free of harmful alleles by selection in which the coefficients are relative and not absolute.

Thus, in order to understand the probable composition of the genetic variation present in the gene pool, one must consider the relative likelihood that different classes of variation, once they have arisen, will be maintained. When this is taken into account, the picture of genetic variation that emerges is strikingly different from that in which these relative likelihoods are not considered.

(b) Selection in which the selection coefficients are dependent on circumstance can affect a much larger fraction of the genome than selection in which they are not

Much of the book is concerned with the consequences of a particular class of circumstance-dependent selection—*rank-order* selection. In this type of selection the organisms in the population are ranked according to some property or properties, and then a truncation or quasi-truncation is imposed on the population, the severity of which is dependent on the environment. The chance of survival of a particular organism is thus dependent on its properties, on the properties of the rest of the population, and on the environment. As many authors have noted, such selection can lead to very large per-locus selection coefficients at many different loci simultaneously. Despite the pleasing properties of this type of selection, acceptance of the idea that it may be acting has been slow among population geneticists. As will be pointed out, however, new molecular evidence almost demands this type of selection, and potentially fruitful experimental approaches can be taken to determine its extent and importance. I hope to show, in this admittedly tentative exploration of the consequences of rank-order circumstance-dependent selection, that natural selection has the potential to be a much more powerful shaper of the genome than most population-genetics theory would appear to permit.

Plan of the book

The first chapter deals with the historical importance of the question of the nature of genetic variability, and why it is essential to our understanding of the evolutionary process to know how much there is and how effectively selection can act upon it. Since evolution does not proceed at a constant pace in a given lineage, but may occur very rapidly or slowly, the variation in a particular gene pool must be such that these great changes in rate can be accommodated. I suggest that if a large pool of modifier genes or potential modifier genes is always present, perhaps most easily maintained by balancing rank-order selection, then rapid changes will be much more successfully accommodated than if the gene pool is relatively depauperate of variability.

In Chapters 2 and 3, the controversy between the 'neutralists' and 'selectionists' is examined. This conflict, over whether most of the experimentally detected variation in natural populations is selectively neutral or nearly so, has fragmented the field of population genetics. It has also, I am afraid, been viewed with some bemusement by those outside the immediate storm centre, to the point where it has been compared with the great theological debates of the Middle Ages. Luckily, it has raged in full fury for only little more than a decade, rather than centuries. Like the theological debates, it is impossible to resolve, since it is impossible to prove the absence of something. But it has certainly forced experimental population geneticists, myself included, to learn a lot about theory and to adapt their ideas accordingly.

In Chapter 3 it is suggested that there are powerful selective arguments to counter most of the arguments used by neutralists to explain both present-day variability and the course of the evolutionary process in the past. Some of these counterarguments are based on circumstance-dependent selection. Many nagging questions still remain, however, and these are dealt with later in the book.

In Chapter 4, I turn to new molecular evidence that powerful selective forces are acting to shape most of the genome. It would appear, from a careful consideration of the evidence, that only some form of rank-order truncation selection is sufficiently powerful both to bring about the observed changes, and to maintain the majority of the genome in its current stable and obviously selection-influenced state.

Chapter 5 is a detailed examination of the body of knowledge that has been accumulated concerning the selective forces maintaining genetic polymorphisms. Most of this information, perforce, is drawn from man. A great variety of mechanisms is revealed, some of them highly circumstance-dependent, and some, but not all, dependent on selective response to disease. A contrast between this extensive body of information and the primitive state of our knowledge concerning the maintenance of polymorphisms in other organisms is instructive on two counts: first because it suggests that we may be missing many important selective pressures acting on other organisms, and second because it reveals that there are still uncertainties about even the most well-studied human polymorphisms, and that many pieces of the puzzle are still missing.

This information on the nature of genetic variability is essential to an understanding of Chapter 6, in which the probable types of selection acting on different classes of genetic variability are examined. This chapter introduces the concepts of selected functional polymorphisms and conditional subvital genes, both of which are most probably acted on by circumstance-dependent truncation selection. A number of interesting relationships emerge when genetic variability is subdivided in this way. For example, it transpires that lethal and unconditionally subvital genes are more effectively removed from a population by selection in which the fitnesses at individual loci are invariant and multiplied together, while conditionally subvital genes are more effectively removed by truncation or quasi-truncation selection.

Chapter 7 wrestles with a multifaceted problem: Why, if many polymorphisms are maintained by balancing selection, are not more of them observed at intermediate frequencies? These are least likely to be lost and should accumulate on a selective model. The neutralist model, in contrast, provides a ready explanation of the observed distribution.

The conclusion reached is that a preponderance of intermediate frequencies will be observed if polymorphisms are maintained by unconditional heterozygote advantage, but may not if they are maintained by other types of balancing selection. The consequences of different types of balancing selection, particularly marginal overdominance and conditional heterozygote advantage, are examined in detail. Evidence is discussed that bears on the question of whether these are, in fact, the likely types of balancing selection acting on populations, and suggestions are made for further experimentation to illuminate this question. The point is made that only carefully designed experiments can distinguish between balancing selection of the marginal overdominance or conditional heterozygote advantage type and the most commonly proposed alternative, which is selection against recessive or partially dominant alleles introduced by mutation.

The last two chapters of the book deal with the most complex set of issues. They are an attempt to construct a consistent theory of multilocus finite Mendelian populations, based on the premise that different classes of variability are being acted on by different types of selection. Two major conclusions are drawn: first, it is highly unlikely that 'crystallized' supergenes will appear in outbred populations regardless of the selective regime, and second, balancing truncation or quasi-truncation selection provides a powerful and continuously acting mechanism for the maintenance of a high level of genetic recombination. This latter finding seems to me to provide powerful circumstantial evidence that this type of selection is operating, since a 50-year search by theorists has failed to reveal a satisfactory mechanism giving the advantage to recombination except for special cases.

Central to the elaboration of this theory is the prediction that the *strength,* though not necessarily the *sign,* of the linkage disequilibrium to be expected between a pair of loci in outbred populations is a function not of selection but of the sampling process, dependent only on the effective population size and the recombination between the loci. The theory on which this conclusion is based can be used to accurately predict the outcome of published computer simulations by many authors. Further, it predicts the pattern of disequilibrium seen at the thoroughly-studied MNSs and Rh complex loci in humans, and provides independent support for Fisher's model of the Rh locus.

If this theory is correct, there are easily realized conditions under which the gene will be the unit of selection. The ancient 'bugbear' of the experimental population geneticist, that the effect seen at a locus that can be measured may be due to loci closely linked to it, can thus be circumvented. Experiments can be devised to test whether truncation or quasi-truncation selection act at the level of individual polymorphisms. Such approaches are essential if the theories elaborated in this book are to be put to the test.

Many people have made this book possible. James F. Crow and Warren Ewens have read the entire book in manuscript and made innumerable detailed and trenchant criticisms. Roger Milkman has given great support and advice over the years, and has been most instrumental in the development of the ideas set forth here. I have not always followed their advice, and errors in the book are solely my responsibility. I have benefited greatly from discussions with my colleagues Michael Soulé and Michael Gilpin. I would like particularly to thank Arthur Mourant for allowing me to see his compilations of the association of human polymorphism with disease and Theodore Friedmann for providing the DNA sequence of polyoma virus, both in advance of publication. Many other workers have also provided me with manuscripts and information about work in progress. I am indebted to the patient staff of the Bayswater Branch of the British Library. The writing was aided by a fellowship from the John Simon Guggenheim Memorial Foundation. Finally, I must pay tribute to the memory of two fine teachers and scientists whose views underlie every part of this book, Theodosius Dobzhansky and I. Michael Lerner.

CW

La Jolla
August, 1979

Contents

Genetic variability

1. The role of genetic variability in evolution

Problems raised by Darwin and his contemporaries about the nature of inherited variation remain unanswered at the present time

In this book we will be concerned with the nature of genetic variability, which was the greatest problem facing Charles Darwin after his realization that evolutionary change is due to the process of natural selection. It is generally assumed that with the rediscovery of Mendel's work and the realization that inheritance is particulate, the difficulties with Darwin's theory disappeared and the neo-Darwinian synthesis made sense out of the whole process. This is, of course, far from the case. We have no idea even now of the function, if any, of over 90 percent of our genetic material. We are not sure, any more than was Darwin, of the true magnitude of the role of natural selection. We are not certain whether all adaptive evolutionary change occurs in a gradual manner or whether at least small leaps (saltations) are permitted.

Darwin and variability

In 1959, a century after the publication of Darwin's *The origin of species,* Theodosius Dobzhansky remarked that 1859 had marked the beginning of the post-Darwinian era, just as it marked the end of the pre-Darwinian era (Dobzhansky 1959b). By this he meant that while 1859 is generally conceded to be a watershed year in our understanding of the natural world, and many of the scientific problems raised by Darwin have been solved, many still remain. Indeed, some can still be phrased in much the same way as they would have been in 1859, and we are still as far from solving them.

Although a number of writers (see Eiseley 1961, for a review) claimed precedence for the idea of natural selection, Darwin was the first to articulate it clearly with copious examples, and the first to divorce it from any teleological principles. His most famous forerunner, Lamarck, had concerned himself primarily with the question of direction in evolution, postulating that this direction was provided by the needs of the animals themselves and assuming that the variability was generated by the inheritance of acquired characters. Darwin rejected the first assumption but was forced to embrace the second, and while it is implicit in the first edition of *The origin of species* it becomes explicit in later editions, as can be seen clearly in the variorum text of Peckham (1959).

Darwin was aware that variability, somehow passed from one generation to the next, was essential to this theory, since it was on such variability that natural selection had to act. Selection acting on variation which had no heritable component would, of course, accomplish nothing. As had been clearly shown by Jenkin (1867), the generally accepted mechanism of blending inheritance was incompatible with natural selection unless there were some means for constantly generating new variability. Jenkin explicitly admitted that if characters were to be transmitted unchanged, rather than being averaged in succeeding generations, then natural selection could operate effectively. But he could not imagine a mechanism for this, and remarked that it would be 'contrary to general experience.'

Darwin had already thought about the problem. His theory of pangenesis was put forth in 1868, in the second volume of *The variation of animals and plants under domestication,* and was an attempt to explain the source of new inherited variability. In the theory, 'gemmules' migrated from various parts of the body to the gonads and were passed on to the next generation, their numbers and perhaps characters altered by environmental circumstances. The theory also attempted to encompass observations that could most easily be explained by particulate inheritance:

> . . . it does not seem improbable that certain gemmules, under favourable conditions, should be retained and go on multiplying for a longer period than others. . . . We certainly gain some clear insight into the wonderful fact that the child may depart from the type of both its parents, and resemble its grandparent, or ancestors removed by many generations.

So far as Darwin was concerned, the problem of inheritance was a highly complex one and unlikely to be due to single causes. It is amusing to speculate, as de Beer (1962) has done, on the probable consequences of Darwin's being made aware of Mendel's results, which demonstrated particulate inheritance in peas. With his grab-bag memory, Darwin could undoubtedly have come up with a hundred counterexamples!

The contribution of Darwin's theory of pangenesis was that it made explicit what had, until then, only been implicit; that the inheritance of acquired characters underlay all contemporary thinking about heredity. Once this underlying assumption was revealed, it could be questioned.

The transfusion experiments of Galton, and the famous experiment of Weissman, who played the role of the farmer's wife to generations of hapless mice, began to cast doubt on both Darwin's pangenesis theory and the underlying premise of the inheritance of acquired characters. The alternative was a theory based on particulate inheritance, and especially, as Weissman (1893) perceived, on a separation between the germ plasm and the rest of the organism. The rediscovery of Mendel's work in 1900 accelerated the process which eventually laid to rest the inheritance of acquired characters, even for the small variations which Darwin thought provided the most likely source of evolutionary change. There were, however, a number of difficulties still ahead.

Simple models of population structure espoused by the early Mendelians

The history of the early impact of Mendelism on biology has been ably covered by Dunn (1965) and Provine (1971), but some points are worth emphasizing here. One central argument about how evolution proceeded was the controversy about the relative importance of large and small heritable changes. William Bateson, in his long battle with the biometricians Karl Pearson and W. F. R. Weldon, championed both particulate inheritance and large changes. Natural selection was played down by the early Mendelians, dazzled as they were by the realization that inheritance was particulate. Further, the particles could apparently alter, giving rise to sudden heritable changes. Thus, the mutation theory of de Vries (1901–3) seemed to his Mendelian contemporaries to provide the most likely mechanism for evolution. The inability of Johannsen (1907) to alter inbred lines of beans by selection was initially interpreted to mean that populations were clusters of pure lines with little, if any, heritable variability; evolution must, therefore, proceed through a series of mutational leaps.

The biometricians, on the other hand, ignored Mendelism and stuck to the point of view that natural selection acted on small heritable changes which they inferred to be present in variable natural populations. Continuous variability could be observed in natural populations, and it appeared to them that it was inherited in a blending fashion.

The growth of experimentation in genetics led, perhaps inevitably, to an espousal of the idea of genetic uniformity of populations by the early geneticists. It is remarkable to think that for almost two decades of Drosophila genetics, before the discovery by Muller and Stadler that X-rays induce mutations, virtually every mutation was isolated by painstaking searches through hundreds of thousands of flies by dedicated workers on the lookout for something which was different and would, they hoped, prove genetically interesting. Indeed, many mutations are still found by accident. The variety of mutations encountered in this fashion, ranging from point mutations to duplications, deletions, attached chromosomes, inversions, and translocations, might have given them pause to wonder whether all this variation, much of it concealed in heterozygotes, indicated that there was a store of other kinds of variability as well. But the pace of discovery was so brisk and the genetic story that emerged so exciting (Sturtevant 1965) that no one in the Columbia fly room appears to have thought very much about the matter. Dobzhansky became the leading proponent of highly nonuniform populations during later years. But he recalled in 1959 that, as a disciple of Morgan, he made the same assumption as did his coworkers, that evolution proceeded by the occasional appearance and fixation of an advantageous mutation in a population that was genetically homogeneous except for a collection of deleterious mutations at low frequency. The great labor involved in finding visible mutations gave these workers the impression that the majority of flies were wild type and that variants were few and far between.

The theoretical consequences of Mendelian inheritance of continuous variation were laid by Fisher (1918), following the demonstration by East (1910) that this type of inheritance was, in fact, operating in corn. This did a great deal to close the gap between the Mendelians and the biometricians, but it was still difficult for many workers to imagine that evolution did not somehow proceed by leaps.

Does evolution proceed by saltations?

The early stages of this controversy have been dealt with by Provine (1971). I would like to suggest here that the problem may eventually be resolvable, if only because we are now better able to understand the extent of the difficulty.

Darwin felt constrained to assume that the variations acted on by natural selection in a positive fashion—that is, such that they increased in frequency in the population—would have to be small in effect. The only large changes he was aware of were sports, suddenly appearing variants like the Ancon (short-legged) sheep which so impressed his contemporaries because it was a 'useful' mutation (could not jump over fences or walls). Darwin quite rightly perceived that natural selection would not see such changes as useful, and he felt strongly that large changes were much more likely to be deleterious. Indeed, the Ancon sheep has not proved to be a successful variety.

In the fourth edition of *The origin of species,* Darwin commented on instances of some domestic pigs being born with elongated probosces like those of tapirs. He remarked that it would seem unlikely that such a character would be established in nature without much modification.

Darwin's contemporary T. H. Huxley put much more emphasis on the problem of apparent discontinuities in evolution than did Darwin, who simply assumed that the intermediate stages had disappeared. Huxley was concerned with the fact that most species differed quite clearly from each other, and was also worried about how new organs and functions could have arisen. In a letter to Darwin dated June 25, 1859, in which he emphasized the possible importance of sports in the evolutionary process, Huxley wrote:

> I may illustrate what I mean by a chemical example. In an organic compound, having a precise and definite composition, you may effect all sorts of transmutations by substituting an atom of one element for an atom of another element.

> You may in this way produce a vast series of modifications—but each modification is definite in its composition, and there are no transitional or intermediate steps between one definite compound and another. I have a sort of notion that similar laws of definite combination rule over the modifications of organic bodies, and that in passing from species to species 'Natura fecit saltum.'

It is interesting that a similar comparison occurred much later to Haldane (1932, p. 104) when he wrestled with the same problem:

> . . . the atomic nature of Mendelian inheritance suggests very strongly that even when variation is apparently continuous this appearance is deceptive. On any chemical theory of the nature of genes this must be so.

The problem was certainly a real one, and Darwin penetrated to its essence when he worried about how large changes could possibly be adaptive. Darwin, with his keen observational sense and his enormous interest in every aspect of living things, was perhaps more attuned than most of his contemporaries to the kinds of variation that would be selected for rather than against. He was very much aware of the struggle for existence, and continually stressed that very small changes in the appearance or behavior of an organism could make tremendous differences to its survival. Concerned as he was with demonstrating the validity of the concept of natural selection, he did not want it to founder on the necessity of postulating large changes that were somehow adaptive when all his observation told him that this was very unlikely to happen.

Darwin was also able to imagine at least some of the intermediate steps that might be necessary if organisms with very different forms were to arise from a common ancestor. This is essential for an evolutionist, as pointed out by Fisher (1954):

> The call on the imaginative faculty required to conceive of the intermediate states, which must have connected widely diverse groups of animals and plants, must have been, and must always be, an immense one to minds not already prepared by a knowledge of the diversity and intricate adaptations of analogous forms.

We realize now, that in dealing with the problem there are essentially two questions that must be answered.

(a) How large must the jumps be to bring about the observed discontinuities between species, and to produce new structures and functions?

(b) If a jump results in an organism with both adaptive and maladaptive qualities, is it possible that the adaptive qualities can rapidly be enhanced and the maladaptive ones diminished in order to permit this alteration with large effects to survive in the population?

The first question, how large jumps must be to bring about the observed discontinuities, has essentially been answered: they need not be large. When Punnett (1915) proposed that mimicry in butterflies could have arisen only by a sudden change, Fisher (1927) pointed out that it need not have done so, but could have arisen through a gradual series of steps, each one conferring a slightly higher fitness. Indeed, it had been demonstrated theoretically that small fitness differences between genotypes were enough to bring about large shifts in allele frequencies, if carried out over many generations. These arithmetical consequences of long-continued selection were first worked out by H. J. T. Norton, and included in the very book by Punnett (1915) to which Fisher had objected. As Provine (1971) notes, the theory undermining Punnett's ideas on the evolution of mimicry was actually present in nascent form in the same book—one of the smaller ironies of the history of science.

Further, as more paleontological evidence accumulated, it became apparent that the large gaps in the fossil record were not as large as had been thought. Forms are continually being discovered, among both fossil and living organisms, that provide information about their probable ancestral relations, though for a variety of reasons these forms rarely appear as precisely intermediate 'missing links.' This evidence was summarized in considerable detail by Rensch (1959), who concluded that paleontological evolution could indeed be explained on the basis of relatively small changes cumulatively bringing about large phenotypic alterations.

Still, Simpson (1944) had shown that the early stages of a particular line, which are nearly always missing in the fossil record, were of much shorter duration than the later stages, for which there is fossil evidence. This rapid evolution, which Simpson referred to as tachytelic as opposed to slow or bradytelic evolution, could only have occurred through a rapid accumulation of phenotypic changes. This brings us to the second question. While there appear to be no immense, sudden changes in the fossil record in terms of phenotypic alteration, could changes of smaller but still pronounced effect have occurred, and how could they have survived?

The immediate answer to this is straightforward: it depends. In a well-reasoned argument, Mayr (1959*a*) reexamined the question of discontinuous changes in terms of 'the emergence of evolutionary novelties.' He began by noting that the establishment of the gradual nature of the *microevolutionary* process had tended to push questions of the nature of *macroevolution* into the background. He then set out to reexamine some of these questions.

His central point was that evolution is not infinitely resourceful. It is not possible for organisms to evolve in any direction, since their evolution is constrained by existing structures and functions. The seeds of the new structure or function must already be present. He referred to this (perhaps unwisely, in view of its teleological implications) as 'preadaptation.'

He then made the important point that some morphological changes, which we perceive to be large and presumably important, may not have a great impact on the fitness of the organism—the pattern of branching in plants or corals, for example, or the direction of coiling in snails. Here I will take the liberty of extending his argument. Such large phenotypic reorganizations, were they to have some beneficial aspects, might be established in the population because their deleterious effects were small or could be modified. Thus, the balance between the advantageous and deleterious aspects of a particular phenotypic change will depend not only on the change itself and on the circumstances in which it has arisen, but also on the degree to which the deleterious aspects of the change can be reduced or eliminated as a result of subsequent reorganization of the genotype through recombination and selection.

It has been realized for some time that the expression of single gene characters can be modified with remarkable rapidity by selection. The first theoretical approach to this evolutionarily important process was made by Fisher (1928), who was intrigued by the fact that deleterious mutants tend to be recessive. Making the reasonable assumption that partially dominant harmful genes would be modified towards recessiveness, he investigated the rate at which a modifier of the heterozygous effect of a partially dominant recurrent deleterious mutation would build up in frequency.

The modification is a second-order phenomenon. If modifier and mutant are unlinked, and if the mutant is rare and the modifier has no other effect, then the initial selection for the modifier should be extremely small. Wright (1929*a*, *b*) criticized Fisher's approach by pointing out that the modifier might be expected to have some slight effect on the expression of the wild type allele as well as on the mutant allele. At the outset, this effect would almost certainly overwhelm any effect on the relatively rare heterozygotes for the deleterious allele, so that the frequency change of the modifier would be determined primarily by its effect on the common wild type allele.

Yet it appears that dominance can be modified dramatically. Ford (1937, 1940) investigated a color polymorphism in the moth *Abraxas grossulariata*, in which the colors white, yellow, and deep yellow were controlled by two alleles at a single locus. Three generations of selection were enough to alter the heterozygotes significantly towards either white or deep yellow. The *medionigra* form of the moth *Panaxia dominula* is a heterozygote intermediate in morphology between the homozygotes *dominula* and *bimaculata*. Rapid alteration of this form in both directions was accomplished by selection over four generations (Ford 1959). Modifications of the expression of *medionigra* in both directions were also observed in different natural populations (Ford and Sheppard 1969).

The interpretation of these experiments is difficult because the rapid selective response could have been due to one or both of two processes.

(a) Selection might have been acting on multiple alleles at the locus, each conferring a slightly different degree of dominance (Haldane 1939*a*).

(b) Selection might have been acting on modifier loci elsewhere in the genome. The most effective selection in terms of a rapid change in dominance would be on modifiers that were already polymorphic—that is, had more than one allele at high frequency. The change in allele frequency for a given amount of selection is greater under such circumstances than it is if the selected allele is initially rare, and Wright's argument is also circumvented.

Either hypothesis implies a great deal of hidden variation that can modify dominance. Harlan and Martini (1938) showed that both mechanisms were operating in selection experiments involving the degree of dominance in cotton. We are in need of carefully-controlled experiments to determine just how widespread is the modification of dominance and other single-gene characteristics, and how much can be attributed to mechanisms (a) and (b) in different organisms and for different characteristics.

Frequency shifts and recombination at polymorphic modifier loci can quickly bring about large alterations in phenotypic expression at the locus being selected. In Chapters 6 and 8 it will be shown how it is possible for very strong selection to be applied to many loci simultaneously, so as to bring about rapid shifts in allele frequency. This, coupled with the likelihood that many genes interact in complex and sometimes unexpected ways, as detailed by Wright (1968, 1969), can be invoked to explain changes as rapid as those observed in the moth populations.

Most evolutionary changes are likely to be small, and even saltations will for the most part be small in effect. But sudden and profound mutational alterations such as polyploidy must occasionally become fixed in the course of evolution. It is also difficult to see how drastic developmental changes such as paedogenesis, the sexual muturation of the larval stage that may have occurred during the early evolution of the chordates, could have arisen gradually. The deleterious effects of large, sudden changes must be rapidly modifiable if the altered organisms are to survive more than a generation or two. By analogy with the swiftness with which dominance can be modified, it is possible that this process could also be brought about by the recombination and selection of existing polymorphic modifier genes.

How can potential modifier genes be present as polymorphisms in a population prior to a saltational change? Again there are several, by no means exclusive, possibilities.

(a) The potential modifier alleles were, before the advent of the saltation, neutral or nearly so, so that their frequencies were determined by mutation and subsequent drift. If this were the predominant mechanism, it would imply that the selective importance of a substantial class of alleles could change dramatically as a result of a change elsewhere in the genome.

(b) The potential modifiers were maintained as a result of mutation pressure counterbalanced by selection against the mutant alleles. If selection were sufficiently weak, or if dominance and epistatic interactions diluted the effectiveness of selection, a high frequency of mutant alleles could indeed be achieved by this process. The mutant alleles, since they will be arising by recurrent mutation, will tend to be numerous but individually rare in such a case. However, if selection against the mutant alleles is weak, so that they can achieve substantial frequencies, this would again imply a large change in the strength and even the sign of the selection acting on these alleles with the appearance of a saltation.

(c) The potential modifiers were maintained as a result of some form of balancing selection, leading to their maintenance at substantial frequencies even before the appearance of the saltation. If the selection pressures acting on these polymorphisms were substantial even before the appearance of the saltation, then large changes in the strength of selection need not be postulated.

All of these possibilities are real ones, and all may contribute substantially to the modification of those newly arisen genes which have major effect. Of the three, the third possibility would be considered the least likely from conventional population-genetics theory, since the substantial balancing selection needed to maintain such a large number of polymorphic loci would have too great an effect on the fitness of the population.

A central thesis of this book is that, despite theoretical objections and apparent evidence to the contrary, there are immediate short-term benefits accruing to an organism if it is heterozygous at many loci for alleles maintained by balancing selection, particularly if the environment is complex. Further, there is no impediment to the existence of a large number of such balanced polymorphisms. A long-term benefit of this genetic diversity is that the evolutionary process can bring about more rapid alterations in the phenotype than would be possible if the evolving populations were genetically highly uniform. Saltations are more likely to survive in a highly polymorphic population.

The role of natural selection

This book is primarily concerned with genetic variability that contributes positively or negatively to adaptation, rather than variability that is essentially invisible to the selective process. Arguments will be developed that a population of organisms can be an enormous depository of selected genetic variation, and indeed that all the variation we can currently detect, using even the most sophisticated techniques, could be subject to selection. As was suggested in the preceding section, such a pool of variation would greatly facilitate the evolutionary process. Whatever the quantity of selectively neutral variation in a population may be, it will have no bearing on these arguments so long as it remains neutral.

Nonetheless, chance does play an enormous role in the process of adaptive evolution. The complex of genes that comes together by recombination in a particular organism in a critical place and time is, of course, a matter of chance, and subsequent developments may be greatly influenced by this momentary and unique combination.

The question of the importance of chance is one that goes back to the historical beginnings of the study of evolution. Darwin himself was uncertain of it. Of course, like scripture, Darwin can be quoted to support almost any point of view. This was partly because his own ideas were in a constant flux, indicating his great flexibility of mind and disinclination to adhere to dogma—even the dogmas about evolution that were rapidly created by his contemporaries and attributed to him. He was, in all his later writings, constantly attempting to counter every criticism and difficulty raised by his opponents. In part, too, he fully realized that his theory was in no way provable as a theorem in mathematics is provable.

What, indeed, was the role of natural selection in the evolutionary process? A paramount one, undoubtedly, but even here his lack of dogmatic certainty is apparent:

> I probably attributed too much to natural selection or the survival of the fittest. . . . I had not formerly sufficiently considered the existence of many structures which appear to be, so far as we can judge, neither beneficial nor injurious; and this I believe to be one of the greatest oversights as yet detected in my work. (*The descent of man and selection in relation to sex,* Vol. 1, p. 152, 1871.)

Darwin himself attributed this apparent overemphasis on selection to his strenuous efforts to prove to himself that each species had not been 'purposely created.'

The subsequent battle between the proponents of selection and chance in evolution will be briefly considered in the next chapter. It is difficult to get too excited about these matters at this stage, since it is by now abundantly clear that adaptive evolution cannot proceed without both. But there is a fundamental and important question here that should be kept in mind during the arguments that follow. It is: if both chance and selection play a role in adaptive evolution, what sort of population structure is likely to maximize the rate of adaptation of organisms to new and unexpected environmental conditions?

This is precisely the question addressed so ably by Wright, who in explaining his 'shifting balance' theory of evolution points out that recombination between epistatically interacting complexes of genes will often produce very unexpected results by chance. A subdivided population will, in this view, provide the maximum possible opportunity for evolution to occur, since the effect of drift in such a population will be to increase the likelihood of such unusual recombinants. We are in great need of experiments to test this hypothesis.

What I hope will be the primary contribution of this book is to suggest ways in which large quantities of variation can be retained in natural populations. Such variation can aid in the fixation of saltations, as suggested in the previous section. It will also maximize the probability that random drift and recombination can together produce unexpected, and perhaps highly adaptive combinations of genes. It will even be seen, in Chapter 9, that the retention of large quantities of polymorphic variation may facilitate selection for high levels of recombination and thus the ability to generate new and more highly adapted genotypes. If population structures are of the sort that I suggest, it is remarkable (though of course not unexpected) that they seem superbly adapted to facilitate the process of evolution itself.

2. Neutrality *v.* selection

The controversy between 'neutralists' and 'selectionists' is an old one, and is unlikely to be resolved in the near future

The consequences of having many neutral alleles in a population

As indicated in the previous chapter, the idea that some characteristics in the population are selectively neutral goes back at least to Darwin. But the early workers in the field never imagined that a majority, much less an overwhelming majority, of the detectable variation in a population might be neutral. In this chapter I will set forth the history of the emergence of this idea, and in the next the reasons why this idea has seemed so compelling to many workers. Some counterarguments will also be provided.

Of course, a neutral allele *sensu stricto* must exist only in the minds of population geneticists. Crow (1972*b*) has been careful to distinguish neutral from nearly neutral alleles, which can be defined as those for which the selection coefficients are small and fluctuate around zero. Even the strictly neutral allele will not always behave as if it were completely neutral during its lifetime in the population, since it will, at various times, be in linkage disequilibrium with selected alleles. However, we can define neutral mutations as those which behave in a way that is essentially indistinguishable from neutrality. Small stochastic fluctuations of selection coefficients around zero can easily be accommodated in the elegant theory of neutral genes.

The history of the conflict concerning the relative importance of strict neutrality and selection is only about a decade old. It was sparked by the discovery that electrophoretically detectable variability was common in natural populations (Lewontin and Hubby 1966; Harris 1966), and this discovery led Kimura (1968) to investigate theoretically whether such variability could in fact be generated entirely by mutation and maintained despite the drift to be expected in a finite population. Even for very low mutation rates and moderate population sizes, the equilibrium heterozygosity for neutral genes in a population could be quite high. This had been shown first by Haldane (1939*b*) and developed further by Kimura and Crow (1964), but the full implications of this finding had not been realized by the majority of population geneticists before Kimura's 1968 paper. The relationship was, ignoring complications of minor effect, a simple one:

$$\overline{H} = \frac{4\,N_e\mu}{1 + 4\,N_e\mu}$$

where $\overline{H}$ is the average heterozygosity, N_e the effective population size (generally smaller and sometimes much smaller than the actual population size), and μ the mutation rate to neutral alleles. The model assumes that all alleles are distinguishable, and that all arise at the same mutation rate. For a population size as small as 10^5, quite likely to be achieved in most populations, especially over long periods of time, and a mutation rate to neutral alleles of only 10^{-7}, a hundred times smaller than mutation rates to detectable alleles in higher organisms, the heterozygosity at equilibrium would be about 4%, comparable to those heterozygosities that are actually observed. Of course, for larger population sizes the heterozygosity would be much greater. Heterozygosities close to 1 could easily be achieved. As Lewontin (1974*a*) pointed out, heterozygosities greater than 0.2 are rarely seen, even among tropical insects in which the population size may be expected to be very large. This problem could be neatly circumvented (Ohta 1973) by assuming that even neutral genes are only quasi-neutral; if they have very small but constant negative selection coefficients associated with them, then as the population size increases these very small selection coefficients would eventually have an influence on allele frequency and these variants would be purged by selection from the population. This would tend to reduce the equilibrium amount of heterozygosity.

Further, it will be noted that the theoretical expectation of heterozygosity is for populations in equilibrium between mutation and drift. A population accumulating neutral mutations requires a number of generations approximately equal to the reciprocal of the mutation rate, in order to reach equilibrium (Nei, Maruyama, and Chakraborty 1975). If natural populations go through periodic bottlenecks, particularly during the process of speciation, then it is unlikely that most populations will have reached equilibrium, which is another possible 'neutralist' explanation for the deficiency of heterozygosity noted above.

This postulated large class of alleles, essentially neutral in their effect on fitness, caused a great deal of disturbance among many evolutionary biologists. These workers were used to thinking of alleles as being subject to strong and continually shifting selection pressures, both as a result of changing conditions affecting the locus itself and of recombination which generated new complexes of epistatically interacting loci. Could the existence of alleles with such remarkably constant properties—or constant lack of properties—be proved or disproved? Of course not; it is impossible to prove the absence of something, particularly of something as hard to measure convincingly as selection coefficients. But many workers felt instinctively that there was something unlikely about the existence of a huge family of alleles with almost invariant properties.

It is instructive to consider the history of the idea of neutral alleles up until the revelation in 1966 of large amounts of electrophoretic variability in natural populations.

Selection and drift before 1966

Selection *v.* drift. The *Panaxia* case

While, as Mayr (1959*b*) points out, the idea of genetic drift had been considered by a variety of naturalists before 1931 as being a probable explanation for morphological differences between isolated or semiisolated populations, it was Wright who in that year provided the solid mathematical underpinnings for the idea. It went very much against the grain for biologists trained in the tenets of Darwinism to accept a role of chance events in the process of evolution (though Darwin himself worried about the problem—see p. 12). In fact the difficulty centered at first on whether such chance events were likely to affect genes which were also subject to selective forces, or whether selection would in most cases simply overwhelm chance. Initially the argument was between *all* selection and a mixture of selection and drift, and only later did the grounds of the conflict shift somewhat to almost *all* drift v. a mixture of drift and selection.

One of the most famous collisions between these two earlier viewpoints occurred between Fisher and Ford, on the selection side, and Wright, who emphasized the importance of drift. While not quarrelling with the mathematical consequences of drift, Fisher and Ford argued that drift was highly unlikely to have much effect on a population of more than a few hundred (Ford 1975), while Wright considered drift to be of extreme importance both during founder events (Mayr 1942) and in large populations, even if genes were quite strongly selected. The existence of some neutral or nearly neutral alleles seemed to him very probable, but of course in order for them to play a role in the process of adaptation they would have to lose their neutrality at some point (Wright 1940).

Since Fisher and Ford were dealing with variations in natural populations, they based their arguments on the consequences of variant alleles having measurably large phenotypic effect. For example, an allele producing increased melanism in the Scarlet Tiger moth, *Panaxia dominula,* also produces behavioral changes (see Ford 1975, for review): the homozygote *bimaculata* is more lethargic than the wild type. The frequency of this undoubtedly strongly-selected gene was followed in an Oxford population for a great many years, though the source of the later conflict centered on interpretation of the data collected from 1939 to 1946 by Ford (Fisher and Ford 1947). The frequency of *medionigra,* the heterozygote, dropped from its highest measured value of 11.1% in 1940 to only 6.8% in 1941, and stayed approximately constant thereafter. It has continued to decline slowly until the present time, but with statistically significant fluctuations.

The fluctuations were attributed by Fisher and Ford to changes in the direction and magnitude of the selection coefficients involved. Wright (1948) analysed the data carefully and showed that the fluctuations might be more easily accounted for by drift. The point he made was that fluctuations in selection coefficients must be very great to produce the same effect as fluctuations due to drift, even in quite large sized populations. For example, given additivity of genotypic effects, a standard deviation of 0.005 in the selective value of the heterozygote (a standard deviation of 0.01 in the homozygote) only produces slightly more pronounced allele-frequency fluctuations in an effectively infinite population than does sampling error in a population of 100 000—a number substantially greater than the population sizes of the moths in question, as measured by mark-recapture studies. Further, the effective population size might have been even smaller than the measured one, again increasing the role of drift.

Indeed, such fluctuations in selection coefficients may not be unreasonable to postulate, as we will see. In retrospect, however, the value of the argument may lie chiefly in the way it illustrates two very different patterns of thinking in sharp conflict with each other. Even seventeen years after the event, Ford (1975) heaped scorn on Wright for not realizing that *Panaxia* lays its eggs over a wide area, thus making it unlikely that density-dependent factors might reduce its effective population size as Wright had suggested. Of course, there is much truth in his contention that one is unlikely to obtain an accurate picture of the magnitude of the selective pressures on an organism without a good idea of its natural history, but this seems to extend still further to a general contempt for that very large group of biologists who are disinclined to get their feet wet.

The resolution of the conflict is simple, though in fact it resolves nothing in the last analysis. Both selection and drift must be acting on the population. Selection against this gene is very strong, and has been estimated at a 10% disadvantage (Sheppard 1951*b*). It is certainly strong enough to overcome the effects of drift, except in a very small population size, but since true effective population size could not be estimated, but rather only the number of imagines in the population, the true role of drift must remain problematical. The argument serves to point up in miniature, however, the gap between the points of view of observers of natural populations and of more theoretically inclined biologists. The former tend to work with large phenotypic differences, in which the influence of selection is so overwhelming as to be detectable by the observer and therefore theoretically rather uninteresting. The latter deal with the consequences of genes of very small or even zero effect, and are quite capable of blithely considering spans of time running into millions of generations, during which the properties of these genes do not vary or vary so little and so randomly as to have no measurable effect on the outcome. The first class of genes certainly does exist, but does the second—at least in numbers sufficiently great to make them evolutionarily interesting?

The attempt by Fisher and Ford to equate the controversy with an either-or position, *either* selection *or* drift, was not responded to by Wright, who claimed quite reasonably that each had an important role to play and that drift would become progressively more important as the strength of selection diminished. But his view of the evolutionary process was considered by many to be somewhat static, too dependent on the mathematically convenient concept of constant selection coefficients. Wright has often repeated that it seemed to him that the most favorable conditions for evolution would be provided by a subdivided population, since stochastic fluctuations in allele frequency would be preserved in such a situation. Wright suggested that unusually fit genotypes arising in such a subdivided population could survive longer, before being broken up by recombination, than if the population were panmictic. Such unusual genotypes would be available to 'climb' new adaptive peaks on an 'adaptive landscape' as they appeared. Wright (1970, 1975) has expanded on his earlier classification of evolutionary processes in his 'shifting balance' theory, which explicitly takes into account the presence in populations of large numbers of pleiotropically interacting loci with small individual effect. Shifts in the frequencies of these loci due to large environmental changes, or selection for different combinations in patchy environments, would occur rather slowly were it not for the effect of drift in increasing the likelihood of survival of new combinations of alleles. While not explicitly dealing with fluctuating selection coefficients, the model could be expanded to embrace them without difficulty. It has become much closer conceptually to a model (or perhaps more properly a metaphor) proposed by Ford (1975) as an alternative to what he considered to be Wright's too-static model. Ford considers the adaptive landscape of Wright to be more of a seascape, with constantly shifting peaks and valleys. Perhaps these two workers are not so far apart conceptually as their writings sometimes make them appear.

Selection *v.* drift. The case of *Cepaea*

Cepaea nemoralis, a highly polymorphic snail with a wide distribution over western Europe, provides a particularly revealing case study of the difficulty of disentangling the relative roles of selection and drift. The genetics of its color and banding pattern is quite well understood. While modifiers make the picture complex, major genes controlling color and the number and intensity of shell bands are organized into a single 'supergene' and can show marked linkage disequilibrium.

Since the alleles result in such striking morphological changes, it might be thought that selection acting on them would be easy to detect. However, this is not the case. There are certain advantages to the study of these organisms not possessed by *Drosophila,* for example. The shell is easily examined and scored, and outlasts the creature itself, so that the banding pattern (though not the color) is available for scoring fossil specimens. As noted by Diver (1929), the polymorphism of these snails has been present since the Paleolithic. Populations are large, and the organisms are (to say the least) easily captured. The snail has a good many predators, including the song thrush *Turdus ericetorum,* which uses stones as anvils in order to break open the shells, thus providing an excellent record of the morphological types of snails most susceptible to this particular type of predation in each population. And finally, the snail is sufficiently large for physiological measurements such as body temperature to be made on it. The disadvantages are that the snail reproduces slowly (one generation every four years) and is cytologically unfavorable. Although Diver (1940) found differences in color and banding in several populations of *C. nemoralis* and the closely related *C. hortensis,* he was inclined to attribute these differences to random events. He did not attempt to correlate them in detail with the type of vegetation in which they lived. Sheppard (1951*a*, 1952) conducted more detailed surveys of populations around Oxford, England, and was able to show that both banding and intensity of color could be correlated with the habitat. In woods, brown and pink unbanded shells tended to be found, while in hedgerows more yellow banded shells were found. In open regions with short turf, shells tended (though not consistently) to be unbanded. But the correlation was not such as to be obvious—it was interesting, for example, that populations found in woods and in short exposed turf were rather similar in appearance. There were apparently many unknown factors tending to mask a clearcut correlation of pattern with environment.

Extensive studies of relative predation of the different morphs were carried out by Sheppard (1951*b*) and Carter (1968*a, b*). They were able to demonstrate that thrushes do not carry their prey for long distances, so that the fragmented shells near an anvil come from the immediate population. Sheppard showed a trend (based on very small numbers, but marginally significant) from a high to a lower proportion of yellow shells in the total predated by thrushes in an Oxford population as spring advanced (the yellow shells actually appear greenish when the snail is inside them). This suggested that they were being eliminated disproportionately early in the year when the background is less green. This could also be shown with marked snails, though again the numbers were small owing to the technical difficulty of the experiments. In the marked populations, for example, the proportion of yellow shells dropped from 16 out of 21 in early May to 11 out of 25 in late May–early June. None of the observations or experiments individually is significant, but taken together they show a trend. The tendency of polymorphisms to reestablish themselves was shown by Carter (1968*a*), who found that when the number of yellow morphs in a population was artificially increased, it had returned to its original proportion after a lapse of 12 years.

The situation in the diverse countryside around Oxford, broken into clearly defined habitats because of the farming practices of the area, was not duplicated elsewhere. Cain and Currey (1963*a*, *b*) found that in rolling downs in the south of England there was little correlation of color and banding type with habitat, though the populations continued to be highly polymorphic. This lack of obvious correlation was also found in extensive surveys by Lamotte (1952, 1959) in French populations, and indeed Lamotte was at first inclined to attribute the polymorphism entirely to drift and a very high mutation rate (1952). Later (1959) he did claim to find a correlation 'greater than zero' (though how derived he does not say) between habitat and various morphs. It is amusing to note that Lamotte attributed this general lack of a clear correlation to the fact that the French countryside is 'more diverse' than the English (1959), while Cain, King, and Sheppard (1960) considered it to be less diverse and more like the uniform chalk downs of England. A list of the variables which a reviewer of this literature must take into account includes, it would appear, chauvinism.

Indeed, the situation in different parts of Europe seems to be strikingly different from that surrounding Oxford. Jones and Irving (1975) showed that the proportions of banded and unbanded shells in populations of this snail varied strikingly from one valley of the Pyrenees to another. In particular, the correlation of unbanded shells with altitude in the Ter valley was almost the inverse of that found in the valley of the Garonne not far away. Thrush predation was not a factor in these populations.

Since field studies have such diverse outcomes, can selection be demonstrated to be acting on this organism in the laboratory? Not without controversy. Attempts to show differential mortality of different morphs caused by temperature differentials (summarized by Arnold 1969) have been strongly criticized by Jones and Irving (1975) as being, for the most part, statistically nonsignificant.

A reading of the literature leaves one confused (Jones et al. 1977). Just what is going on here? The polymorphism of these snails has persisted, though not in its current proportions, for many thousands of years (Cain 1971; Currey and Cain 1968; Diver 1929). There are tantalizing clines of frequencies, weak correlations in different parts of the species range with a variety of presumed environmental influences such as predation and altitude, and a general feeling from a survey of the data that something quite complex and probably nonrandom is going on. But the details escape us, perhaps simply because too little is known about the interaction of these snails with their environment. What does emerge is that the selection operating in these populations appears to be neither constant nor overwhelmingly strong. As will be emphasized later (pp. 160–82), selection does not need to be overwhelmingly strong or constant to maintain a polymorphism, and perhaps what we are seeing here is simply an expression of that fact. But there is certainly plenty of room in these populations for the action of drift, particularly the founder effect. At least, in its details, the pattern of polymorphism is likely to be partly due to this influence.

The *Cepaea* story has been commented on at some length here, not because it is unique but because its features can be found generally in many different studies of the effects of selection. In very few instances indeed can the action of selection be pinned down as being due to the action of a single gene subject to a single environmental influence. One thinks most particularly of the superb investigation of industrial melanism (Kettlewell 1961). But in most cases, particularly those not due to disturbance by man, the situation is tantalizingly and frustratingly complex. Rather than being a cause of despair among selectionists, however, this should be viewed as a clue to how selection acts on genetic variability in general. It is complex and subtle in its manifestations, and will require all our effort to understand even a small portion of its underlying mechanisms.

In fact, such subtle interactions of polymorphism with selection are exactly what one would expect to be selected for. If a very large proportion of a given genotype were to die as a result of some selective process, then modifiers of that genotype which reduce the amount of selection against it should rapidly increase in frequency. It would therefore be expected that the effects of selection would be much easier to observe in recently-disturbed environments, such as that leading to the appearance of industrial melanism. In complex environments, the organisms have had many generations to accumulate such modifiers—as appears to have happened in the case of *Cepaea*. In a well-reasoned discussion, Sheppard (1967) argues that the evolution of modifiers is relatively easy to accomplish if the modifiers affect an already polymorphic allele in the population, and he cites considerable experimental and observational evidence to show that modifiers can in fact be selected for very readily. The importance of such modifiers in the evolutionary process has already been emphasized in Chapter 1. Thus, it is not at all surprising that the interaction of genes and selection is complex and difficult to measure if there are many polymorphic alleles with attendant polymorphic modifiers present in the population.

Famous 'neutral' genes of the past

A recurrent pattern in the literature of population genetics is the following: a polymorphism is discovered; it is presumed, in the discussion about it, to have no obvious function and therefore to be neutral; and later more detailed investigation shows it to be subject to selection—sometimes quite strong selection. The example of Diver's investigation of the *Cepaea* populations was mentioned above, though the reader will be quite correct in pointing out that the case for selection in this organism has not been proved. But other more vivid examples abound. In a single survey paper written in 1940, Wright remarked that blood group polymorphisms in humans and inversion polymorphisms in *Drosophila* both appeared to be quite neutral. In that same year, Rh incompatibility was reported, and a few years later Wright himself played a part in showing that the inversion polymorphisms are subject to strong selection. The discoverers of the inversion polymorphisms in *D. pseudoobscura,* Dobzhansky and Sturtevant (1938), also assumed that they were neutral, and this belief persisted through the first extensive surveys of these inversions in natural populations (Dobzhansky and Queal 1938).

Wright wrote his paper just before the discovery of the Rh polymorphism by Landsteiner and Wiener in 1940 as a direct result of their investigation of a baby with erythroblastosis. In fact, the possibility of ABO incompatibility was realized by Hirszfeld (1928) and it was only because of its relative rarity among the newborn that he missed it. As will be noted later (pp. 113–18), ABO incompatibility is much more common than had previously been thought, and the web of interactions among the blood groups, diseases, and other polymorphisms that is beginning to emerge is highly complex. Little of this was realized or even imagined at the time, however.

More recently, the Duffy blood group polymorphism, for which no obvious selective raison d'être had been found, has been shown to be involved with resistance to *vivax* malaria (pp. 132–33).

In view of this recurrent pattern, it might be wise, as future polymorphisms of unknown function are discovered, to bring in the Scottish verdict of 'not proven,' rather than presuming that they are neutral at the outset.

3. Arguments for and against postulating a large class of neutral genes

It is shown that the need for postulating a large class of selectively neutral genes is, at least in part, a function of the assumptions made about the nature of genetic variability

When Kimura (1968*a, b*) suggested that there is a very large class of selectively neutral genes that remain neutral, on average, for very long periods of time, there appeared to be strong theoretical reasons for doing so. Taken together, these reasons strongly supported this 'neutralist' hypothesis. An appreciable part of this chapter will be spent examining both the reasons and the premises which underlie them. I hope to demonstrate that there are selectionist counters to each argument. Recent molecular evidence (Chapter 4) suggests that selection may be more powerful and more universal than has hitherto been suspected. It now appears that it can affect different parts of the genome for which the functions are not yet known. Thus, while a neutralist position continues and will continue to be feasible, the selectionist is now confronted by the necessity of explaining unexpectedly intense selection at a large number of points in the genome. My conclusion will be that there is in the final analysis no *need* to postulate a large class of neutral mutations—which is not, of course, the same thing as saying that this class does not exist. It now seems safe to conclude that, in the mathematical sense, neutralism is sufficient to explain most genetic variability, but not necessary.

I will deal with the more common neutralist arguments in turn in the subsections of this chapter.

The magnitude of the substitutional load

'Haldane's dilemma'

In a famous paper, Haldane (1957) calculated that the cost in terms of selection-caused deaths of substituting one allele for another could be prohibitively high if too many alleles were being substituted at once. He suggested that the rate of substitution in evolution was probably such as to allow for only about one gene substitution every 300 generations, using the following argument:

First, natural selection could not, in his opinion, be very intense, since even artificial selection cannot be made as intense as the breeder might like:

> . . . in slowly breeding animals such as cattle, one cannot cull even half the females, even though only one in a hundred of them combines the various qualities desired.

And he extended this argument to natural (or at least seminatural) populations by considering the case of the moth *Biston betularia*. If, as Kettlewell observed using experiments with marked moths, the frequency of the most conspicuous phenotype can be halved in a day, then if there were ten independent characters being selected with the same intensity only one moth in 1024 would survive at the end of a day. Such an intensity of selection would be insupportable.

The argument is a very clear expression of the concept of *multiplicative selection coefficients*. This concept derives directly from the single locus model which says that the reduction in fitness of a population due to the effect of genotypes at a single locus is simply the product of the frequencies of the deleterious genotypes and the selection coefficients associated with each. (It should be noted that even the single locus model does not normally take into account increases in fitness.) The multiplicative model extends this assumption to more than one locus, so that the proportionate reduction in fitness caused by each additional locus affects the organisms remaining in the population. If the population is a finite one, the model makes the assumption that a particular maximum number of organisms will survive if all have the best available genotype, and that any deviation from this genotype will result in the survival of fewer organisms.

Haldane had previously (1954) derived a formula for the intensity of selection, based firmly on this multiplicative view. If selection at the ith locus is responsible for d_i deaths, then the survival to adulthood of the species will simply be the product of all these deficits:

$$W = \Pi\,(1 - d_i) \qquad (3.1)$$

When each d_i is small, this can be closely approximated by $e^{-\Sigma d_i}$

Haldane defined the intensity of selection as the natural logarithm of the ratio of the survival of the optimum phenotype to that of the population as a whole. This had the pleasing property that in the case of multiplicative fitness with very small effects contributed by each gene, it could be taken simply as Σd_i, without introducing too great an error.

Haldane then assumed that alleles which were to be substituted for old alleles in a population would mostly be present in the population in low frequency because they would be initially disadvantageous. They would therefore be held in the population entirely by mutation pressure. At the time circumstances changed to give the mutant alleles an advantage, they might therefore be found at initial frequencies of about 10^{-4}. He used this low value as a starting point for the substitution of one allele for the other.

Dealing first with a haploid system, Haldane determined by integration the approximate number of deaths resulting from selection necessary to change the frequency of the newly-advantageous allele by a given amount. This turned out to be

$$D \cong \sum_{n=0}^{\infty} s_n q_n \qquad (3.2)$$

where D is the number of selective deaths, s_n is the selection coefficient associated with the deleterious allele, and q_n is the frequency of the allele in the nth generation.

When this sum was evaluated, it turned out that the number of selective deaths was remarkably invariant over quite a range of per-locus selection coefficients, so long as the coefficients were not too large. Of course, the deaths due to selection could be spread out over a relatively small or a large number of generations. A rapid replacement can indeed reduce the number of genetic deaths, but only at the cost of a tremendous loss of numbers in the population in a given generation.

The same calculations were extended to a diploid population, and Haldane found that for a variety of selection coefficients and partial dominance values D varied only between about 10 and 100. He took 30 as a representative value, and then asked how many substitutions could be tolerated by the population if the total intensity of the selection were not too great. He took that intensity to be 0.1.

The genetic deaths are spread out over a number of generations. Thus, for one gene substitution every n generations, the fitness of the population W will be reduced by approximately $30/n$. But if a number of substitutions are going on simultaneously, as must certainly be happening in natural populations, the fitness of the population will be depressed even further. The number of loci being substituted at a particular point in time can be obtained by $\Sigma d_i/D$. The average number of generations needed to bring about the substitution is the inverse of this fraction, $n = 30/\Sigma d_i$ (using Haldane's estimate for D). For these multiple substitutions, the fitness $W = e^{-\Sigma d_i} = e^{-30/n}$. Therefore, the intensity of selection $I = 30/n$.

If the intensity must be 0.1 or less, and from his earlier argument about how much selection can occur each generation Haldane considers this likely, then it will take about 300 generations to effect a single substitution. On the assumption that a typical mammalian species persists for about half a million generations (Simpson 1953), then only about 1700 substitutions could take place during this period, and closely related species should differ by about this number of loci.

It will be seen that this argument depends very heavily on its assumptions.

One is the assumption of a multiplicative model. The second is that the favorable allele is very rare at the beginning of its period of replacement in the population. In fact, if the allele is already at an appreciable frequency in the population, then even given Haldane's assumptions it takes fewer genetic deaths to accomplish the replacement.

Such a rate of substitution is quite respectable if it is assumed, as can be estimated from the rate of mutation to lethals in Drosophila, that the number of genes with indispensable function is about 1000 (Alikhanian 1937). This is in approximate agreement with the more recent estimate of the number of indispensable functions in Drosophila by Judd et al. (1972). Judd and Young (1973) suggest that this number can be safely multiplied by 10 to give 10 000 cistrons coding for proteins in Drosophila, not all of which are indispensable to the functioning of the organism.

But there is much more DNA in higher eukaryotes than is needed to code for proteins. In man, for example, there is enough DNA in a haploid genome to code for about one million genes of 3000 nucleotides each. Further, the rate of nucleotide substitution observed in the course of molecular evolution is very high, and has been calculated to be as high as 2.5×10^{-9} per year in the genes coding for the fibrinopeptides and 4×10^{-9} per year for the third position of the codons in the gene for histone H4 mRNA (Kimura 1977). If we assume that only a small minority of bases in the genome are not subject to allelic substitution because of selective constraints, then most of the three billion bases in the primate line are available for substitution. Taking the fibrinopeptide rate as a reasonable one (Kimura would argue that the rate may be even higher on average, because of all the untranslated DNA which is not subject to selective constraint), this should result in about 15 nucleotide substitutions per year or (assuming the average primate generation to be eight years) about 60 million substitutions during the lifetime of a primate species. This is just about forty thousand times as great as the rate considered by Haldane to be bearable for a species without producing too much of a substitutional load! It is this discrepancy of four orders of magnitude or more between the rate Haldane's calculations appear to permit, and the observed rate which provides such a powerful argument for the neutralist viewpoint. The substitution of selectively equivalent alleles through drift, while regulated by the mutation rate, will have no effect on the fitness of the population.

Of course, all these arguments are based on the assumption of a multiplicative model. The model was explicitly criticized by Mayr (1959*b*, p. 5), just two years after the publication of Haldane's paper:

> Haldane (1957) has recently pointed out that selection places a considerable strain upon populations. Too rapid a rate of simultaneous selection against too many genes might eliminate the entire population. By these considerations, Haldane has called attention to a neglected aspect of selection. But is this really a serious problem? Is this not a situation where the new genetic theory of relativity should be applied? Haldane himself emphasizes correctly that we are dealing

> with relative values and that phenotypes are exposed to selection, not genes. Those individuals who have the greatest number of minus genes have the lowest reproductive potential. Those with the greatest number of plus genes have, quite obviously, the highest reproductive potential. Natural selection thus can deal simultaneously with a great number of variable loci. The 'goodness' or 'badness' of a gene will be relative depending on the fitness of the averages of all the individuals in the population. And as the population density drops, even rather poor phenotypes may survive by not being exposed to competition from better ones.

Regardless of the counterarguments that have been made to Haldane's dilemma, and there have been several, though few quite so cogent as Mayr's, it can easily be seen that the sheer numbers of substitutions which must occur in a natural population can be quite daunting to someone used to thinking in terms of multiplicative selection coefficients. It can even be daunting to someone attempting to think in terms of relative selection coefficients that are dependent, as Mayr suggested, on the density of the population or more generally on the severity of the environment. Indeed, can substitutions on this scale take place in a selectionist framework? Perhaps not all of them can, but as we will see numbers of substitutions thousands of times the upper limit postulated by Haldane can be accomplished without imposing too heavy a genetic burden on the population. The discrepancy may be closed even further by the likelihood that not all of the genome may be evolving as a result of single gene substitutions. The sudden appearance and disappearance of whole classes of satellite DNA in the course of evolution (pp. 86–90) provides evidence that the process of substitution can be greatly accelerated in some parts of the genome. This and other recent discoveries will probably make a mockery of all our calculations.

Multiplicative *v.* truncation models

A number of quantitative arguments have been made that lead to conclusions unlike those of Haldane. These are based on the assumption that at least some of the selection acting on natural populations is not of the multiplicative type, but rather occurs as a result of a *truncation* of the population. Because the differences between multiplicative and truncation selection models will form a *leitmotiv* of this book, let me take some time to compare the two models. The differing predictions of the two models, and variations of them, will be followed in detail in this and subsequent chapters. I will use the multiplicative model as a paradigm of those population-genetics models in which one starts with the genotype and uses it to determine the fitness, and the truncation model as a paradigm of models in which the relative position of the organism in the population with respect to some rankable character will determine its survival, regardless of its precise genetic makeup.

Truncation selection has received far more attention from quantitative geneticists than from theoretical population geneticists, primarily because domesticated plants and animals are selected for desirable qualities by the breeder using a truncation method. The least desirable organisms are culled from the population by the breeder, on the assumption that the heritability of the desirable characters is high enough that this culling will make a difference in the expression of the characters in the next generation.

The first important distinction between truncation and multiplicative models is that the former can easily accommodate a low heritability for the character being selected, while the latter *as it is normally applied* does not. In almost all cases, the multiplicative model is applied as if the variance of the selected character is entirely genotypic. A given genotype results in a given fitness. This is not the case with truncation selection, however, This is because the determination of where the truncation occurs must be based on the phenotype, and this may in turn be only the dimmest reflection of the underlying genotype.

Second, truncation models have in common a most important property, which is that they permit a great *range* of fitnesses among the organisms of a population. The existence of such a range greatly enhances the effectiveness of a selection scheme resulting in a given number of genetic deaths.

The process of truncation

When the experimenter truncates a population, he does so in two stages. In the first, the organisms are ranked according to the phenotypic expression that the experimenter is interested in. Then the truncation is made, at a point in the ranking which is determined by a complex of factors. Too large a truncation may reduce the population so much that it results in excessive inbreeding or will not permit the population to recover by the next generation. Too small a truncation may result in little phenotypic change in the next generation, particularly if the range of the desirable characters in the population is not very great. The experimenter will also be influenced by what he knows of the heritability of the character. If it is high, then a small truncation might be quite sufficient to achieve a steady generation-by-generation increase without the attendant danger of exhausting genetic variability. If it is low, then a much larger truncation may be necessary. Thus, the fitnesses assigned by the experimenter to members of the population will be only an indirect reflection of their genotypes and will be very much dependent on circumstances.

How might truncation occur in natural populations? Again, it can be visualized as a two-step process. In any population, whether or not it is genetically heterogeneous, organisms will exhibit a variety of phenotypes which will ultimately, in conjunction with environmental conditions, determine their fitness. If, as in the case of lethal genes, the fitnesses of the organisms carrying them are directly determined by their genotypes, then multiplicative selection or some variant of it will operate. If, as will often be the case, the genotype does not have an obvious and directly proportional effect on fitness, then it is possible for truncation selection based on the phenotypes to occur. The truncation need not be a sharp one, of course, and it is unlikely to be sharp in nature.

The phenotype which will ultimately, in conjunction with the environment, determine the organism's fitness, has been termed by Milkman (1973) the *fitness potential.* It is the partial divorcing of fitness potential from the ultimate fitness of the organism that distinguishes truncation from multiplicative selection. Depending on the circumstances, a fitness potential need not be fully expressed. A person with diabetes living in the year 1900 had the same fitness potential, in the sense of having the same phenotype, as a person with diabetes living in 1979. But their fitnesses (assuming appropriate medical care in the latter case) would be very different.

There is no difficulty in envisioning a vast multiplicity of types of truncation selection. The shape and severity of the truncation will depend on the actual phenotypic qualities that make up fitness potential, on the distribution of these qualities, and on the severity of the environment.

Fitness potential may depend on a restricted set of phenotypic characteristics, influenced by a small number of genes. Alternatively, but by no means exclusively, it may depend on an accumulation of many small phenotypic effects that are influenced, at least in part, by more general properties of the genome. It may be, for example, that (leaving aside genes of major effect) much of fitness potential may be determined by the average heterozygosity of the organism, or by the average number of deleterious genes of small effect that the organism is homozygous or heterozygous for. We will spend some time exploring these latter, and more general, cases.

I will defer until later a more precise quantification of the effects of different kinds of truncation selection, including a discussion of the effects of a variety of shapes of selective curve (pp. 141–55). The question of the relative effectiveness of truncation and multiplicative selection on different classes of genetic variability will also be reviewed (pp. 150–58). But I will begin by putting matters in historical perspective by considering the approaches various workers have taken to truncation models in the past. The reader should keep firmly in mind the distinction between models in which fitness is a direct and those in which it is an indirect function of the genotype.

Truncation and the substitutional load

Sved (1968*a*) and Maynard Smith (1968) objected to Haldane's approach to measuring the substitution load on the grounds that the multiplicative model was unrealistic. I will follow and expand upon Maynard Smith's argument here, because it is more subject to quantification, although Sved's approach is very similar.

Maynard Smith began by assuming alternative alleles A and a at a large number of unlinked loci, producing a Gaussian distribution of fitnesses in the population. Calculating the number of loci that can be acted upon by selection coefficients of a given size, he then applied two types of selection: Haldane's type based on the multiplicative model, and a truncation selection in which the lower half of the population is simply prevented from reproducing. Although he did not remark on this explicitly, these two modes of selection result in very different fitness distributions in the population. In the multiplicative model, the burden of selection falls with a surprising degree of equality on all the organisms in the population, while with the truncation model it falls on only some of them. This is because of a kind of 'moving average' effect of the multiplicative model, which is easily seen in Fig. 3.1. We assume that organisms in a population are segregating for L loci, at each of which A is replacing a. Each locus is at a frequency 0.5, and selection against the aa homozygotes is s at each locus. The average organism will be homozygous aa at 25 loci, with a standard deviation of 4.33 loci. The average fitness $\overline{W}$ of the population is given by $(1 - sq^2)^L$, where q is the frequency of the a gene at each locus. The fitness of a given organism W_t is given by $(1 - s)^{L_{aa_t}}$, where L_{aa_t} is the number of loci carrying aa in that particular organism. The fitness distribution for $L = 100$

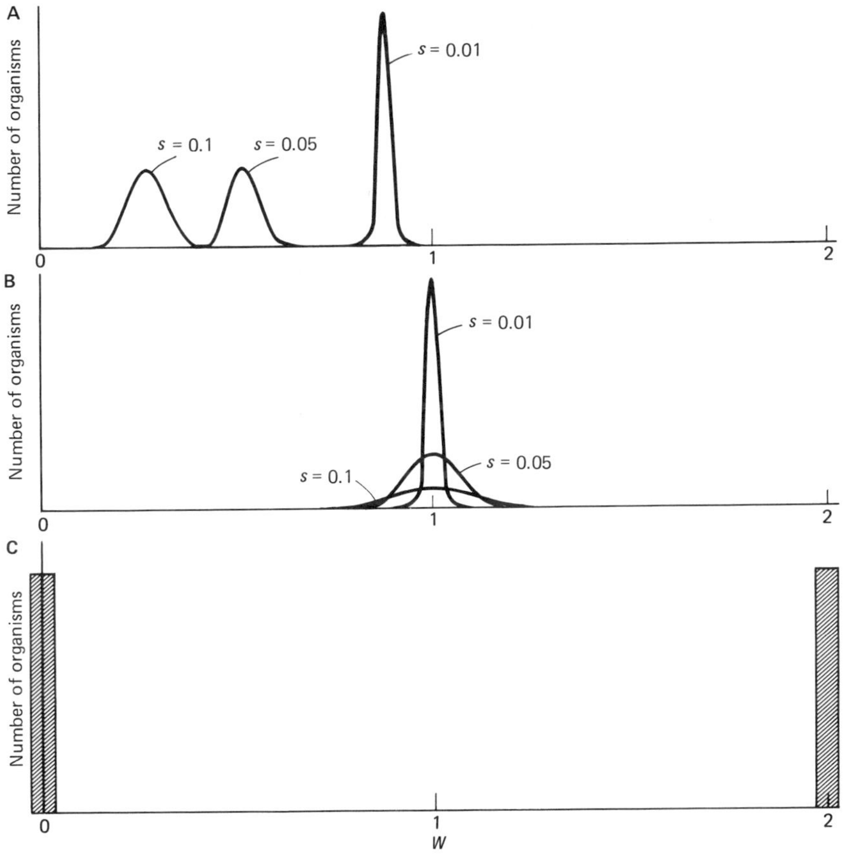

Fig. 3.1. The distribution of fitnesses in a population in which $L = 100$, $p = 0.5$ at each locus and s is the selection coefficient associated with the mutant homozygotes at each locus. **A.** Multiplicative model, before normalization. **B.** Multiplicative model, after normalization. **C.** Truncation model of the Maynard Smith type. Half the organisms have a fitness of zero or a fitness of 2.

and different values of s are plotted first (Fig. 3.1(a)) on the same scale. It will be seen that there is a narrow distribution of fitnesses and little depression of average fitness when s for each locus is small. When s is larger, the average fitness drops dramatically but the range of fitnesses in the population is still small.

The effect is seen even when the fitnesses of all populations are normalized to an average of 1.0 (Fig. 3.1(b)). In order to obtain even a moderate spread of fitnesses in the multiplicative model, the mean fitness must be lowered to such an extent that a very appreciable fraction of *all* genotypes must die or be prevented from reproducing each generation.

Even when the fitnesses are normalized each generation, as they would have to be if the population size were to remain constant, the spread of fitnesses is not as great as that obtained through truncation. Such a spread of fitnesses, normalized to keep the population size constant, is shown in the third part of the figure for truncation selection of the type investigated by Maynard Smith. Here half the organisms have a fitness of zero and half have a fitness of two, a distribution quite unattainable by the multiplicative model. Thus the intensity of selection per locus can be seen to be vastly greater in the truncation model, because some organisms have a high fitness and some have a fitness of zero. Fitnesses of zero can only be obtained with the multiplicative model when $s = 1$. In this particular model with $q = 0.5$ and $L = 100$, this would lead to an average fitness of 3×10^{-13}! Organisms with both high and zero fitness can undoubtedly occur in natural populations. It is this spread of fitnesses which constitutes the advantage of truncation models (Wills 1978), provided that the variance of fitness potentials is large, as it is in this case.

Maynard Smith used an ingenious argument to determine the effect of the different models on the rate of which alleles can be substituted. He assumed, in his multilocus haploid population carrying two alleles A and a at each locus, that the fitness of A is one and that of a is $1 - s$. The A allele is therefore replacing the a allele at each locus. All A alleles have the same frequency p.

The mean of the number of favorable alleles is therefore np, where n is the number of loci involved, and the variance is npq. First, Maynard Smith considered the simple truncation case. If the less fit half of the population is eliminated each generation, then the mean number of favorable alleles in the remaining fraction is $np + \sqrt{2npq/\pi}$. At each locus, the new frequency p' is $(np + \sqrt{2npq/\pi})/n$, which reduces to $p + \sqrt{2pq/n\pi}$. So Δp, the change in one generation, is $\sqrt{2pq/n\pi}$.

It is possible to derive Δp in another way. After one generation, $p' = p/[p + q(1 - s)]$, which is approximately $p(1 + qs)$ when s is small. So $\Delta p = p' - p = pqs$. Solving for n, we find $n = 2/pqs^2\pi$.

Maynard Smith now considered the multiplicative model. From Haldane's theory, for 50% selection, $I = \ln 2 = 0.69$. This is approximately equivalent to Σd_i. Since $\Sigma d_i = nqs$, then $n \cong 0.69/sq$.

The numbers of loci acted upon by a given selection coefficient s differ sharply when calculated by these two formulas, as one might expect. If s = 0.01, then if all the less favorable alleles are at a frequency of 0.9 only 77 loci can be acted upon by that intensity of selection, while 71 000 can be acted on by a truncation of the population to the same size.

The calculations of Maynard Smith for the truncation model, rough though they are, can be seen to reconcile the numbers of substitutions seen during molecular evolution, and the numbers that can be imagined to take place as a result of selection. Interestingly, the two modes of selection differ even more in their properties than these few figures indicate. Because truncation selection is a function of the variance of fitness in the population, as will be developed at greater length later, a tenfold decrease in the size of the selection coefficients results in a hundredfold increase in the number of loci that can be acted on by that intensity of selection. By contrast, the multiplicative model is, at least for a narrow range of selection intensities, approximately linear in its efficiency. The effect of this difference can be seen in Table 3.1, where the selection intensity per locus is dropped to 0.001, tenfold less than that employed by Maynard Smith. With the multiplicative model, the number of loci that can be acted upon by selection increases a hundredfold as the per-locus selection is reduced by a factor of 100, but with truncation selection the number of loci that can be acted on by selection increases by a factor of 10 000.

Arguments of this type would seem to go some way towards eliminating the problem of the substitution load. They depend, however, on certain assumptions about the properties of loci that are selected by truncation. It must be borne in mind that only certain classes of genes are likely to be subject to trun-

Table 3.1. A comparison of the number of loci that can be acted upon by multiplicative selection sufficiently strong to reduce the mean fitness of the population to 50%, and by 50% truncation selection. At each locus the frequency of the more favorable allele is 0.9. Modified from the data of Maynard Smith (1968).

s	No. of loci acted upon by s
Multiplicative model	
0.1	69
0.01	690
0.001	6960
Truncation model	
0.1	707
0.01	70 700
0.001	7 070 000

cation selection. The properties of these classes, and how prevalent they are, will be dealt with in Chapters 6 and 7.

The magnitude of the segregational load

The problem posed by the segregational load

The introduction of truncation selection provides a way of escaping from Haldane's dilemma. For classes of genes subject to truncation selection, the rate of substitution may be very rapid and the selection coefficients associated with each may be very high. But other types of genetic load are present in Mendelian populations. For example, if many genetic polymorphisms are being held in a population by balancing selection, then the homozygotes at these loci will be on average less fit than the heterozygotes. While this load may be quite bearable for a small number of loci, it becomes very large even with weak per-locus selection if the number of segregating loci is so great that every member of the population is homozygous for a large number of deleterious alleles.

Another aspect of this problem quickly emerged. When Kimura and Crow (1964) calculated the number of alleles that could be maintained in a population under conditions of balancing selection, they made two assumptions, one explicit and the other implicit in the mathematics. The explicit assumption was that all heterozygotes for all the alleles at the locus in question had a fitness of one, and that all homozygotes had the same selection coefficient associated with them. These conditions do in fact lead to a stable multiallelic polymorphism, but slight deviations from this highly symmetrical arrangement can easily lead to instability and the loss of alleles (Kimura 1956). This difficulty must have influenced their thinking about the probability that such a multiallelic polymorphism would occur. The problem with the stability of a multiallelic polymorphism is one to which we will return in Chapter 7; suffice to say now that the difficulty of establishing such a stable polymorphism can be circumvented by abandoning the assumption that populations are panmictic. Considerable theoretical difficulties remain, however.

The implicit assumption in their calculations was that of multiplicative selection coefficients. The problem in the multiallelic case dealt with by Kimura and Crow was that as the number of alleles increased, the relative proportion of the population that was homozygous and was therefore selected against, decreased. For two equally frequent alleles this proportion is 1/2, but for three it decreases to 1/3, for four to 1/4 and so on. If selection coefficients remain constant, the ability of selection to maintain polymorphisms decreases as the number of alleles increases. Kimura and Crow showed the apparently inevitable tradeoff between the number of polymorphisms that could be maintained, and the magnitude of the load. As the number of alleles increased, so did the probability of fixation of one or more of them. An equilibrium could be established between alleles lost through drift and introduced through mutation, and this equilibrium number of alleles could only be raised by increasing the population size or by increasing the strength of selection against the homozygous genotypes.

It was possible, using their assumptions, to compare the heterozygosity at an average locus expected under the neutrality model, with that obtained by a heterotic model. It was quite easy to increase the equilibrium heterozygosity *at a single locus* to a level far above the neutral one by application of strong heterotic selection. But it was apparent to them that attempting to extend such selection to many loci simultaneously would quickly result either in an unacceptably high segregational genetic load or (if the population were to survive) in the reduction of selection coefficients to the point at which the behavior of 'selected' alleles would be indistinguishable from that of neutral ones.

How has this problem of increasing load been circumvented? Again, as with the substitutional load, by the application of truncation rather than multiplicative selection. A number of thinkers in the field were stimulated by the particularly explicit expression of the multiplicative selection-coefficient model proposed by Lewontin and Hubby (1966), when they were confronted by immense quantities of genetic variability revealed by electrophoresis of enzymes and other proteins from a random sample of Drosophila. When these authors applied the multiplicative model to the situation revealed by electrophoresis, in which for the first time a statistical estimate could be obtained of the amount of variability to be found among structural genes, the discrepancy between selection theory and the facts became very obvious indeed.

For many organisms, the proportion of polymorphic loci appears to be about 30%. Extrapolating from this subsample of genes to the entire genome of—say—Drosophila, would mean about 3000 polymorphic structural genes. This leaves out of consideration polymorphisms in parts of the genome that are not translated, which may greatly increase this number. Suppose the total were 10 000 two-allele polymorphisms with frequencies at 0.5. Then the average fly would be homozygous at 5000 of these loci. If the selection against homozygotes at each locus were 0.001, then the fitness of the average fly would be 0.999^{5000} or only 0.6%.

Again, the problem is essentially with the number of organisms which have to die—not, in this case, to bring about a given number of gene substitutions, but to support a given number of polymorphisms. Given the multiplicative model, one of two things can be done. Substantial selection coefficients can be postulated, for example, of the order of 0.01, a size which tends to appear continually in theoretical discussions of selection. In that case, the average fitness of the population quickly becomes low as the number of segregating polymorphisms increases. Or, one can use much smaller selection coefficients, which do not result in a substantial reduction in fitness unless quite large numbers of polymorphisms are segregating in the population. For example, if s at each locus is 10^{-4}, and all polymorphic loci have two alleles each at frequencies of

0.5, the average fitness of a population segregating for 1000 polymorphisms would be 0.9, which is quite a reasonable number. But it drops to 0.37 if 10 000 polymorphisms are segregating.

The difficulty with such small selection coefficients is that they are not that different from neutrality. In a series of papers, Kimura and Crow considered the problem of just how small selection coefficients should be before the behavior of selected alleles would become essentially indistinguishable from that of neutral ones. Obviously this criterion varies somewhat, depending on how one defines indistinguishable, but the consensus of these papers was that selection coefficients should be of the order of $1/N_e$, the effective population size, before they could be considered to be effectively neutral. [A particularly rigid criterion was proposed by Kimura (1971), who suggested $1/4N_e$.] Later work by Ohta (1973) on the population behavior of nearly neutral mutations showed, however, that small amounts of selection could affect the outcome. Completely recessive deleterious alleles, at equilibrium between mutation and selection, are fixed by chance at 43% of the rate of neutral alleles if $s = 1/N_e$, at 94% of the rate if $s = 1/10N_e$, and at 99% if $s = 1/100\ N_e$, Even for these minuscule selection coefficients, the behavior of selected and neutral alleles is detectably different.

If we take the rather conservative limit of $1/N_e$, then for the example in which $s = 10^{-4}$ the effective population size would have to be 10 000 or more before the alleles begin to behave appreciably differently from neutral ones. Even so, many polymorphisms would be lost by drift. We are faced with a dilemma: if 10 000 polymorphisms are segregating at frequencies of 0.5, then even with weak selection at each locus the genetic load is high. And despite all the load, even in a population of 10 000, polymorphisms will be lost by fixation at a rate not greatly different from that if they were neutral. On top of this, selection coefficients of 10^{-4} or less are not exactly inspiring—particularly if one is faced with the prospect of looking for them experimentally. To ensure a .95 probability of seeing a difference of that magnitude between two genotypes of Drosophila segregating 1:1, for example, one would have to count approximately a hundred million flies. Such a task might give pause to even some of the more indefatigable Drosophila population geneticists.

As we will see, however, the selection coefficients associated with truncation selection or some variant of it are likely to be substantially higher than $1/N_e$ even for large numbers of segregating loci. They may still be difficult to detect, and some discussion (pp. 248–53) will be devoted to the most likely methods for going about their detection.

The first attempts to reconcile the observed amount of electrophoretic polymorphism with some sort of truncation selective model continued to be haunted by the ghost of the multiplicative model. The three papers, by Sved, Reed, and Bodmer (1967), King (1967), and Milkman (1967), were published together in a single issue of *Genetics,* and all are usually quoted together. Yet each took a different approach to the problem, and it is instructive to examine the underlying assumptions of each paper to see how their models differ.

The Sved, Reed, and Bodmer model

The paper by Sved, Reed, and Bodmer is simply a modification of the multiplicative model. As was seen above, one problem with this model is that, as the numbers of loci or the sizes of the selection coefficients increase, the mean fitness of the population decreases. Eventually it becomes obvious that size of the population should decrease to zero.

Simply normalizing the fitnesses each generation, which has to be done if the population is to remain the same size from one generation to the next, as Sved et al. (1967) realized, merely begs the question: how might such a normalization occur? If the selection takes place as the organisms grow to adulthood, then the only way that the population can be normalized is to have each organism produce a very large number of offspring. If the multiplicative model is taken to its extreme, this quickly becomes physiologically impossible, as Lewontin and Hubby (1966) showed.

Sved et al. took the approach that the problem with the multiplicative model was that it compared all fitnesses with an ideal genotype, a genotype in which all the loci conferred maximum fitness (in a simple heterotic model, this would be a genotype in which all the loci were heterozygous). So why not simply stop comparing fitnesses with this ideal genotype? If, for example, heterozygosity beyond a certain level—so many standard deviations above the mean of the population—conferred no additional fitness, then the multiplicative model would still apply to the majority of organisms in the population, but the fitness would no longer drop so dramatically as the number of segregating loci increased.

The cutoff was presented as an *ad hoc* assumption, with no biological justification offered in its support. Their underlying premise was that differences in fitness could be distinguished over a certain range but that the differences became progressively less and less important as the presumptive fitness of the genotype increased. They assumed, though no numbers were given, that this would allow more polymorphisms to be maintained for a given load. In fact, while many polymorphisms might be permitted to segregate in such a population, all alleles in organisms above the cutoff would be neutral and behave as such.

There is another problem with the model. The populations dealt with in their paper were assumed either to be random-mating or to be partially inbred, but in either case they were endowed with a given average heterozygosity. But let us see what might happen if that heterozygosity should decrease over time—through a process of bottlenecking or continued inbreeding, for example. We will consider that this process occurs under two types of circumstance, one in which the cutoff remains at a certain level and one in which the cutoff remains (for some reason) a certain number of standard deviations above the mean of the population. The consequences of these two assumptions as the mean heterozygosity of the population drops are shown in Scenarios I and II of Fig. 3.2.

If the cutoff remains at a particular point, then it will eventually no longer be relevant to the population in question since only a few organisms on the upper tail of the distribution will be affected by it. Thus the fitness of the population will drop as it converts from the model of Sved et al. to the simple multiplicative model. By this time the fitnesses of the organisms in the population are being compared with that of the very few organisms just below the cutoff, and one should see a very marked degree of inbreeding depression. Highly homozygous populations should be extremely unfit—which is, inciden-

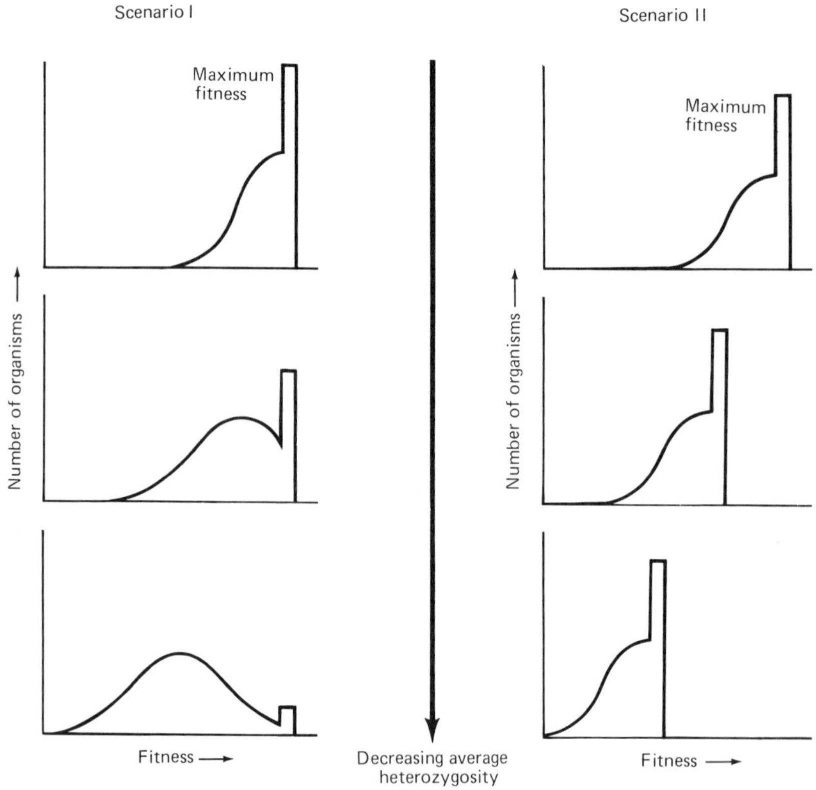

Fig. 3.2. The effects of reduced heterozygosity on the fitnesses of the organisms in a population, given the assumption—as in Sved, Reed, and Bodmer (1967)—that heterozygosity above a certain cutoff level has no effect on fitness. Scenario I: a fixed cutoff level. Scenario II: a moving cutoff level. See text for details.

tally, contrary to observation. Once they have been 'cleansed' of lethals and subvital genes, highly inbred strains of Drosophila and mice have been established that seem perfectly fit, at least in the laboratory.

It was this consequence of the model which worried Sved et al. The model presupposed, though they did not make this point explicitly, that the cutoff stayed at a certain point on the heterozygosity axis regardless of the average heterozygosity of the population in question. But suppose that we introduce a moving cutoff (Scenario II) without, of course, making any suppositions about the mechanism of such a cutoff. Now things improve from the standpoint of inbreeding depression, because as the heterozygosity of the population drops the fitness of the organisms in the population are always being compared, not with some almost unattainable genotype, but rather with a genotype which is just below the cutoff.

Such a system, however, is unlikely to maintain polymorphisms. As soon as a polymorphism is lost through drift or inbreeding, the mean fitness of the population remains relatively unchanged, because the cutoff moves down. New polymorphisms appear by mutation or immigration. Suppose that these are slower processes than their loss through drift. Since there is no disadvantage to the population in losing variability, there should be nothing under these circumstances to prevent it from eventually becoming virtually homozygous. Further, one runs into all kinds of conceptual problems with this moving cutoff. Suppose a population were highly homozygous, and an immigrant from a highly heterozygous population arrived on the scene. Would this single immigrant be enough to shift the cutoff upwards again? Since the population is no longer normally distributed, where might the cutoff lie?

Sved et al. introduced a cutoff or truncation into the discussion of the load problem, but I believe their model founders because it is a hybrid. It attempts to retain multiplicative selection for polymorphisms and as a consequence poses more questions than it answers.

King's model

The model of King (1967) has certain points of similarity with that of Sved et al. in that it is a hybrid model. But it illustrates certain features of cutoffs or truncations of normally distributed characters that are not found in the previous paper, although they are presented incidentally and King does not make a large point of them.

King postulates a threshold, specified at the outset, below which no organism survives. He suggests that each homozygous locus contributes one unit to what he calls a 'survival factor parameter.' This parameter has the property that organisms with a factor greater than some value t do not survive, and it can be seen then that the distribution of fitnesses in such a population will resemble that in the population of Maynard Smith (p. 29), in that some organisms will have a fitness of zero. The remainder will have fitnesses greater than one, the precise values depending on the placement of the threshold t.

Once again, however, this threshold does not move but remains at the same place while the population, depending on the effects of selection and drift, moves back and forth under it. At some times more of the population will be subject to truncation, at other times less.

Suppose the population were segregating for some number of polymorphisms, L, and the average homozygosity at each locus were a value h. Then the variance of numbers of homozygous loci per organism would be $Lh\,(1-\mathrm{h})$, and this would in turn be some small proportion r of the total variance of fitness in the population. Thus σ^2, the total fitness variance, is equal to $Nh\,(1-h)/r$.

Further, suppose loci are selected such that the selective advantage of heterozygosity is a given value s. Now, the proportion of a population segregating for L loci which survives to reproduce is W_0, that proportion of the normal curve lying to the left of the threshold t. One can consider two subgroups of that population, one homozygous and one heterozygous at a given locus (King assumes that all loci have the same effect). The proportion of the group homozygous at the given locus that survives is W_a, while the proportion of the subgroup heterozygous at that locus that survives is W_b. By definition, s, the selective advantage of heterozygosity at that locus, is $1 - W_a/W_b$. If s were 0.001, then W_a/W_b would be 0.999 or very close to unity.

Assuming that the main population and its two subpopulations are all normally distributed, one can ask the following interesting question. Is the variance of fitness in the population large enough to permit L such subpopulations to exist? That is, if W_a/W_b is some number less than unity, can L such different subpopulations, normally distributed around W_0, exist in the population? It is this question which King sought to answer by evaluating W_0, W_a and W_b for different values of L, s and h.

The three Ws were evaluated by integrating over the normal distribution, from minus infinity to the cutoff, given that the mean homozygosity in the population was M. Using King's notation, we have:

$$W_0 = 1/\sqrt{2\pi\sigma^2}\int_{x=-\infty}^{x=t} \mathrm{e}^{-[(M-x)^2/2\sigma^2]}\mathrm{d}x = 1/\sqrt{2\pi}\int_{x=-\infty}^{x=z} \mathrm{e}^{-(x^2/2)}\,\mathrm{d}x, \qquad (3.3)$$

where $z = (t - M)/\sigma$.

$$W_a = \frac{1}{\sqrt{2\pi\sigma^2}}\int_{x=-\infty}^{x=t} \mathrm{e}^{-[(M+1-h-x)^2/2\sigma^2]}\mathrm{dx}$$

$$= \frac{1}{\sqrt{2\pi}}\int_{x=-\infty}^{x=z-(1-h/\sigma)} \mathrm{e}^{-x^2/2}\mathrm{d}x \qquad (3.4)$$

since homozygosity at the locus adds $1 - h$ units to the mean homozygosity.

$$W_b = \frac{1}{\sqrt{2\pi\sigma^2}} \int_{x=-\infty}^{x=t} e^{[(M-h-x)^2/2\sigma^2]} dx = \frac{1}{\sqrt{2\pi}} \int_{x=-\infty}^{x=z+(h/\sigma)} e^{-x^2/2} dx \quad (3.5)$$

since heterozygosity at the locus adds the fraction h to the mean homozygosity.

W_a and W_b can be evaluated straightforwardly from tables of the cumulative normal distribution, assuming that the difference between them reflects the selective difference at the locus in question, plus the remaining variance in the population. The total variance is taken to be unity. Suppose, as in King's first example, $N = 1000$, $s = 0.001$, $W_0 = 0.5$ and $h = 0.7$. Then W_a will be 0.4997 and W_b will be 0.5007. From the table, the result of one such substitution will be to shift t relative to the mean of the subpopulation by 0.00125, a very small value. The total variance due to all sources is the reciprocal of the square of this value, $\sigma^2 = 640\,000$. But the variance due to the heterozygosity is only $Nh\,(1 - h) = 210$. The amount of the variance that is attributed to heterotic selection is the ratio of these two figures and also turns out to be a very small value, $r = 0.000\,33$.

What these manipulations mean is that the difference between W_a and W_b is slight, because of the small difference between the fitnesses conferred by a single gene substitution. This difference is a very small fraction of one standard deviation in the subpopulation, so that many such loci could be selected simultaneously with the same intensity.

In fact, King was able to show that 1000 polymorphisms could be acted upon by very strong selective pressures per polymorphism, without coming anywhere near to exhausting the variance in the population. For example, each could be subject to selection pressures of 0.01 and even so r, the proportion of the variance due to balanced heterosis, would only be about 3% of the total variance (see Table 3.2). By implication, very large numbers of polymorphisms could be acted on by smaller selective pressures without exhausting the variance. In a way analogous to Maynard Smith's model, the number of polymorphisms that can be maintained for a given r goes up as the square of the reciprocal of the selection coefficient.

In addition to the bimodal fitness distributions which can be obtained, truncation selection has, as was pointed out earlier, the further advantage that selection is acting on the standard deviation of the character, while the number of loci being selected determines the variance. Thus, if a tenfold decrease in per-locus selection coefficients can be permitted, the number of loci acted upon by selection will increase by a factor of 100. This advantageous aspect of truncation models will be more fully explored in a later section (pp. 150–58).

King's model, however, with its rigid truncation, is also a hybrid and suffers from the same fault which Sved et al. (1967) perceived in theirs. Once the majority of the population, through loss of heterozygosity, moves below the truncation, then the multiplicative model comes into play again and one expects strong effects of inbreeding. This can be illustrated rather vividly by

Table 3.2. The consequences of King's truncation model. Figures taken from King (1967), with a column added for 50% inbreeding.

No. of polymorphic loci (N)	Frequency of heterozygosis ($1-h$)	Selective advantage of heterozygosity (s)	Proportion of zygotes surviving (W_o)	Effect of one substitution ($1/\sigma$)	Proportion of variance due to overdominance (L)	Survival with 25% inbreeding (W_F)	Inbreeding depression with 25% inbreeding ($1 - W_F/W_0$)	Survival with 50% inbreeding (W_{F1})	Inbreeding depression with 50% inbreeding ($1 - W_{F1}/W_0$)
1000	0.3	0.001	0.5	0.00125	0.0003	0.4627	0.075	0.33	0.34
1000	0.3	0.0025	0.5	0.00313	0.0021	0.4075	0.185	0.3192	0.36
1000	0.3	0.005	0.5	0.00627	0.0083	0.3202	0.36	0.1736	0.65
1000	0.3	0.01	0.5	0.01253	0.0330	0.1711	0.66	0.0301	0.94
1000	0.3	0.005	0.2	0.00361	0.0021	0.1525	0.25	0.1178	0.41
1000	0.1	0.005	0.5	0.00627	0.0035	0.4397	0.12	0.3783	0.38
500	0.3	0.0025	0.5	0.00313	0.0010	0.4533	0.09	0.4052	0.19
500	0.3	0.005	0.5	0.00627	0.0041	0.4121	0.18	0.32	0.36
500	0.1	0.01	0.5	0.01253	0.0071	0.4404	0.12	0.3133	0.37

extending King's table of values. The homozygosity added by inbreeding depression subtracts a fixed amount from the outbred population, so that W_F, the proportion surviving after inbreeding, is given by:

$$W_F = \frac{1}{\sqrt{2\pi\sigma^2}} \int_{x=-\infty}^{x=t} e^{-[(M+NF-NFh-x)^2/\sigma]} dx$$

$$= 2\pi^{-1/2} \int_{x=-\infty}^{x=z-(NF-NFh)/\sigma} e^{-x^2/2} dx \quad (3.6)$$

King evaluated this integral for various values, using the value of σ already obtained, for 25% inbreeding. He did not carry out his calculations of inbreeding any further, but the results of taking them to 50% inbreeding are seen in the last column of Table 3.2. The loss of fitness due to inbreeding is very great once an appreciable fraction of the population moves below the truncation, and this effect is enhanced as the proportion of variance due to balanced heterosis increases. Note that the greatest inbreeding depression occurs in the cases in which the greatest proportion of the variance is due to overdominance.

Still, despite the difficulty with inbreeding which also plagued the model of Sved et al., King was able to show that a truncation model was markedly more efficient at maintaining polymorphisms than was a multiplicative one. Indeed, such is the power of truncation selection that many polymorphisms can be maintained when only a tiny fraction of the variance in fitness in the population is due to these polymorphisms. Regardless of the details of the model, the old pattern of thinking had been broken.

The model of Milkman

Milkman's paper, the third of the series and the shortest, postulated a model that was purely truncation, but in this case a truncation in which a fixed proportion of the organisms was removed from the population each generation regardless of their genotypes. His model was essentially based on the comment of Mayr quoted above (p. 25). Because selection coefficients in such a model are entirely relative, there is of course no difficulty with inbreeding. A highly homozygous population should be able to survive quite well in the absence of competition, but its members might do quite badly in the presence of competition from organisms with a higher level of heterozygosity. Because, in this model, there are differences in fitness between organisms, regardless of their levels of heterozygosity, there is no tendency for such populations to move towards monomorphism, as there is in the model of Sved et al. when the truncation is made movable. Milkman made the point that such a system should be extremely efficient at the maintenance of polymorphism, but did not quantify his remarks further.

The size of selection coefficients

Each of these papers, in its own way, examined a portion of the problem which Kimura and Ohta (1971) stated was uppermost in Kimura's mind when he proposed the idea of large numbers of neutral alleles in 1968. Although they did not remark on it specifically, King's and Maynard Smith's papers also had a direct bearing on another related problem mentioned by Kimura and Ohta, that of the magnitude of selection coefficients. Haldane (1927) had shown that the probability of fixation of an advantageous allele is only 2 s, where s is the selective advantage of the heterozygote (this is because of the high probability of loss of even an advantageous allele in the first few generations). Kimura and Ohta (1971) showed that Haldane's formula should be modified to take the population size into account:

$$k = 4N_e s\,\mu \tag{3.7}$$

where k is the number of loci fixed and μ the mutation rate.

From the molecular-evolution data available at the time, it appeared that there were about six substitutions per generation in the average mammalian genome. So, assuming $k = 6$, a fairly small effective population size of 5×10^4, and s of 0.001, they obtained the answer that advantageous genes would have to arise in 3% of the gametes each generation to account for this substitution rate—a figure as high as that observed by Mukai (1964) for the accumulation of lethal and subvital genes in *Drosophila.* If we take the more recently estimated figures of 120 substitutions per generation (fifteen substitutions per year and 8 years per generation), then advantageous mutants would have to arise in 60% of the gametes each generation.

Of course, this argument depends on the assumption that advantageous selection coefficients are indeed this small, which circles around once again to the problem of the segregational load. Kimura and Ohta dared not postulate selection coefficients that were too large because of the substitutional load that would ensue—but we have seen that even King's hybrid model can allow selection coefficients many times greater than 0.001, so long as the truncation falls somewhere near the middle of the fitness distribution. Maynard Smith's calculations show that millions of alleles can be substituted at average selection coefficients of 0.001.

This rather lengthy discussion has been intended to show, in some detail, how many apparently compelling arguments that neutral alleles should be the most numerous, are the result of the properties of the multiplicative selective model. Once at least some selection is assumed to be acting in a way that produces a large range of fitnesses in the same population, these difficulties can be seen to disappear. Further, such truncation-selected genes have interesting properties that make their existence in large numbers highly probable, as will become apparent. One thing that this discussion of the early models shows rather vividly is how tenacious the grip of the multiplicative way of thinking has been and how difficult it was for population geneticists to escape from it.

The decrease of fitness from inbreeding

It was noted in the previous section that inbreeding can result in a considerable decline in fitness in a truncated population when the truncation threshold remains constant (the model of King 1967), but essentially disappears if the truncation involves an approximately constant proportion of the population (the model of Milkman 1967). In fact, the problem with inbreeding is severe when loci are selected in multiplicative fashion. Such selection places strong constraints on the number of genetic loci that can mutate to multiplicatively selected alleles.

It was exactly this problem that was addressed by Morton et al. (1956). Their reasoning was stimulated by their estimate of the number of lethal equivalents carried in the average human genome. The number turns out to be small—3 to 5 per genome as calculated from older data on the effects of inbreeding in human populations, the results of which are probably closer to 'natural' rates of mortality than more recent studies would be. With such a relatively small number of lethal genes, the chance of survival of a mildly inbred child would be good.

A very large load of lethals and severe subvitals would not harm a population, so long as it remained outbreeding. If it were to undergo a severe size bottleneck, however, with consequent inbreeding, the average fitness would decrease so rapidly that the population would be threatened with extinction.

Some examples, on the assumption of multiplicative selection, are given in Tables 3.3 and 3.4. In all cases the mutation rate is assumed to be 10^{-5} per gene per generation, and L is the number of loci that can mutate. In Table 3.3,

Table 3.3. Population fitness $\overline{W}$ to be expected when outbred populations, at equilibrium between mutation and selection for lethal alleles, are subject to sudden inbreeding.

	F			
L	0	1/16	1/4	lethals/organism before inbreeding
$h = 0, \hat{p} = 3.16 \times 10^{-3}$				
1 000	0.99	0.81	0.45	6
10 000	0.90	0.13	0.0003	60
20 000	0.82	0.016	1.1×10^{-7}	120
$h = 0.05, \hat{p} = 10^{-4}$				
10 000	0.90	0.80	0.55	4
20 000	0.82	0.64	0.30	8
50 000	0.61	0.32	0.05	20
$h = 0.2, \hat{p} = 2.5 \times 10^{-5}$				
50 000	0.61	0.52	0.32	5
100 000	0.37	0.27	0.11	10
200 000	0.135	0.072	0.011	20

Table 3.4. Population fitness ($\overline{W}$) for loci mutating to subvital alleles ($s = 0.1$) at 10^{-5}/gene/generation.

	F		
L	0	1/16	1/4
$h = 0, \hat{p} = 0.01$			
1 000	0.99	0.93	0.77
10 000	0.9	0.48	0.074
20 000	0.82	0.234	0.006
$h = 0.05, \hat{p} = 10^{-3}$			
10 000	0.9	0.802	0.56
20 000	0.82	0.64	0.31
50 000	0.61	0.33	0.055
$h = 0.2, \hat{p} = 2.5 \times 10^{-4}$			
50 000	0.61	0.52	0.32
100 000	0.37	0.27	0.11
200 000	0.135	0.072	0.011

a selection coefficient of 1 is assumed. For complete recessives, the fitness on moderate inbreeding ($F = 1/16$) drops rapidly with L much above 1000. L can be greater with partial dominance, but low partial dominance seems to be the rule for recessive lethals (Mukai 1968; Wills 1966). The number of loci that can mutate to lethal genes is unlikely to exceed 20 000, since with higher L the population would begin to run into the triple constraints of too much reduction in fitness on inbreeding, too much reduction even without inbreeding, and more lethals per organism before inbreeding than are actually observed.

Table 3.4 shows similar figures for subvitals of moderate effect ($s = 0.1$). While the average fitnesses of populations segregating for completely recessive subvitals are somewhat different from those segregating for completely recessive lethals, the introduction of a small amount of partial dominance results in average fitnesses that are essentially the same for subvitals as for lethals. This is because, with partial dominance, most selection takes place against the heterozygotes.

There is no doubt that lethal genes are selected multiplicatively. It is possible that, since mutation rates cannot be reduced greatly without very strong selection (Leigh 1973), the effects of inbreeding have placed constraints on the number of loci of *indispensable* function that the organism can possess. The neutralist view would be that the rest of the genome is the abode of neutral or nearly-neutral mutations. But truncation selection with a movable threshold may reduce or eliminate the effect of inbreeding. Thus, if mutations of conditionally harmful effect can take place in the rest of the genome, they should produce no or little inbreeding depression. Evidence that conditionally harmful mutations can occur in loci or regions on the genome of both dispensable and indispensable function will be considered in Chapter 4. The effect of truncation selection against such mutations will be considered in Chapter 6. Suffice to say at this point that truncation selection provides a way in which mutations of this type, if they exist, can be dealt with.

The constancy of rates of molecular evolution

The rate of substitution of neutral alleles

Kimura's 1968 paper, in which he raised the possibility of a very large class of effectively neutral alleles, followed logically from the paper by Kimura and Crow (1964) which demonstrated and worked out some consequences of the mathematically very pleasing properties of these alleles. Because of the unassuming air of Kimura's paper, it might have passed without much notice, were it not for the publication of a paper by King and Jukes in the following year (1969) with the inflammatory title *Non-Darwinian Evolution.* It was this which caught the attention of every Darwinian, and did perhaps the most to trigger the entire controversy between the selectionists and the neutralists. In fact it is ironic that, extrapolating their calculations, King and Jukes suggested that the substitution of selectively neutral alleles might account for something like 10% of the allelic substitutions observed during evolution, a figure with which most selectionists might find themselves in agreement. But the neutralist camp quickly staked out a much larger portion of the genome for itself. Kimura (1968) had already suggested that 40% of all mutations might be neutral, based on synonymous or chemically synonymous changes in the genetic code. In a later paper it was suggested that '. . . molecular evolution has mainly been carried out through random fixation of selectively neutral mutations . . .' (Kimura and Ohta 1971).

A major point made by Kimura (1968) and King and Jukes (1969) was the overall constancy of rates of molecular evolution. This constancy is precisely what one would expect from the neutralist model, provided that average generation time remains constant, for the fixation rate per generation is equivalent to the mutation rate to neutral alleles, and is completely independent of the population size.

That this is so can be seen from the following simple argument (paraphrased from Crow and Kimura 1970, pp. 368–9):

Assume first that the mutation rate at a locus is μ. Then in each generation the number of mutants arising in that generation will be $2N\mu$, where N is the population size in that generation.

Now, suppose that our population is at generation $t = 0$. If we go back far enough, we will find a generation $t = -a$ at which a single gene present in that generation has given rise to all the genes in the present generation. If a gene from that generation were picked at random, then the probability that it would be this ancestral gene will be $1/2N_{t=-a}$, where $N_{t=-a}$ is the population size in that generation.

Let us now consider the present generation, in which the population size is

$N_{t=0}$. The probability that a mutant gene arising in *this* generation will be the ancestor of *all* the genes at some point in the future is $1/2N_{t=0} \times 2N_{t=0} \times \mu$, which is simply the mutation rate μ. Provided that the genes remain neutral for a sufficiently long period of time, and that N stays constant during this period, the rate of fixation of neutral genes is simply equal to the mutation rate and is independent of population size.

Because the mutation rate is expressed in mutations per generation in these calculations, however, the rate of fixation of neutral mutations will be an inverse function of generation time. If populations with the same mutation rate per generation have different average-generation times, the rates of fixation will be different. Thus, the neutralist model predicts that the rate of substitutions for equivalent populations should be the same, but the neutralist is not surprised by different rates in different organisms if their generation times differ or if the mutation rate per generation can be inferred to be different. Further, the fraction of the gene itself in which substitutions are permitted is likely to vary from one gene to another and from one period of time to another.

Indeed, it is quite possible to imagine situations in which neutral genes do not exhibit a constant rate of substitution. If a population should remain large for many generations and then undergo a bottleneck—during the process of speciation, for example—the neutral genes which have accumulated during the period of large N will be fixed in greater numbers during the bottleneck. Thus neutral alleles can show 'bursts' of evolutionary activity just as selected ones might be expected to do. J. F. Crow (private communication) points out that the 'burst effect' will be enhanced if the alleles are only approximately neutral, and their s values fluctuate around zero. Once a population has passed through a severe bottleneck and expanded again, as was suggested earlier (p. 14), it may take a very long time for the neutral mutations to accumulate sufficiently so that their rate of fixation again equals the mutation rate. The number of generations during which the rate is slower than average is approximately equal to the population size. For a population of a few million, the return to the equilibrium situation could easily take longer than the lifetime of the average species.

Thus, the constancy of substitutions is unlikely to be observed over any period of time shorter than a geological one. If our ability to discriminate is finer than that, we should expect large fluctuations in the rate of fixation of neutral alleles.

Given all these caveats, it is hardly surprising that the neutralist model fits the molecular-evolution data so well.

The molecular 'clock' and its consequences

The idea of using the growing family of protein sequences as a probe to reveal the probable nature of past sequences and the course of protein evolution had been suggested by a number of workers, but was first worked out in detail by

Zuckerkandl and Pauling (1962) using hemoglobins, and Margoliash and Smith (1965) using cytochrome *c*. Globular proteins which can be isolated in quantity and which, because of their structure, have a nonrepetitive sequence of amino acids, are obvious candidates for such a molecular probe.

Heme-containing globins have another advantage in that they are relatively small proteins all of approximately the same subunit size. This makes the task of aligning the sequences of amino acids less difficult, since relatively few insertions and deletions of genetic information have taken place at various points during the evolutionary process.

Fig. 3.3 shows the current 'family tree' of the globins. It is, of course, quite a straightforward task to calculate the number of amino acid differences between two proteins, and with knowledge of the genetic code to obtain an estimate of the minimum number of base changes that must have occurred (only an estimate can be obtained from protein data, because of the degeneracy of the third position of the codons for most amino acids). But it is rather more difficult to estimate the most probable number of amino acid or base changes that have actually taken place. This is because, while one can always employ a simple stochastic model to estimate the number of base changes which are

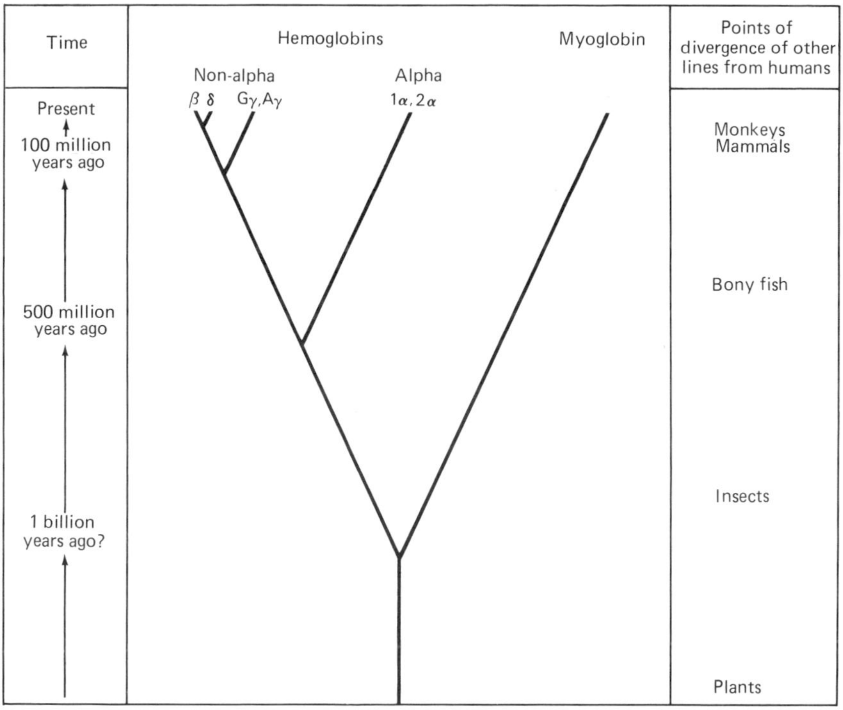

Fig. 3.3. The evolution of genes for the fetal- and adult-human hemoglobins, and for myoglobin. The order, and approximate times of the gene duplications are given. From Dayhoff (1972).

'hidden'—either because they took place in the third codon or because they took place at a position which has changed several times during the divergence of the two known sequences—there is no guarantee that the permitted rate of change for different parts of the molecule has remained the same, or that the amount of the molecule free to evolve has always remained constant. These difficulties continue to haunt workers in the field, and as we shall see are not susceptible of easy solutions.

There is also a difficulty with the time scale. In order to provide a time scale for the evolutionary process, Zuckerkandl and Pauling assumed a date of 80 million years ago (mya) for the divergence of relatively unrelated placental mammals. This is a reasonable estimate based on the fossil record. The divergence of placental mammals can be dated to the beginning of the Cenozoic, about 70 mya, and it seems reasonable to include an extra ten million years just in case. But in fact fragmentary jawbones of rodentlike mammals have been found in early Jurassic deposits of 180 mya (Datta et al. 1978), and it is possible that at least some of the divergence of placental mammals took place between that time and the beginning of the Cenozoic. Thus, even this date, which is normally taken by molecular evolutionists to be one of the best-established, depends on a fossil record which is fragmentary and on which it may be dangerous to place too much reliance.

Extrapolations from the fossil record have been used to establish other divergence dates: monotreme–marsupial, monotreme–placental, bird–mammal, and so on. The fossil record is free of obviously eukaryotic or multicellular organisms until about 800 mya, and it is presumed on this evidence that the great divergence of eukaryotic phyla must have taken place during the late Precambrian and the Cambrian periods. This also may not be correct, and by the time one has gone that far into the past, the possible errors are measured not in tens but in hundreds of millions of years.

On the assumption of these tentative dates, and that double and triple replacements were distributed randomly in time and over the entire molecule, Zuckerkandl and Pauling estimated the rate of evolution of the hemoglobin molecule to be about one amino acid replacement every 7 million years. Such a period of time was certainly longer than the average lifespan of a species, and so it was not surprising that when human and chimpanzee globins were compared, all were identical in sequence except for the delta chain, which exhibited one difference (Boyer et al. 1970).

As other molecules were examined, it quickly became apparent that certain simplifying assumptions did not hold. Cytochrome *c* is a much smaller molecule with a much slower evolutionary rate, and as the first sequences were analysed it was seen that parts of the molecules were far less susceptible to evolutionary change than others. Particularly invariant regions were those involving heme binding and the disulfide links that stabilized the tertiary structure of the molecule. This observation led Fitch and Markowitz (1970) to postulate the idea of the concomitantly variable codon or *covarion*. Covarions are those codons in a particular cistron in which changes are permitted to occur at any particular time. Because covarions can alter in number over time, and are assumed to represent different proportions of different proteins, this concept removed a difficulty which confronted a neutralist explanation of molecular evolution, since it explained the very different rates of evolution of different molecules. Because not all the molecule was free to vary, selective constraints were placed on that portion of it which was directly involved in the molecule's function while the remainder was free to accumulate random variants.

Fitch (1971) calculated that during cytochrome *c* evolution the number of codons free to vary at any given time appears to be very small—between four and ten out of approximately 104, with an average near the lower end of that range. Further, there is a very high probability, approximating 75%, that one of these covarions will lose its status because of a change elsewhere in the genome, with an equal probability that on average another covarion would appear elsewhere in the molecule to take its place. Thus, very different sets of codons were free to vary in different lineages.

The picture that emerged for cytochrome *c* evolution, according to the neutralist view, was of a tightly controlled molecule in which mutations changing most amino acids did not survive, but in which there were small regions where mutations were permitted. These mutations themselves then changed the 'ground rules' for the molecule, opening up new regions in which mutations were permitted and closing down other regions. If Fitch's model is correct, it is unclear whether such mutations can be considered neutral except in the formal sense, since substitutions in one part of the molecule have effects on other parts.

The addition of selective constraints poses problems for the neutralist view. Suppose a neutral allele, destined for eventual fixation, arises in a gene, such as that for cytochrome *c*, in which the number of covarions is limited. During the time that the neutral allele is polymorphic in the population, one set of covarions will be available elsewhere in the cistron from which subsequent neutral alleles can arise. But in cistrons not carrying this neutral allele, different sets of covarions will be available. A further mutation occurring in a covarion of the neutral allele might have different selective consequences from the same mutation occuring in another allele where this position was not a covarion. This difference would be apparent until this second allele and the neutral allele reached linkage equilibrium. Under the circumstances, can such a neutral allele be considered neutral? It could, for example, be lowered in frequency in the population by 'hitchhiking' if a second allele deleterious only in the context of this first allele were to arise.

Regardless of the conceptual problems for the neutral-allele theory, the imposition of selective constraints was necessary to explain unequal evolutionary rates. Further, as shown above in connection with cytochrome *c*, those portions of the molecule directly concerned with some obvious functions such as binding a ligand were the least altered during evolution, just as the neutralist model would predict. Some of the most rapid protein evolution rates were found in the fibrinopeptides, small peptides nine to fifteen amino acids long that are cleaved from fibrinogen to produce fibrin. This might be expected if it is assumed that these molecules simply serve to 'cap' fibrin. It will be seen (p. 78) that the functions of even these small molecules may be complex, however, and that there is evidence for selective constraints operating on them.

This rate is not the maximum that has been observed. With the advent of rapid techniques for sequencing nucleic acids such as those of Maxam and Gilbert (1977) and Sanger et al. (1977), it has proved possible to determine the precise use of codons for quite long segments of the genome. When Grunstein et al. (1976) compared the H4 histone messenger RNA from two species of sea urchin, *Lytechinus pictus* and *Strongylocentrotus purpuratus,* they could unambiguously determine 28 third-position codons and found substitutions at eleven of them. Since this histone has the same amino acid sequence in both organisms, the only evolution that has gone on is in the 'unconstrained' portion of the gene, and Kimura (1977) calculated that this must have involved $3.7 \pm 1.4 \times 10^{-9}$ nucleotide substitutions per site per year. This is half again as high as the rate observed for fibrinopeptides, and Kimura suggested that it might represent an upper limit to the rate of neutral mutation.

As we will see later (pp. 79–86) there are strong selective constraints acting even in these positions, at least for a number of completely sequenced genes from a great range of organisms. But if we leave aside these recent observations, this apparently unconstrained evolution constitutes the greatest strength of the neutralist argument: those portions of the molecule that matter the least are the fastest-evolving.

At first sight there seem to be no exceptions to this rule. The selectionist has only two ways to reconcile selection with the observations. The first and less convincing argument is to propose that certain portions of the protein molecule, notably its surface, interact strongly with its cellular environment and are therefore subject to rapid change as the cellular environment changes. Very little is known about the sort of cellular *milieu* polar-surfaced globular proteins might find themselves in, and so these arguments, unsupported by evidence, remained speculative. There is, as a general rule, nothing less convincing than an attempt to explain on the basis of selection, a structure for which the function is not thoroughly understood.

However, a second approach to the problem is possible. It does seem reasonable that the portions of the molecule not directly concerned with vital functions might be able to respond to weak selective pressures, just as they might be subject to neutral substitution. This is simply because changes in less critical parts of the molecule are not instantly fatal to their carriers—as, for instance, would be an alteration in the ability of cytochrome *c* to bind heme.

Thus it may be that there is a semantic difficulty at the root of this problem. The selectionist viewpoint is perhaps not fully understood by those molecular evolutionists who make the argument for the prevalence of neutral mutations.

Whether wittingly or not, the assumption may be that selection is considered to be an all-or-nothing affair, so that for selection to be effective each amino acid alteration must be a life-or-death matter for at least some of the organisms carrying it. But this is not necessarily so. A selection differential of 0.01 can mean that either one organism in every hundred carrying the gene actually dies of its effects, or (more probably) that only 99 offspring carrying the less fit genes survive to adulthood on the average for every 100 of the less fit genotype in the previous generation. The effect of a particular genotype might therefore be very difficult to trace. Thus, no particular organism need be 'condemned to death' by its possession of this almost imperceptibly disadvantageous genotype, and most selection is likely to be quite undetectable other than statistically.

What this means in terms of molecular evolution, therefore, is not necessarily that every isoleucine–valine substitution is critical in determining the survival of the organism, but rather that it may have a small cumulative effect which is different from zero. It is, therefore, exactly such alterations which are likely to occur most freely in the course of molecular evolution, since their effect on any particular organism is so slight. And they are most likely to occur in the least 'essential' parts of the molecule, since small advantages associated with substitutions would be more likely to occur there.

To summarize, the neutralist picture is one in which the least important portions of proteins and the very large untranslated portions of the genome are the most free to evolve at something approaching the maximum rate dictated by the mutation rate for neutral alleles. As shown earlier, this will result in a constant rate of evolution provided that the generation time, mutation rate and the effective population size do not show secular trends during the period being considered. Of course, a lack of constancy in molecular evolution does not disprove the neutralist view.

The fine structure of molecular evolution

As more information has accumulated about protein sequences, and as the statistical techniques used to interpret it have become more sophisticated, tantalizing irregularities have appeared in what has, until recently, seemed to be a fairly constant 'clock' for each molecule. Because of the fragmentary nature of the data, and the fact that ancestral events have to be inferred from present-day sequences, there has been a good deal of disagreement about the best way to proceed. Three methods, all yielding similar results, have been used by three groups of workers.

(1) The maximum parsimony tree. Fitch (1976) has taken the view that the most accurate picture of the events that have occurred in the lineages of these molecules can, at the moment, be obtained only by using the sequences themselves, without attempting to fill in the details of the tree by some statistical

method. Consequently, when Langley and Fitch (1974) examined the question of the constancy of the evolutionary clock, they did so using the trees constructed by the 'maximum parsimony' method of Fitch and Margoliash (1967). That is, what is the smallest number of changes which can bring about the observed substitutions? They then used a maximum likelihood estimator to find the evolutionary rate for each branch of the tree which would give a rate of substitution as close as possible to the observed rate. This gave rates of evolution sufficiently different that a χ^2 test showed a significant amount of heterogeneity between rates. The variance of the clock for hemoglobins, cytochrome *c* and fibrinopeptide A was about twice that expected for a constant clock.

The chief source of the heterogeneity was the slowing down of the rate of evolution, which Wilson et al. (1977) strongly argue may be an artifact of the precision with which divergence times are known from the fossil record, but there were some indications of heterogeneity within particular lineages as well.

(2) Random evolutionary hits. Jukes and Holmquist (1972) attempted to fill in the details of the trees by a stochastic method. This approach has been criticized by Fitch (1976) as not adding information but merely adding noise, but it seems to me that a proper application of a stochastic method might enable one to obtain a closer estimate of the *number* of alterations that have occurred. Since it is the relative numbers in different branches of the tree that are of concern here, presumably the more information that is available about them, the better will be our estimates of the rate of the molecular clock.

Unfortunately, this methodology is highly dependent on its assumptions, as Jukes and Holmquist fully realized. It had early been remarked by Margoliash and Smith (1965), that different portions of the cytochrome *c* molecule evolved at different rates, and Fitch and Margoliash (1967) were led to postulate three portions of the molecule evolving at different rates to explain their data adequately. Jukes and Holmquist were able to show that an excellent fit to the observed data for cytochrome *c* and for globins could be obtained if *for any particular lineage* they postulated two regions, one constrained from evolving and one free to accumulate random mutations along its length. The numbers of these random mutations were determined by counting the number of single, double and triple substitutions in each codon, and comparing these proportions to proportions expected on the basis of corrections for degeneracy of the code and multiple fixations at the same site, including back mutations. Thus, the total number of mutational events at the level of the DNA could be estimated for each lineage.

For a particular lineage, it was possible to postulate very different rates of evolutionary change for the different parts of the molecule, as is shown in their Table 10 for the carp and warm-blooded-mammal α lineages, which has been adapted in Table 3.5. An excellent fit to the observed data could be obtained by postulating a relatively small number of hits on more than half the molecule

Table 3.5. Data from α-hemoglobin chains of carp and warm-blooded mammals expressed as number of changes per codon. The table shows that divergent values for the number of hits per codon and the proportion of codons subject to these hits can be used to generate distributions very similar to the actual data. Adapted from Table 10 of Jukes and Holmquist (1972).

	0	1	2	3	Average No. of
Actual data	51.0	29.9	19.1	0	'hits' per codon
Average hits/codon generating two Poisson distributions					
4.0 on 57% 0 on 43%	50.4	30.0	18.5	0.7	2.28
8.0 on 41% 0.25 on 59%	52.0	28.7	18.4	0.8	3.43
6.0 on 46% 0.2 on 54%	50.4	29.9	18.6	0.8	2.87
5.0 on 49% 0.15 on 51%	51.0	30.0	18.3	0.8	2.53

and none on the remainder, which was their method of choice, or a large number of hits per codon on a smaller portion of the molecule and a small number on the remainder. The data are open to a number of statistical interpretations—indeed, there is nothing in the data to argue against the division of the molecule into three or more sections, each with different probabilities of random base changes.

An interpretation of the data based on randomness is rendered less likely by a remarkable observation documented in Moore et al. (1976) and Holmquist et al. (1976). This is that the number of observed and predicted three-base changes in the cytochrome *c* lineage were reasonably close to each other (585 vs. 636), but that there was a striking and extremely significant deficit of three-base changes in the intrafamily globin lineages (20 observed and 108 predicted). This extended to the interfamily lineages. Though it was proportionately not quite so pronounced (548 observed and 1178 predicted), it was even more significant. What difference could there possibly be to account for this deficit? They examined the possibility that three-base changes are in some qualitative way different from one- and two-base changes, but the average charge properties of amino acid pairs separated by three-base changes are not noticeably different from those separated by two- and one-base changes. In any case, unique properties of three-base changes could not account for the fact that the cytochrome data showed the expected number of such changes while the hemoglobin data did not. Thus there are obvious nonrandom forces at work in the globin lineage which make the application of a stochastic model suspect.

(3) The augmented maximum parsimony method. This method, presented by Moore et al. (1976), works out the maximum parsimony tree and then uses the information in that portion of the tree which is based on the most data to estimate the probable number of mutational substitutions that occurred in less dense portions. These authors suggest that on the whole, this method would tend to equalize the rates in the known and unknown parts of the tree, and therefore be a conservative measure. If one is looking for inequalities in the rate of the evolutionary clock, therefore, this method should underestimate them. Any differences observed in the rates would, as a result, be that much more believable.

Their augmentation algorithm compares the number of known or inferred links between two points in the tree with the number of base substitutions observed. Suppose there are two pairs of points in the tree, and each pair is separated by five known or inferred links. Five base substitutions are known to have occurred during the evolution of the first pair, but only three during the evolution of the second. Then the number of substitutions occurring in the second pair is augmented by two on the assumption that there is a good probability that at least two additional ones occurred but were due to a variety of circumstances invisible.

In two joint papers (Moore et al. 1976; Holmquist et al. 1976), the two groups of workers compared the augmented parsimony method and the random evolutionary-hit method and found that statistically there was an excellent correlation between them. Further, the variances of the two measures were comparable. There were, however, some differences. The augmented-distance method gave values only about two thirds those of the random evolutionary hit values between closely related vertebrate taxa in the cytochrome *c* lineage, but gave good agreement when the taxa were more distantly related. The same pattern was seen for the globins, but was even more pronounced; the random evolutionary-hit method resulted in more apparent mutational steps for every distant comparison except those between the α and β globins.

Both these methods revealed what these authors considered to be significant 'fine structure' in the evolutionary tree—though the two methods did not always give concordant results. I have examined 26 comparisons of the rate of globin evolution in different branches from a common ancestor (Holmquist et al., Fig. 5, p. 56), and find that seventeen are concordant for the two methods (that is, one branch shows more mutational steps than the other by both methods), seven are discordant and two cannot be compared because one method gives the same number of substitutions in the two branches. These differences between lineages are not dependent on assumptions about the time of branching, since the two branches are determined relative to a third reference sequence in each case. Using the sign test, I calculate that the concordance between these two methods is not quite significant at the .05 level.

Thus all three methods have their difficulties, and it may not be possible to obtain a universally agreed-upon approach to the question of uniformity of the evolutionary rate. But it seems that all the workers in the field now agree that over short periods of time and in different lineages the rates of evolution can be markedly different—though the level of significance in each case is problematical and more data are needed. Fitch (1976) and Wilson et al. (1977), however, continue to argue that over very long periods of time the rate of the molecular clock is fairly constant, and here they disagree with the other two groups of workers. By extrapolating back, particularly from the cytochrome *c* data, these groups find that the very early eukaryotic ancestral divergences—between plants and fungi, metazoans and protozoans, etc., took place, according to both the stochastic and the augmented parsimony methods, between two and three billion years ago—even though the fossil record appears quite barren of eukaryote remains before about 800 mya according to the reassessment of the evidence by Knoll and Barghoorn (1975). Is there a possibility that eukaryote lineages are much more ancient—more than twice as ancient—than we have records for, during most of which time they left no trace in the fossil record, or is there an alternative possibility that just before the Cambrian period there was a tremendous burst of molecular evolution which explains why so many substitutions have to be postulated in these distant lineages? The fact that we can now formulate this question means that answers to it may not be far off.

The influence of generation time

Despite the evidence of some irregularities, the consensus seems to be that the *average* clock is remarkably constant—at least among the amphibians, reptiles and mammals. Wilson et al. (1977) have recently summarized the evidence that the evolutionary clock is constant among divergent groups of organisms and dependent for its rate not on generations but on sidereal time. Their Fig. 5 (p. 596) summarizes information on the rate of evolution of eleven proteins. The rates for long-generation organisms are plotted on one axis, and those for short-generation organisms on the other. Despite a considerable scatter, there is no significant deviation from a slope of one, suggesting that generation time has no consistent effect.

This is remarkable on two counts. The first is that generation times do vary widely among organisms, with the smaller rodents having several generations a year and the larger animals such as primates, elephants and whales having generation times of several years. Further, it seems likely that population sizes in the rodent lineage will have been on the average considerably greater than sizes in the insectivore–primate lineage. In view of the strong dependence of the rate of fixation of neutral genes on mutation rate and secular changes in population size, why should the clock be apparently constant in these very different groups of organisms?

The problem is compounded by the fact that the rate of spontaneous mutation is (so far as it can be measured accurately) a function of generation time rather than of absolute time. The rates for spontaneous forward mutation per locus per generation have been summarized by Propping (1972) and by Vogel and Rathenberg (1974). They range from 10^{-5} to 10^{-6} for both *Drosophila* and mouse, and from 10^{-4} to 10^{-6} for man, even though the human generation time is 75 times as long as that of the mouse, and perhaps 2–300 times as long as that of *Drosophila*.

The increased mutation rate and short generation time of rodents compared with primates should lead to an increase in the rate of fixation of mutational changes if they are neutral. The fact that the effect, if there is one, is very small suggests rather that there is a strong tendency for the rate of molecular evolution to be stabilized despite order-of-magnitude changes in the variables that might be expected to determine the rate.

The estimates of constancy of molecular evolution rates are only as good as the estimated times of divergence. Consider the time back to a common ancester for man and chimpanzee, which Romero-Herrera et al. (1973) note have been estimated to lie anywhere between 5 and 38 mya. It should be remembered, in assessing the relative likelihoods of these figures, that even the smaller estimate is not impossible. There is no proof, for example, that *Homo erectus* is really our direct ancestor, and no fossil record has yet been found for the great apes. These authors placed the chordate–polychaete divergence somewhere in the Cambrian period, 516–590 mya, but the recent discovery of vertebrate remains in the upper Cambrian (Repetski 1978) casts some doubt on this estimate, too. In view of these very large uncertainties, which in some cases are almost as large in absolute magnitude for recent events as for those in the more distant past, one is driven more and more to the conclusion that any observed constancy in evolutionary rates is likely to be a statistical accident. As Stebbins and Lewontin (1972) have noted, even a neutralist model must take into account the uncertainties in dating, generation time and mutation rate, compounded by the impossibility of determining the effect of secular changes in average population size.

If we consider very long periods of time, the best selectionist explanation of approximately constant evolutionary rates may be some variation of Van Valen's Red Queen hypothesis (1974). Van Valen suggests that much evolution may result from a constant deterioration of the environment with respect to the population of organisms—so, like the Red Queen in Looking-Glass Land, the species must move as quickly as it can just to stay in the same place. An elaboration was proposed by Zuckerkandl (1976). In his view, proteins are always slightly suboptimal for one or more functions. An improvement in one part of the molecule renders other parts less optimal in function. There is a seesaw approach to, but never quite an achievement of, optimal function. Such mechanisms seem much more likely to generate a constant evolutionary rate than does neutral substitution, with its strong dependence on mutation rate and generation time.

As information on protein sequences accumulates at an ever increasing rate, we are now beginning to see, at least in outline, how selection has shaped particular protein molecules. It comes as no surprise that the perceived role of selection has continued to grow. Some of this evidence will be considered in Chapter 4.

The distribution of allele frequencies as an indicator of selection or neutrality

The examination of distributions of allele frequencies in populations will by itself never provide evidence for or against the classical or balance models of population structure. This conclusion was painfully arrived at by Lewontin (1974*a*), who showed that drift, migration and mutation can, with a few appropriate assumptions, produce any array of allele frequencies without the need for postulating any selection. Some of the assumptions one is forced to make in order to do this may seem unlikely, but that is, of course, a purely subjective judgement.

Polymorphism frequency and heterozygosity

One of the strongest neutralist arguments, and the most difficult to counter, was produced by Yamazaki and Maruyama (1972). These authors showed elegantly that neutral theory demands that the fraction of the total heterozygosity contributed by alleles of a particular frequency class will be a constant. Most neutral alleles will be rare, and each will contribute a small amount to total heterozygosity. The few that are at intermediate frequency will each contribute far more to total heterozygosity. On a neutral model or a frequency-dependent one in which the rarer allele has an advantage, the distribution of allele frequencies is U-shaped, with a few near the middle and many near fixation. Selection against harmful alleles produces a further dip at intermediate frequencies. Selection for overdominant alleles should produce a bulge in the center of the distribution, since alleles at intermediate frequencies will tend to accumulate. The data available at the time followed the neutral expectation with remarkable exactitude. Later, Yamazaki and Maruyama (1974) found that blood-group data from man did not follow this expectation. Blood groups showed a distribution to be expected from an overdominant model (Fig. 3.4).

In fact, the expectation on the basis of an overdominant model is extremely different from what is observed, much more so than Yamazaki and Maruyama suggested, since they did not take finite populations into account. In finite populations, balanced polymorphisms with intermediate equilibrium frequencies, say at about 0.5, should accumulate to very high levels, orders of magnitude greater than those with equilibrium frequencies near fixation. This problem will be commented on extensively in Chapter 7. When the data are plotted in this way they indicate only the mildest amount of heterotic selection, and only in a restricted subset, the human blood groups. In Chapter 5, the possible

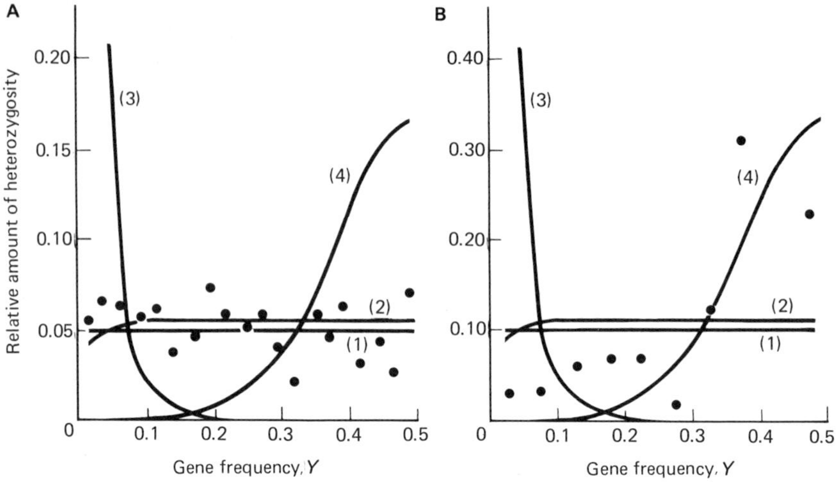

Fig. 3.4.A. Enzyme-polymorphism-frequency data taken from a large number of population surveys of a great variety of organisms. The data are grouped into gene-frequency intervals of 0.025, and the relative amount of heterozygosity contributed by the polymorphisms within each interval is plotted. The solid curves show the theoretical expectations calculated by Yamazaki and Maruyama (1974), which assume infinite population size for (1) neutral, (2) advantageous, (3) deleterious, and (4) overdominant polymorphisms. **B.** The same type of plot is given for human blood-group polymorphisms.

mechanisms for the maintenance of human blood groups will be discussed at length. It is interesting, and highlights the difficulties inherent in the study of selection, that no evidence for heterozygote advantage exists at those loci other than the Yamazaki and Maruyama plot.

The proponents of the neutral view have found it necessary to postulate that another type of selection may be important in order to fit the data. This was done in order to provide an explanation for the fact that the average heterozygosity in very large populations is lower than it should be on the assumption of neutrality (Ohta 1973; see p. 34. This type of selection is *purifying* rather than *balancing*. That is, it is selection against harmful mutants rather than selection for intermediate frequencies as a result of a balance between selective forces.

We have seen how purifying selection might operate in our examination of molecular evolution; parts of molecules are not available for evolution at the present time simply because any changes would critically affect their functions, so that any mutant is rapidly removed from the population. This undoubtedly happens, and indeed entire molecules, like some of the histones, are thus sequestered from evolution at the amino-acid level. Other regions of the genome show evidence of strong selective pressures to preserve particular sequences, or particular patterns such as those observed in nonrandom codon utilization (see Chapter 4). But is such purifying selection the whole story?

The next great battle of population genetics will be to decide whether most selection is purifying with the occasional substitution of a favorable allele, or whether there is still room for an appreciable role to be played by balancing selection. This important question will be dealt with in Chapter 7.

The role of cryptic alleles

Much intellectual effort has been expended in attempting to fit the patterns of variability observed in natural populations to neutralist or selectionist models. As has been seen, the average heterozygosity of a population accumulating neutral mutations can be quite high, and for a product $4N_e\mu$ greater than three or four the great majority of loci in the population will be heterozygous for neutral alleles—provided the population is at equilibrium for mutation and drift. There is thus no problem, at least in populations of large size, in accounting for all the observed heterozygosity. Once the product $4N_e\mu$ becomes too large, however, an equilibrium population will have more heterozygosity than has ever been observed. For large populations such as the Neotropical Drosophila so extensively investigated by Ayala (1975, for a survey), something must be limiting heterozygosity to the observed levels of about 20% rather than the levels close to 100% that might be expected on the neutral model.

It seems unlikely that recent advances in the technique of electrophoresis or eventual gene sequencing will do more than double the estimated average heterozygosity for a species, particularly since 'cryptic' alleles which can be detected only by refinements in electrophoretic methods, appear not to be present at some loci. Perhaps the observed level may be raised to 40% rather than the 100% expected for very large populations. Singh et al. (1976) and Coyne (1976) have surveyed the xanthine dehydrogenase loci of *D. pseudoobscura* and *D. persimilis,* using slight modifications of the technique of acrylamide gel-electrophoresis in an attempt to determine at least the majority of the 'cryptic' alleles that may be masquerading as the same electrophoretic alleles. The modified techniques, which consist of alterations in pH and gel concentration first employed in a population survey of Colias butterflies by Johnson (1976), have met with what appears to be great success. A large number of distinguishable alleles, most of them rare, have been found, though not all have been demonstrated to be true alleles at the locus in question. Because most of the new alleles are rare, however, the impact on levels of heterozygosity at this locus is relatively minor, resulting in less than a doubling. For *D. pseudoobscura* the heterozygosity at this locus is raised from 0.439 using 'standard' techniques that reveal eight alleles to 0.628 for extended techniques that reveal 37 alleles. The fascinating information to emerge from this survey is that *D. pseudoobscura* and *D. persimilis* appear to share only three rare alleles at this locus, and that an isolated population *D. pseudoobscura* from Bogota, Colombia (raised to subspecies status by Ayala and Dobzhansky 1974) is also highly differentiated from the main population.

More recent work has shown that the xanthine dehydrogenase situation may not be the rule for all loci. While Cobbs and Prakash (1977) showed that the esterase-5 loci of *D. pseudoobscura* and its relatives have a large number of cryptic alleles, Beckenbach and Prakash (1977) found that application of the same range of techniques to the hexokinase loci of *D. pseudoobscura* and *D. persimilis* revealed no detectable cryptic variation. This tremendous difference between loci might at first sight be attributable to selection (but see below). It would also appear at first that balancing selection maintaining large numbers of rare alleles is unlikely; this is because of the difficulty of maintaining a large number of alleles through heterozygote advantage. There are ways of circumventing this difficulty, however, as will be seen on pp. 160–82. Further, there are two well-documented cases in man (GGPD deficiency and the β-thalassemias) in which large numbers of alleles are maintained by known selective forces (see Chapter 5).

The new extensions of electrophoretic techniques are being used to investigate phenomena at the very limits of detectability, with consequent problems of interpretation. It is not clear, because it has not yet been possible to perform genetic tests on a large number of these new alleles, whether some or even a majority of them may not be the result of variation at other loci responsible for post-transcriptional modifications of the enzyme that is being examined. Dominant alleles at modifier loci should result in an apparent deficiency of heterozygotes in population surveys, a phenomenon found by Johnson (1977) in surveys of Colias butterflies. It is quite possible to imagine unlinked modifier loci which covalently alter amino acids that are on the surface of some proteins but not on others, explaining without recourse to selective differences the very different patterns of variability emerging at the xanthine dehydrogenase and hexokinase loci of *Drosophila.*

The ultimate answer will come from sequencing a large number of genes from a single population, a process which can be expected to reveal the extent of variability at nearby regulatory sites and in intervening untranslated sequences as well as in the structural gene. Inherited variation has already been found by mapping the sites of cleavage of restriction enzymes in the spacer regions of ribosomal DNA in *Xenopus* (Reeder et al. 1976). In the meantime, it should be noted that it is possible for a locus to be monomorphic at the amino acid level. Doolittle et al. (1970) have examined the A and B fibrinopeptides of 125 humans in a survey of variability, and while these peptides are very short (16 and 14 amino acids, respectively) it is instructive that they found none. At these regions of the fibronogen locus, in humans at least, there appears not to be a wealth of hidden variability at the amino acid level. Fibrinopeptide B in gibbons, however (13 amino acids), has been shown to be polymorphic at one position in a small sample size (Mross et al. 1970).

Watterson (1978) has applied a statistical technique of Ewens (1972) to the data of Coyne (1976) and of Singh et al. (1976). Ewens (personal communication) had been reluctant to do so until extensive and reliable data became available, because his technique is based on the requirement that essentially all mutations at the locus produce uniquely detectable alleles. Any cryptic variability, particularly if it tends to be associated more with certain easily detectable alleles than with others, will greatly reduce the power of the test.

The test is simply an examination of the distribution of alleles at a locus. Given a certain number of alleles, neutral theory demands that they be distributed in a particular way. The probability that a given distribution of alleles could have been generated by random processes is given by:

$$P(n_1^{\alpha_1}, n_2^{\alpha_2}.\ .\ .\ . / K,n) = \frac{n!}{n_1^{\alpha_1} n_2^{\alpha_2}.\ .\ .\ .\alpha_1!\ \alpha_2!\ .\ .\ .\ .\ |S_n^{(k)}|}, \qquad (3.8)$$

where α_i is the number of alleles each of which is represented n_i times, K is the sum of the αs and $S_n^{(k)}$ is a Stirling number of the first kind and can be evaluated from tables (e.g. Abramowitz and Stegun 1965).

The probabilities that the alleles are distributed in neutral fashion for the Coyne data and for the data of Singh et al. are 6×10^{-7} and 2.3×10^{-9}, respectively. Watterson showed, using a variety of further analyses, that these values for individual probabilities do in fact reflect a very low probability that the two distributions are the result of mutation and drift alone. In both cases, the most common allele is too common, and the number of rare alleles in the Coyne data is also too great.

Watterson is at pains to show that the discovery of a few more alleles, subdividing the most common allele, could make this distribution indistinguishable from that expected from neutrality. On the other hand, small modifications of the neutral model may bring about large changes in the distribution of alleles expected at equilibrium. Because the distribution derived by Ewens is dependent for its generation entirely on mutation and drift, if there are any differences in mutation rate among the different alleles they will be directly reflected in the final allele distribution. An increase of the mutation rate to a particular allele will result in that allele being represented more often at equilibrium than it otherwise would. Now, very large differences in mutation rate within a single gene are already known in microorganisms (see Drake 1970, for review). For example, depending on the surrounding base composition, reversion rates of similar T_4 *rII* mutants can vary over more than six orders of magnitude (Drake 1970, Table 5-1). Such mutation rate differences would, if they persisted, be quite sufficient to account for any sort of allele distribution that is likely to occur in natural populations. Further, Strobeck and Morgan (1978) have pointed out that Ewens' model ignores the possibility of intragenic recombination. This can be thought of as a kind of mutation rate which is dependent on the distribution of frequencies of alleles in the population. If two pseudoalleles become sufficiently common in the population, new alleles can be generated by recombination between them at a high rate. For instance, for a two-site model Strobeck and Morgan find that if $4N\mu = 4$ and the recombination rate is ten times the mutation rate, both the mean number and the variance of alleles will be doubled at equilibrium, compared to the case without recombination.

In short, one is stymied at every turn. Since we are quite unaware of the sizes of the variables defining the population—effective population sizes and mutation rates to neutral alleles are both quite unknown—and because variables such as mutation rate are likely to differ greatly from locus to locus and even within loci, the neutralist model can account for anything—*except* direct evidence for selection.

The constancy of some allele frequencies

An early puzzle emerging from surveys of natural populations, yet to be resolved except on a neutralist basis, continues to haunt the proponents of selection. Prakash et al. (1969) surveyed four natural populations of *Drosophila pseudoobscura* for polymorphic loci. All except those on the third chromosome had allele frequencies so similar to each other that all the samples except those from the isolated South American population at Bogota could have come from the same population. The more recent surveys of cryptic alleles are not, as yet, sufficiently extensive to provide contrary evidence. The most common cryptic allele at the XDH locus tends to be common in all populations except Bogota, although the large number of alleles obscures any other pattern. Prakash (1973) and Prakash and Levitan (1973) found quite a different pattern among easily detectable alleles in different samples of *D. robusta,* with only a few of the loci as invariant in frequency as those of *D. pseudoobscura.* Alleles at the majority of loci showed complex clinal patterns or were geographically restricted in their occurrence. Some of these complex patterns can be traced to linkage with inversions found on every long chromosome arm of this fly, but some are quite uncorrelated with inversion frequency.

It is striking that, as Carson (1955) initially observed, chromosome polymorphism tends to decline considerably at the periphery of a species distribution—perhaps, as he suggested, because of the need to release genetic variability on the periphery to adapt to rapidly changing conditions, or perhaps simply because of repeated size-bottlenecks on the periphery of the distribution. But no such decline at the periphery has been observed for enzyme polymorphisms in any species examined (Soulé 1973).

The absence of such a decline, coupled with the observed decline for chromosomal polymorphism, strongly argues that the chromosomal polymorphism decline is not being brought about simply by bottlenecks, but rather by some selective mechanism. It begs the question, however, of why in some species allele frequencies should be relatively invariant throughout the species range.

Suppose, as seems likely, that the variants easily detected by electrophoresis involve discrete charge changes on the surface of the molecule, and that variants more difficult to detect involve amino acid substitutions of smaller effect. If the majority of the latter kind of variant are deleterious and held in the population by mutation pressure, it might be expected that in different regions of a species ranging over a large geographical area as *D. pseudoobscura* does, there would be different rare alleles found in different areas.

No single explanation seems to be satisfactory. If rare alleles are selected against and thus localized in space, then why are the electrophoretic classes due to charge differences not similarly localized? Surely, if small differences are selected against, then large differences involving charge changes on the molecule should be even more subject to selection. If the latter are being selected by balancing selection, then why should such selection be constant in its effects in different parts of the species range? Questions such as these are unlikely to be answered by population surveys, no matter how detailed or extensive.

The variability of other allele frequencies—the case of the Yanomama

Arguments about the amount of genetic differentiation to be expected in different parts of a population depend very much on the time factor. There is simply no way of knowing with assurance how many generations the present populations of *Drosophila pseudoobscura* have persisted—indeed, we do not even have a reliable estimate of the number of generations this organism passes through in a year. There are, however, firmer estimates to be derived from a few human populations, particularly from the Yanomama Indians of South America.

These Indians live in scattered forest villages (though they are now moving to the river banks to take advantage of trade routes) in a region of rain forest about 200 by 300 miles in extent on the border between Brazil and Venezuela (see maps, Fig. 3.5). This region straddles a watershed, and as a consequence movement across this barrier has been somewhat restricted.

There are probably about 10 000 individuals in the tribe as a whole, scattered among roughly 125 villages, some of which have not been visited by outsiders. The fascinating story of their social characteristics, particularly their extreme quarrelsomeness culminating in ritualized fighting, has been recounted by Chagnon (1968). As a result, the village groups have continually divided along tribal and family lines, establishing new villages at a relatively rapid rate. Verbal historical records can be used to reconstruct the sequence in which villages were established, and the approximate times at which the division took place can be gleaned from the recollections of older tribal members and from oral history (Ward 1972). Intensive genetic investigation by Neel and his co-workers has established that this social structure has led to extremely rapid genetic differentiation. (For a recent survey, see Neel et al., 1977.)

It was soon apparent that such differentiation is precisely the sort visualized by Wright (1929*c*) to be conducive to rapid evolution, and indeed Fitch and Neel (1969) calculated that the rate of evolution in the Yanomama and nearby tribes was very high indeed. Assuming that the Indians arrived in this region about 15 000 years ago, they calculated a maximum rate of gene substitution of 130 000 years per substitution per locus. This is about 100 times faster than an average rate based on polypeptides from a variety of mammalian species.

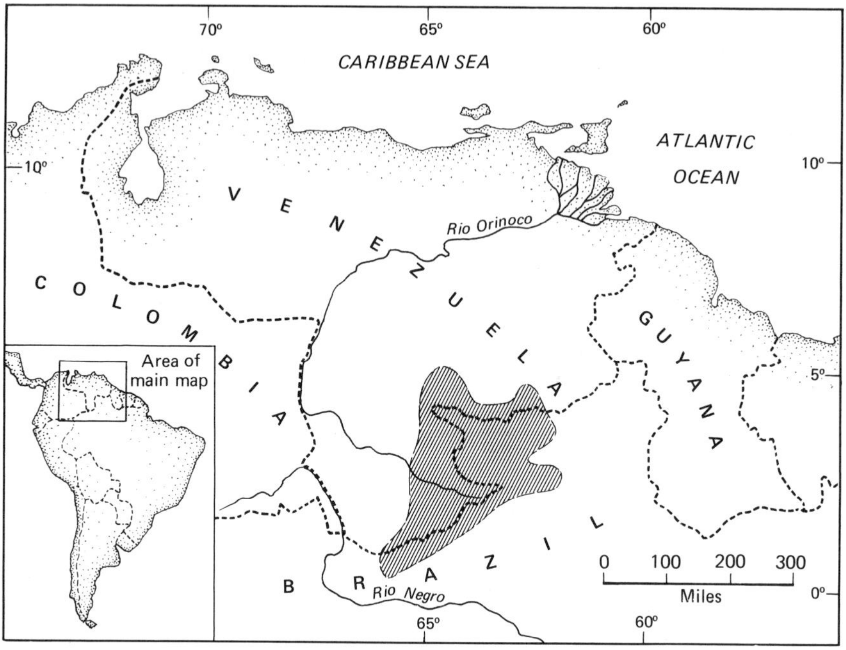

Fig. 3.5.A. Location of the Yanomama Indians in South America. From Chagnon (1968). **B.** Distribution of Yanomama villages in the Venezuelan part of the tribe's range. From Smole (1976, p. 53).

The comparison is presumably fair, since electrophoretic and blood group surveys in the Yanomama also measure differences between polypeptides.

Quite obviously, evolution is simply not proceeding at this rate. The amount of diversity due to drift in this tribe is not a measure of the rate at which evolution is proceeding in terms of allele substitution.

This discrepancy can be visualized as follows. A dispersion index calculated for the Yanomama villages (Neel and Ward 1970) shows that a great deal of genetic differentiation has occurred, because of repeated founder events as new villages were established by small groups. This index is 90% of that calculated for pooled data from the Yanomama and the adjoining tribe the Makiritare and, perhaps most remarkable, half of the value for the human population as a whole (estimated by Cavalli-Sforza et al. 1964, for 15 world populations, on the basis of a somewhat different sampling of polymorphic loci).

The genetic composition of the Yanomama reflects their complex history. There are some unusual frequencies of common alleles, and nine unique alleles present in frequencies greater than 0.01 (Neel and Thompson 1978). One unique variant of serum albumin is sometimes found at frequencies higher than 0.3. But this tribe still shares most of its alleles with the human race at large.

How long have these alleles been free to drift? Spielman et al. (1974) have

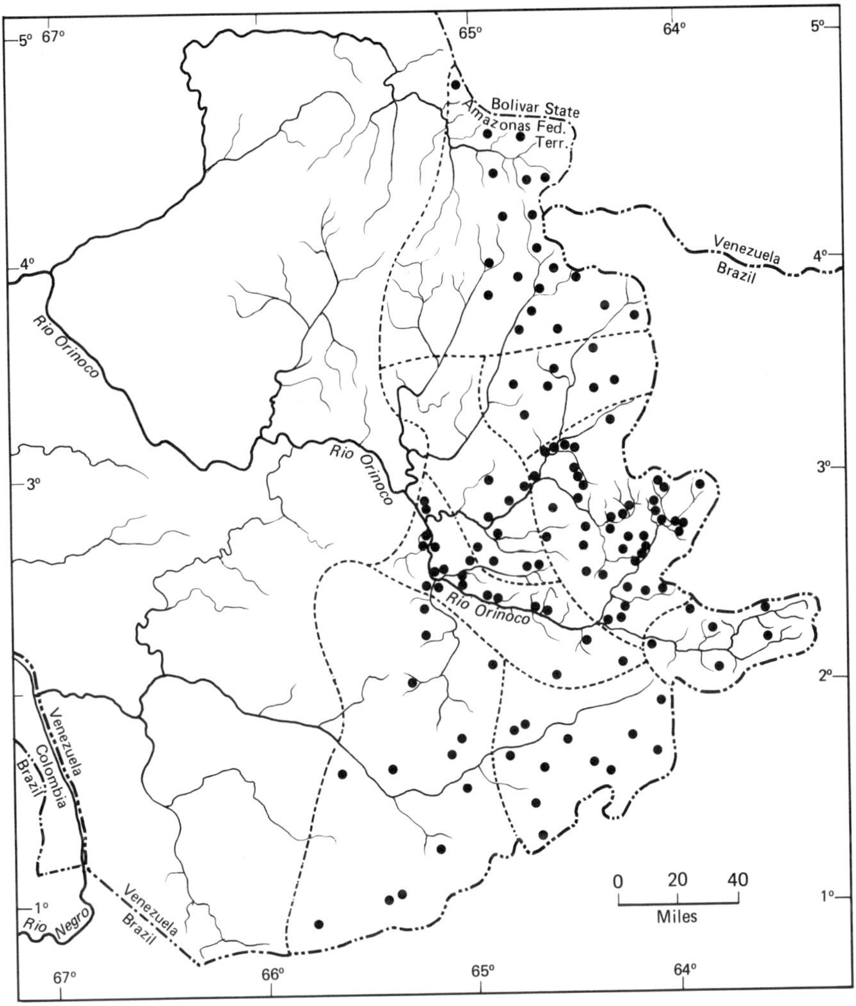

examined language differentiation among the villages, which is correlated (though imperfectly) with genetic distance. There are complications in the analysis: the Indians use a simplified formal *lingua franca* for communication between villages, which must have had a stabilizing effect. And there has been contact with other tribes to the north, reflected in an unusual amount of language differentiation in that direction. But taking these complications into account, and using as a yardstick measures of the amount of differentiation observed among Indo-European languages, these authors estimate that the Yanomama have been differentiating spatially for about 600–1200 years.

Assuming 15 years per generation, this is a mere 40–80 generations. During that time the Yanomama have generated half the variance of gene frequencies observed for the human population as a whole. It is, of course, not clear how much genetic differentiation there was among the founders of the Yanomama, but it seems likely that these consisted initially of one or a small number of groups, and that most of the differentiation occurred subsequent to the founding of the tribe.

Of course, it may be argued that the Yanomama are an unusual case—that their population structure lends itself to extremely rapid genetic differentiation, which is certainly true. But it seems likely that a similar, though perhaps not so extreme, subdivision of population structure characterized the human race throughout much of its history.

There are two possible explanations for this discrepancy in apparent evolutionary rates. The first is that selection is acting to stabilize the allele frequencies at these polymorphic loci—drift is free to generate diversity up to a point, but beyond that point selective forces exert a stronger and stronger influence, presumably through frequency-dependent selection (p. 178), preventing them from straying too far from their selective equilibrium. It is striking, in this connection, that addition of data from the adjoining tribe, the Makiritare, adds very little to the dispersion index, though there is little evidence of an interchange of genes between these two groups during the time that the Yanomama have been drifting. But of course the second interpretation is that such migration has gone on, to a lesser extent (because of local historical reasons) between these two tribes than between outlying members of the human race as a whole, and that this migration has over long periods of time tended to average out the allele frequencies. As pointed out by Wright (1943), it takes very little migration to accomplish this.

Yet we shall see later (pp. 113–33) that some of the polymorphisms investigated in the Yanomama (e.g. Rh and the haptoglobins) are subject to selection. Further, the selection seems likely to be strong enough to overwhelm the effects of drift in a large population.

In this case, as in the others in this section, we are faced with the problem that allele frequencies can be explained in any of a variety of different ways, combining the phenomena of selection, drift and mutation. To put it bluntly, we simply need more hard data concerning the effects of selection. We have models enough.

Neutrality v. selection or purifying v. balancing selection?

This chapter has detailed some of the reasons for postulating a large class of neutral or nearly neutral alleles present at appreciable frequencies in populations. Some of the reasons, as we have seen, are directly related to the consequences of the multiplicative selection model and seem less compelling when other selective models are considered. Others rely on evidence from paleontology and molecular evolution, still others on the distribution of alleles found in present-day populations. None of the neutralist postulates can be proven directly. The selectionist must chip away by direct experiment and observation at the very large class of polymorphic alleles for which no selective differences are known, in the hope that if a sufficiently large number of them, in both the translated and untranslated parts of the genome, can be shown to be affected by selection, then the rest can also be assumed to be affected by default. There seems every likelihood that the selectionist will succeed in the majority of cases as our ignorance diminishes.

In reality, however, the neutralist–selectionist controversy has tended to mask an underlying schism in population genetics. This schism has its roots in philosophical differences between workers in the field about how complex the process of evolution really is.

The extreme positions are simply stated. Do the genetic variants 'seen' by natural selection in a given population behave predictably over long periods because their average properties change very little? Or are the majority of these variants subject to constantly shifting selective pressures which change in sign as well as magnitude, their behavior a complicated function of their frequencies in the population, environmental conditions, and the presence and frequencies of variants at other loci? We can call these the 'simple' and 'complex' models of population structures.

The tide now seems to be turning against a strictly neutralist interpretation of the majority of genetic variability, but it is being replaced by a view that selection is widespread but predominantly purifying rather than balancing. This view still falls in the 'simple' category. Indeed, purifying selection must be widespread and powerful. In the next chapter we will examine new evidence that purifying selection acts on many parts of the genome that were hitherto thought to be selectively unimportant. But the situation appears to be much more complex than that. Chapter 5 examines the great variety of balancing-selection mechanisms that also appear to be operating, deferring until Chapter 7 the fundamental and difficult question of how much selection is strictly purifying, how much is balancing, and how much may be a combination of the two. Even leaving balancing selection aside, the 'simple' model has become progressively more and more complex as strict neutrality is supplemented by purifying selection. To this must be added the likelihood that epistatic interactions will reduce the effectiveness of purifying selection at individual loci, allowing harmful alleles to build up to high frequency. The differences between the two viewpoints will continue to be blurred.

4. New molecular evidence for selection

Selection must be more powerful than conventional population-genetics theory permits if it is to influence all parts of the genome

It was suggested in Chapter 3 that an important observation, leading to the assumption of a large role for neutral mutations, was that only a small part of the genome appeared to code for gene products of indispensable function. Constraints imposed by inbreeding and other factors meant that if selection were acting on the rest of the genome, it must be of a different type from the strictly multiplicative selection that acts on lethal and severe subvital genes. In this chapter I will consider a variety of lines of evidence that selection—and almost certainly some form of truncation selection—is operating in the shaping of both gene products and of other regions of the genome the function of which is not yet fully understood.

The evolution of molecular function

We really know very little about the ways that enzymes and other protein molecules interact with their environment within the cell. In many cases, we are not even clear about the major function of a well-studied enzyme [e.g., liver alcohol dehydrogenase, which has had attributed to it several functions, only one of which is the oxidation of ethanol (Björkhem 1972)]. It is thus altogether too early to presume, as does Nei (1975), that regions of a molecule not directly concerned with a known catalytic function have no importance beyond, perhaps, a general maintenance of the shape of the molecule.

Two groups of workers have recently investigated well-known molecular lineages for evidence that changes may be correlated with function. The molecules investigated are the hemoglobins and insulin, the functions of which are thoroughly—but by no means completely—understood. A survey of some of the known functions of these molecules will illustrate the many ways in which it is possible for selection to act on them.

The hemoglobins

Hemoglobins are globular molecules with a heme prosthetic group. These globins readily multimerize in higher vertebrates to form tetramers made up of two pairs of slightly different chains. Some hemoglobin subunits can also form tetramers made up of four identical chains, but normally do so to an appreciable extent only when in certain pathological conditions there is an excess of one or another subunit. There are, in man, at least ten different globin genes (including at least one for the related muscle protein myoglobin), and all code for sequences that can be shown to be ancestrally related to each other. The most likely genealogy for the globins, therefore, is through a series of duplications, followed in some cases by translocation to different parts of the genome. Polyploidy may also have played a role. The likelihood that this has been the sequence of events is reinforced by the fact that the most closely related genes (those coding for the β, δ, and γ chains) are in man still genetically linked to each other, while this complex is not linked to the more distantly related α genes. The times during which particular genes are translated and the relationship of the different chains to each other are shown in Fig. 4.1.

The primary function of the hemoglobin molecule is to transport oxygen from the lungs to the tissues—from regions of high oxygen tension to those of

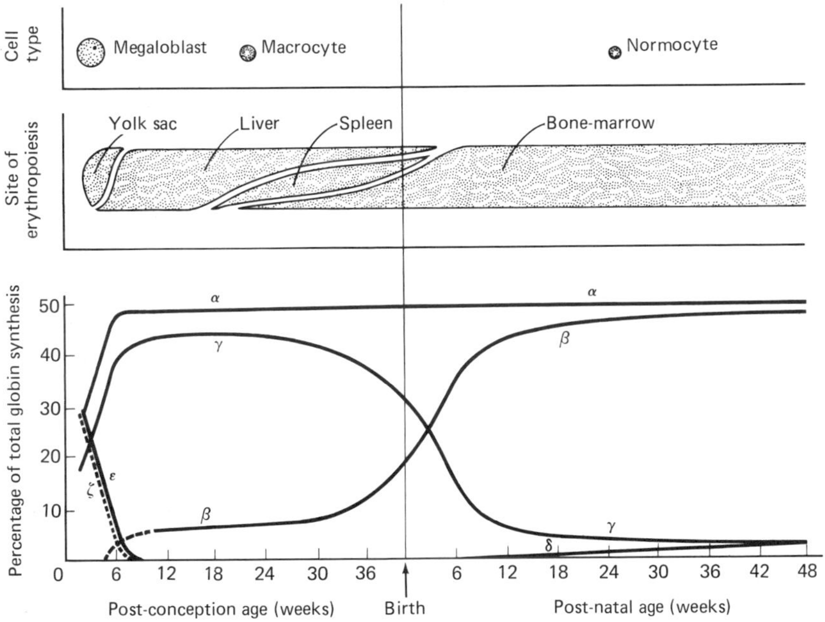

Fig. 4.1. Changes in erythroid cell type, site of erythropoiesis and globin-chain synthesis during human development. This sequence results in the predominant hemoglobin types being Hb Gower-1 ($\zeta_2\epsilon_2$), Gower-2 ($\alpha_2\beta_2$), fetal ($\alpha_2\gamma_2$). and adult ($\alpha_2\beta_2$). From Wood (1976, Fig. 1). After Huehns and Shooter (1965) and Kleihauer (1970).

low tension. It also plays a highly important direct and indirect role in the transport of carbon dioxide from the tissues back to the lungs. Because of the extremely detailed X-ray diffraction studies of both oxygenated and deoxygenated forms of hemoglobin, chiefly on the α and β subunits but also to a more limited extent on the γ subunit of fetal hemoglobin there is an excellent understanding of just how these functions are accomplished. This is not the place for a detailed discussion of the mechanisms—for a very readable review see Lehmann and Huntsman (1974)—but the important point from the standpoint of the present discussion is that many parts of the molecule are involved.

The overall structure of the globin portion of the molecule is generally spherical, with a groove or slot into which a heme moiety is inserted and noncovalently bound. The interior surfaces of this groove are chiefly made up of hydrophobic residues, which provide a 'tunnel' through which the oxygen molecule can pass into the interior of the globin.

The bioconcave disc shape of the red blood cell (RBC) is ideally suited for such diffusion, which must take place during the fraction of a second the cell spends in the capillary bed of the lungs. It has even been suggested by W. H. Crosby (quoted by Lehmann and Huntsman 1974) that the hemoglobin molecules, which are free to rotate inside the cell, might pass the oxygen molecules rapidly into the interior of the cell through this rotational action. This would imply that the external surface of the hemoglobin molecule, governing its ability to rotate freely, is very important in determining the efficiency of diffusion. Once the oxygen molecule has diffused in, it is bound by a reversible coordinate bond, the details of which are not fully understood, to the iron of the heme. During this process the heme is kept in the reduced state by bonding to a histidine. Free heme is easily oxidized irreversibly; a major function of the globin is to keep it in a reduced state so that the coordinate bond can form. Thus, hemoglobin carrying oxygen is oxygenated rather than oxidized.

There are two major properties of hemoglobin molecules, exhibited to different extents by different hemoglobins, which determine their efficiency as an oxygen carrier. These properties are now at least partially understood in terms of structure.

The first is the *Bohr effect*. Dissolved carbonic acid builds up in the red blood cell as it passes into oxygen-poor tissues, and the resultant drop in pH causes oxygen to be released more readily. This is the so-called alkaline Bohr effect, and it continues to operate down to pH 6, which is never reached under physiological conditions. Below pH 6, the affinity of hemoglobin for oxygen starts to rise again, the 'acid' effect. Much of the Bohr effect can be destroyed by mild digestion with protease, and it was found that three-quarters of it can be attributed to two residues: the N-terminal valine of the α subunit and the C-terminal histidine of the β subunit (Perutz et al. 1969). The residue or residues responsible for the fourth quarter of the effect have not been determined with certainty, but a number of other residues in both chains interact with or have the potential of interacting with these sites. The Bohr effect is exhibited only when certain unlike monomers, such as the α and β monomers of the adult hemoglobin, form tetramers.

The second property is the *Hill effect,* or cooperativity. Cooperativity in the hemoglobin molecule is achieved as a result of asymmetry in the binding of the α and β subunits. This cooperativity, which is of the type known as *positive,* can be illustrated by a plot of the binding of oxygen to the molecule as a function of the partial pressure of oxygen (Fig. 4.2). This type of plot is now generally known as a Hill plot (Hill 1910). If the cooperativity is sufficiently strong and uncomplicated by other factors, the degree and direction of cooperativity can be estimated by the size and sign of the exponent to the Hill equation:

$$y = \frac{K_p^x}{1 + K_p^x}$$

where K_p is the concentration of oxygen at which half the sites are bound.

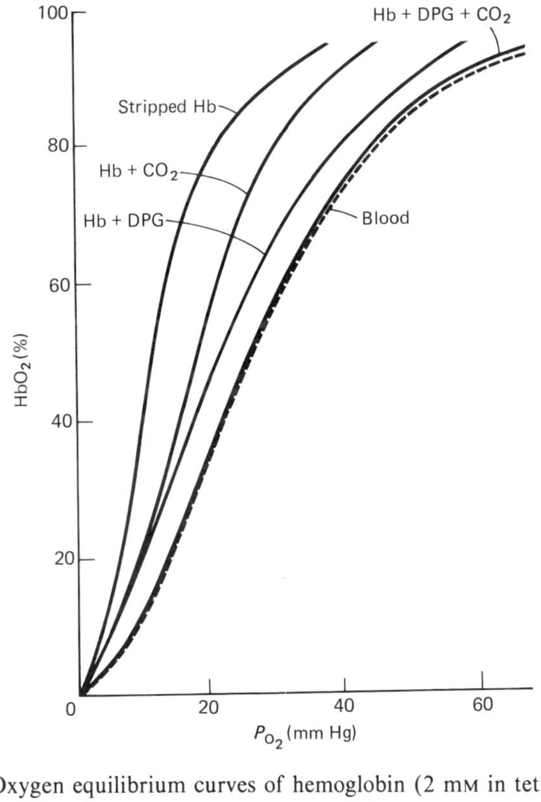

Fig. 4.2. Oxygen equilibrium curves of hemoglobin (2 mM in tetramer) at 37 °C in 0.1 M-KCl. DPG is 2,3-diphosphoglycerate. It will be seen that while the general shape of the cooperativity curve remains unchanged, the binding capacity of the molecule changes dramatically with the addition of CO_2 and DPG. From Kilmartin (1976, Fig. 1). Data from Dahms, et.al. (1972).

A Hill exponent of less than one indicates negative cooperativity, while an exponent greater than one indicates positive cooperativity. For adult human hemoglobin, the exponent is about 2.8, so the cooperativity is very strong. This is a diagnostically useful constant (it varies greatly among different normal and abnormal hemoglobins). Binding of oxygen to one subunit changes the ability of the adjacent subunits to bind to other oxygen molecules. Thus, under ordinary circumstances, the first subunit binds oxygen with a low affinity, but once this binding has taken place, binding to the remaining subunits can occur much more readily. This means that as the hemoglobin moves from regions of low to regions of high oxygen tension it does not immediately begin to pick up oxygen. Otherwise, it would be picking it up in oxygen-poor regions of the capillary bed before it returned to the lungs. On the reverse journey, the high affinity for oxygen exhibited by all the subunits when the molecule is saturated results in oxygen not being given up until the partial pressure of oxygen drops once the RBC reaches the oxygen-poor tissues.

The changes in conformation induced by oxygen binding in adjacent subunits have been investigated in detail with X-ray crystallography by Perutz and his coworkers (summarized in Perutz 1970), and a number of amino acid residues involved in salt bridges have been implicated in these changes of conformation. Another important factor influencing oxygen binding and cooperativity was demonstrated as recently as 1967 (Benesch and Benesch 1967; Chanutin and Curnish 1967). It is the result of the binding of the β subunits of adult hemoglobin to 2,3-diphosphoglycerate (2,3-DPG), which is found in high concentration in mammalian RBCs. The 2,3-DPG molecule is bound between the two β subunits by salt linkages. This in turn has the effect of squeezing together the sides of the heme pockets in these subunits, which effectively prevents oxygenation. The penultimate residue before the carboxyl terminus in the β chain is a tyrosine, and the squeezed state of the molecule is made rigid by the insertion of these residues into pockets between the F and H α-helices of the molecule. When the α subunits, in which the heme pocket remains open, become oxygenated, a conformational change is propagated to the β subunits which increases the probability of loss of 2,3-DPG and the removal of the tyrosine residues from the pockets. Oxygen can then be picked up readily by the β subunits. Once the oxygen begins to be unloaded, however, the original conformation reasserts itself, thus making it very unlikely that the β chain will bind oxygen again once it has been removed, unless the oxygen tension rises again.

In birds, in which the demand for oxygen in the tissues can be extreme, an even more efficient mechanism has evolved. Here the binding molecule is not 2,3-DPG but phytate, the hexaphosphate of *myo*-inositol. This molecule has a very high affinity for avian hemoglobin, greatly aiding the essentially irreversible unloading of oxygen in an oxygen-poor environment.

Cooperativity is a property of multimeric and not monomeric molecules. In hemoglobins it is exhibited by tetramers made up of pairs of unlike subunits.

It appears that such an elaboration of the structure of the molecule is necessary for a reversible reduction of oxygen affinity. Monomeric molecules of hemoglobin or myoglobin have a very high affinity for oxygen, as does tetrameric hemoglobin from which the 2,3-DPG has been removed.

Myoglobin, which is sequestered in muscle tissue and therefore not involved in the physical transport of oxygen, and some of the 'primitive' hemoglobins do not show cooperativity because they are found chiefly as monomers. Lamprey hemoglobin is, when deoxygenated, able to associate as dimers and tetramers, but without detectable cooperativity (Anderson 1971; Anderson and Gibson 1971). The β subunit of adult mammalian hemoglobin can associate as tetramers when synthesized in excess, again without exhibiting cooperativity, but the α subunit can form only dimers and weakly associated trimers. This suggests that the α subunit may have become more highly differentiated during the evolutionary process, since the β subunit seems to have retained more of the properties of a presumptive ancestral homotetramer.

The other extremely important function of the hemoglobin molecule lies in the transport of carbon dioxide from the tissues back to the lungs. It is now realized that, while only about 10% of the CO_2 in the blood is actually involved in this exchange during a given respiratory cycle, the hemoglobin is directly or indirectly involved in about 90% of the exchange which does take place. This transfer is brought about through three general mechanisms.

About half the carbon dioxide available for transport is bound directly to free amino groups of the hemoglobin to form carbamate complexes. It has been shown that more of these can be formed when the molecule is in a deoxygenated state (Roughton and Kendrew 1949).

The remainder is dissolved as carbonic acid in the red blood cell and in the plasma. This would normally have the effect of lowering the pH with disastrous consequences to the organism, but because of the buffering effect of hemoglobin (greatly enhanced because of its high concentration) the pH inside the red cell does not change appreciably. Most, but not all, of the buffering action at physiological pH is brought about by the histidine residues.

Finally, the pH of the plasma is also kept constant because of a shift of chloride ions between plasma and cell. This transfer of negative charges is so pronounced that it actually increases the volume of deoxygenated RBCs and is of course an indirect result of the buffering effect of the hemoglobin molecule. This increase in volume may in turn aid in the diffusion of oxygen among the less closely packed hemoglobin molecules.

When residues involved in stabilizing the tertiary structure, in determining the internal shape of the molecule, and in cooperative and noncooperative interactions between the subunits are counted up, approximately half the amino acids of both the α and β chains have defined functions (69 out of 141 for the α chain and 76 out of 146 for the β chain) (information summarized by Goodman et al. 1975, from work of Perutz and coworkers). Some have more than one function, and in many cases these functions have been confirmed by studies of the effects of particular amino acid substitutions in naturally occurring mutants of the α or β chain (White and Dacie 1971).

It should be apparent from the above that a remarkably large proportion of the molecule is involved in critical functions that we actually know about. This makes it rather surprising that the chief agent of evolutionary change of this molecule has been attributed by some authors to chance—though as detailed in the previous chapter the theoretical reasons for attributing molecular evolution to chance events seemed at the time to be compelling. It is also perhaps reasonable to extrapolate and suggest that an appreciable fraction of the remaining residues in both molecules will turn out to have important functions as the molecules become progressively better understood. Of course, except for a few well-defined cases, the neutralist argument is simply that neutral substitutions can occur freely among residues with the same chemical properties—that an isoleucine can often easily substitute for a valine, for example. Indeed, these substitutions form the majority of those seen during the course of biochemical evolution. But do such substitutions really have no effect? Work has chiefly been done on easily defined properties of the hemoglobin, most of which are so obvious as to be detectable *in vitro*. But how do small changes in the molecules really affect their efficiency? What about the undefined residuum of hemoglobin properties which cannot yet be attributed to particular amino acids? It must be remembered that for most populations of organisms a change in properties sufficient to alter survival on the average by one part in 10^5 or even 10^6 would be quite enough selection to remove a mutant from the neutral category.

Goodman et al. (1975) have considered the detailed knowledge of the structure and function of these molecules in an examination of hemoglobin evolution. Their study is, of course, only preliminary, but as more information becomes available about the details of hemoglobin evolution it is to be expected that evidence for a selective shaping of the hemoglobin molecules will become progressively more convincing.

Interesting relationships are already apparent from their examination of the data. They employ the augmented maximum parsimony tree discussed on pp. 54–55, and ask whether periods of extremely rapid or extremely slow evolution can be correlated with regions of the molecule known to be involved in a particular function. In some cases they cannot. The positions that have undergone the most rapid changes, as in other globular proteins, are the residues on the outside of the molecule. Many but not all of these are polar, and these functions have not for the most part been defined. This therefore seems to be the most likely area for neutral substitution. As we will shortly see, however (pp. 75–78), functions can be attributed to some of the surface residues of the insulin molecule. Perhaps functions as yet unknown will be found for the surface of the hemoglobin molecule.

There are other residues, however, that have been involved in 'bursts' of evolution. In the residues which are involved in heme contacts, there was apparently a seven-times-faster rate of evolution in the preamniotes than in the amniotes, indicating that either this region was more subject to evolutionary pressure or more susceptible to 'noise' before the appearance of amniotes. Also, much evolutionary activity in residues involved in contacts between the unlike chains can be inferred to have taken place at the time of the emergence of the early vertebrates. This certainly corresponds to the period during which the heteropolymer was evolving.

The evolution of the hemoglobins seems to have settled down considerably during the period from the amniote ancestor to the present, except for the external portions of the molecule which continue to evolve at a rate about the average for all residues before that time (about 0.17 nucleotide replacements per position per 100 million years compared with about 0.26 for all positions before that time).

These comparisons appear to be particularly valid because what is being compared is the relative rates of evolution of different parts of the molecule during different periods of time. Thus, while the absolute rates may be open to question because of the uncertainty of the fossil record, the relative rates can be compared because they are not dependent on the absolute time of divergence.

It is interesting to consider these bursts and quiet periods of evolution in the light of the covarion concept of Fitch (1971). The apparently very tight control over which residues are free to vary in a particular lineage, and the changes in the residues free to vary over lineages, may reflect to a greater degree than we currently understand the functional evolution of the molecule.

Insulin

Some 28 complete insulin sequences from a wide variety of vertebrates have now been identified, and the functional relationships of this molecule during evolution have been considered in detail by Blundell and Wood (1975). This story is particularly interesting, partly because correlations between evolutionary activity and the emergence of particular functions have been found for this molecule as well, and partly because the functions of insulin are so totally different from those of hemoglobin. Again, as detailed knowledge of the protein's structure and function becomes available, its evolution becomes progressively more comprehensible.

Insulin is synthesized as a single-peptide precursor, proinsulin, in the B cells of the islets of Langerhans of the pancreas. It is also stored in these cells in the form of granules surrounded by membrane. In most of the animals investigated, the storage form is a hexamer, and in fact free insulin from all mammals except the hystricodonts (e.g. guinea pig, coypu, chinchilla) will dimerize readily and then hexamerize in the presence of zinc. Within the storage granules, a specific endopeptidase removes an internal segment of about 30 residues, converting proinsulin into the active form of insulin. This also enables the insulin to crystallize in the granules, again in the presence of zinc.

Release of insulin from the storage granules is triggered by high blood glucose, and the free insulin then binds to specific receptor sites on a great variety of cells, facilitating transfer of glucose and other metabolites into the cells from the circulation.

Thus, there are a number of functions of the molecule, not all of which have by any means been related to specific features. A general sketch of the insulin molecule taken from the paper by Blundell and Wood (Fig. 4.3), shows the regions of intermonomer, interdimer, and receptor-binding residues, along with the rather small portion of the surface of the molecule that is exposed when it is in a hexameric state. The functions about which something can be inferred include storage, conversion to the active form, and the interaction with a recep-

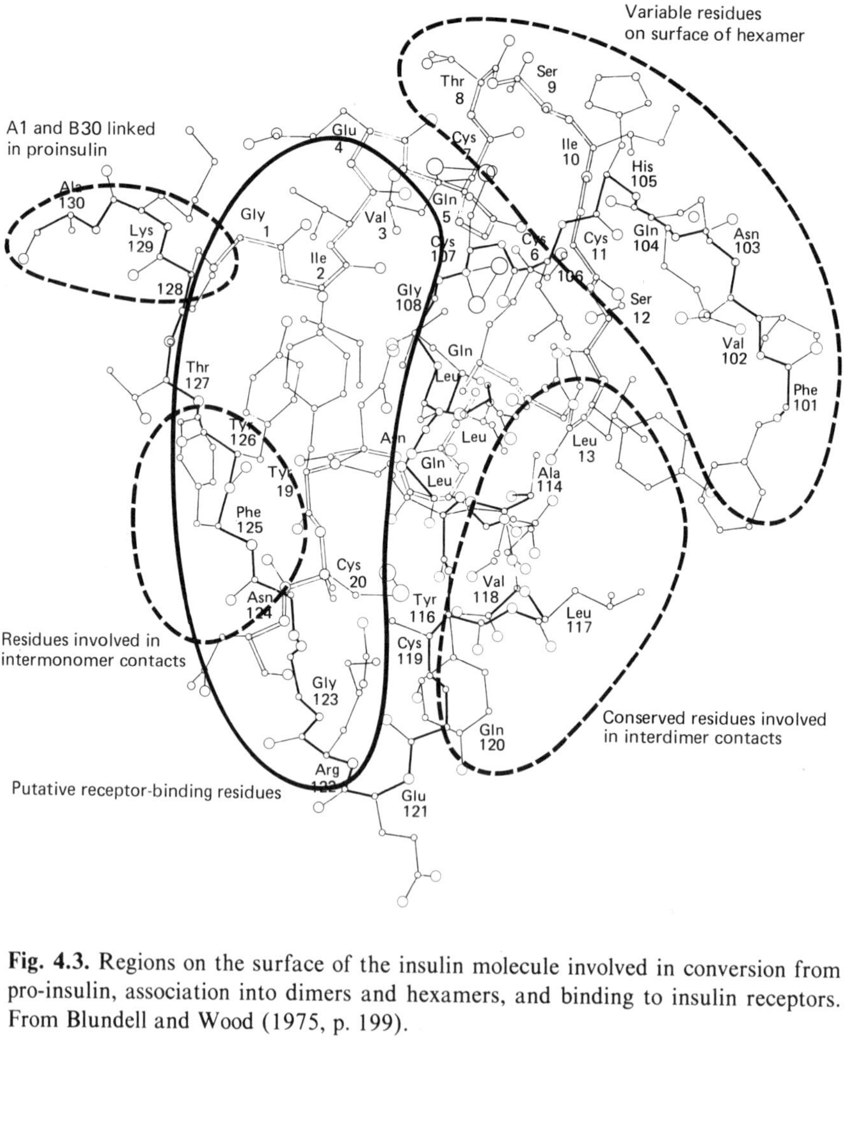

Fig. 4.3. Regions on the surface of the insulin molecule involved in conversion from pro-insulin, association into dimers and hexamers, and binding to insulin receptors. From Blundell and Wood (1975, p. 199).

tor on the membrane surface. The molecule must also be transported to its site of activity, though little is known about how this is accomplished.

As in hemoglobins and cytochrome *c*, the locations of accepted mutations—perhaps corresponding to covarion regions—are very different in different lines of descent. In the fishes, the ability to hexamerize has apparently developed independently of the tetrapod line, since hagfish insulin cannot hexamerize and cod insulin forms hexamers but the resulting crystals have a very different shape from those of mammalian insulin. The change can be traced to a group of related alterations on those surfaces of the molecule which are concerned with the interaction of dimers to form hexamers. The precise effect of all these changes awaits detailed study of hexameric forms.

At some point in the past, hystricomorph insulin has lost its ability to hexamerize. Blundell and Wood speculate that this may have been due to a shortage of zinc in the environment at some point during hystricomorph evolution, though I find this rather difficult to imagine. There are many macromolecules that can multimerize in the absence of zinc, and if hexamerization were essential to insulin's storage function it would surely have been retained even in the absence of zinc. Whatever the reason, these insulins have become more hydrophobic on their surfaces, increasing thermodynamic stability and perhaps making them more resistant to destruction during the storage phase (non-hystricomorph insulin hexamers are far more resistant to tryptic digestion than are monomers or dimers). At the same time they have lost a great deal of biological activity, being only a few percent as active *in vivo* as other insulins. The storage function of insulin must be very important, to permit such a tradeoff of resistance to degradation against loss in activity.

As mentioned earlier, a fascinating aspect of insulin evolution lies in the rapid evolution of surface residues, even though some of these residues can be assigned a definite function and might be expected to be conserved. While much less is known about insulin than hemoglobin, the same sort of story seems to be emerging: the changes in the molecule can begin to be related, no matter how imperfectly, to changes in function. A picture is thus emerging of protein evolution that is very different from the neutralist one, and it is interesting to note that the overall pattern of emerging knowledge is the same as for blood groups and inversion polymorphisms. As more is learned about the systems, selective 'reasons' begin to appear that make the story comprehensible in terms of selection pressures.

Nagging questions still remain, however. Why do different classes of molecules evolve at relatively constant rates over very long periods of evolutionary time? Is there, as Van Valen (1974) suggests, a fairly constant apparent deterioration of the environment which causes organisms to evolve at a particular rate just to 'keep up'? If so, it seems that for hemoglobins, cytochrome *c* and perhaps the dehydrogenases (Rossman et al. 1974) this rate has slowed during mammalian evolution, though not always during primate evolution. And why do molecules with only one obvious function, like the fibrinopeptides, seem to evolve so rapidly? Is this, again, simply a matter of our ignorance? Perhaps we have a lot to find out even about the lowly fibrinopeptides.

The fibrinopeptides

These small molecules comprise the first 20 or fewer residues of the A and B chains of fibrinogen, starting from the amino terminal ends. They are cleaved from the larger molecules by the protease thrombin in the presence of calcium ions. This is one of the first steps in a complex sequence of events leading to the formation of a blood clot. Because thrombin preferentially splits arginine–glycine, the residue at the carboxyl terminus of all fibrinopeptides is arginine. The eight residues adjacent to this arginine are quite highly conserved in fibrinopeptide A, but there is little conservation in the corresponding region of fibrinopeptide B.

These rapidly evolving molecules have proved very useful as a tool in probing the 'fine structure' of molecular evolution, for example in providing molecular evidence concerning the probable evolution of artiodactyls (Mross and Doolittle 1967). Their function, other than simply 'capping' the amino terminals of two of the fibrinogen chains to prevent polymerization, is uncertain. Fibrinopeptide B has been shown to act as a vasoconstrictor in cattle (Colman et al. 1967). The two fibrinopeptides are differentially cleaved by fibrin, with the release of A preceding that of B. Laurent and Blombäck (1958) suggested that removal of A may produce end-to-end linking of the fibrin molecules, and B removal result in crosslinking. Furlar et al. (1976) summarized the evidence that this is so, and showed that thrombin preferentially removes A at pH 6 and B at pH 9 *in vitro*. While A-splitting is essential for clot formation, B-splitting alone will not work.

Despite the rapid rate of evolution of these and other molecules, there are strong indications that the amino acid substitutions seen in the course of this evolution are subject to selective constraint. Bernard et al. (1972) examined the amino acid compositions of hypervariable regions of fibrinopeptide A, ribonuclease and proinsulin C peptide—regions in which all or most of the residues had changed among the six 'species' compared (cow, deer, bison, pig, rat, and turtle or lizard). The expectation on the assumption of neutral changes is that the amino acid composition of these regions should approximate that of 'random' protein generated solely on the basis of the relative frequency of codons for particular amino acids in the code. As Table 4.1, using data taken from their Table 3 (p. 397) shows, this expectation is not fulfilled. The distribution of amino acids in all three hypervariable regions shows highly significant deviations from the neutral expectation. In addition to the deficiency of arginine, which is a general feature of globular proteins (King and Jukes 1969), other highly significant excesses and deficiencies are apparent. There are large excesses of glycine and glutamic acid in the fibrinopeptide A and the proinsulin suquences—which is itself intriguing, since both of these fragments are prod-

Table 4.1. Observed and expected numbers of amino acids in three hypervariable regions of proteins. The expected numbers are generated on the assumption that all sense codewords are equally probable. (From Bernard et al. 1972, p. 397.)

	Fibrinopeptide A	Ribonuclease	Proinsulin *c* peptidase	Expected %
Gly	24 (7)	1 (4.6)	22 (5.9)	7.2
Glu	13 (4.6)	4 (3.0)	13 (3.8)	4.7
Asp	13 (3.5)	3 (2.3)	1 (3.0)	3.6
Ala	9 (5.9)	11 (3.8)	11 (4.9)	6.0
Ser	8 (8.4)	21 (5.5)	2 (7)	8.6
Thr	6 (6.8)	7 (4.4)	0 (5.7)	6.9
Val	6 (6)	0 (3.9)	3 (5.0)	6.1
Phe	6 (2.2)	0 (1.4)	0 (1.8)	2.2
Leu	3 (7.7)	0 (4.5)	14 (6.5)	7.9
Lys	3 (5.4)	2 (3.5)	0 (4.5)	5.5
Pro	3 (4.9)	7 (3.2)	9 (4.1)	5.0
Ile	2 (5.1)	0 (3.3)	0 (4.3)	5.2
Gln	1 (3.8)	1 (2.5)	6 (3.2)	3.9
His	1 (2.9)	0 (1.9)	0 (2.5)	3.0
Asn	0 (4.1)	5 (2.7)	1 (3.4)	4.2
Arg	0 (10.5)	1 (6.8)	0 (8.8)	10.7
Tyr	0 (3.0)	1 (2.0)	0 (2.5)	3.1
Trp, Cys, Met	0 (5.9)	0 (3.8)	0 (4.9)	6.0
$\chi^2_{17\text{ d.f.}}$*	124	94.5	133.4	

*All significant at <0.001 level.

ucts which must be removed from a polypeptide in order for it to function. By contrast, the hypervariable region of ribonuclease shows no excess of either of these amino acids, but rather an excess of serine and possibly alanine. These hypervariable regions, which one might expect to correspond rather closely to neutral expectation, actually show far greater deviations from a random polypeptide than those seen when large amounts of data from globular proteins are tabulated (King and Jukes 1969).

Untranslated and degenerate regions of the genome

The nonrandom use of codons

In the process of being driven from the field of protein evolution, the neutralist has taken up residence in what until very recently has seemed an unassailable citadel. Most of the DNA of higher organisms consists of regions that do not code for gene products or that form a relatively nonspecific part of the transcribed code. If neutral mutations are to be found in large numbers, surely they should be here. Unfortunately, even the limited data now available shows that selection has made its mark in these regions too. At least some of these regions are as conserved during evolution as are proteins.

The genetic code provides an excellent method for determining whether selection has been operating on a particular nucleotide sequence. This is because it is *degenerate*—eighteen out of the 20 amino acids can be coded for by more than one codeword. If all codons for a particular amino acid are equivalent, then in any translated segment of DNA there should not be preferential use of one codon at the expense of another. This prediction is very clear-cut, dependent entirely on the properties of the genetic code, and does not depend on speculations about the function of various parts of the gene product such as those that were made in the previous section. If there are four codons for a particular amino acid, which is the case for seven out of the 20 coded amino acids, then all four should be represented in those proportions expected from the total base composition of the genome, and deviations can legitimately be quantified by χ^2.

Berger (1977, 1978) examined the data available at the time and concluded that the use of codons was indeed nonrandom and not always related to obvious physiological factors. For example, although there was a deficiency of codons ending in A for valine, serine, proline, threonine, alanine and arginine in most of the sequences determined at that time, there is no shortage of transfer RNAs capable of recognizing these code words in most organisms. There is a deficiency of tRNAs recognizing AGA and AGG in *E. coli,* however. Further, contradictory results are actually observed in the same organism. In mammals, the concentration of AUA-recognizing tRNA is low. In the SV40 virus, the AUA codon is common in the A protein and replicate messages, but not found in the coat protein gene (Fiers et al. 1976).

The pattern of use of codons varies dramatically from one organism to another. Messenger RNA's, or entire genomes, have now been sequenced for a variety of organisms with more being added every month. The β-hemoglobin message from rabbits and humans (Efstratiadis et al. 1977) and messages of rat insulin (Ulrich et al. 1977) and the chick ovalbumin (McReynolds et al. 1978) are available. Two especially interesting closely related genes for which the message sequence has been determined are human chorionic somatomammatropin (Shine et al. 1977) and rat growth hormone (Seeberg et al. 1977). More than equaling these in sheer volume of data are the complete sequence of mammalian SV40 DNA (Fiers 1978) and the sequence of polyoma (Friedmann et al. 1979; Deininger et al. 1979). The sequence of the MS2 RNA coliphage, along with partial sequences of the closely related RNA coliphages R17 and f2 (Min Jou and Fiers 1977), are also available. The complete sequence of the two closely related single stranded DNA coliphages ϕX174 (Sanger et al. 1977) and G4 (Godson et al. 1978) round out the list at the time of writing.

I have made comparisons among the pooled mammalian data (excluding the somatomammatropin gene, which is similar in its codon use to the rat growth hormone gene), SV40, MS2, ϕX174, G4 and polyoma. The codon use is summarized in Table 4.2. Expected numbers of codons were then estimated from

Table 4.2. Summary of codon utilization for human chorionic somatomammatropin (HCS), rat growth hormone (RGH), rat insulin (RI), rabbit β hemoglobin (Rab β) and chick ovalbumin (COV) genes, compared with a variety of pro- and eukaryotic viral genomes.

		Eukaryotic					Total (excluding HCS)	Viral				
		HCS	RGH	RI	Rab β	COV		SV40	MS2	ϕX174	G4	Py
Phe	UUU	2	8	1	3	11	21	22	19	71	33	39
	UUC	8	6	2	5	9	22	1	29	37	61	31
Leu	UUA	0	1	0	0	3	4	16	17	32	37	16
	UUG	0	2	1	0	4	7	16	11	50	20	16
	CUU	0	1	2	0	8	11	10	15	67	53	44
	CUC	7	6	3	2	8	17	3	26	22	44	39
	CUA	3	0	0	0	1	1	6	15	11	16	61
	CUG	11	15	8	16	10	49	6	9	35	43	53
Ile	AUU	2	3	1	1	6	11	23	12	75	51	25
	AUC	5	5	1	0	16	22	0	25	22	57	15
	AUA	0	0	0	0	3	3	8	19	7	10	35
Met	AUG	5	6	0	1	17	24	13	20	76	72	53
Val	GUU	0	0	1	4	9	14	24	21	83	51	24
	GUC	1	2	2	2	7	13	1	21	17	36	21
	GUA	0	0	0	0	5	5	11	16	15	26	36
	GUG	4	2	5	12	10	29	18	19	20	30	44
Ser	UCU	2	0	0	3	10	13	16	15	65	86	25
	UCC	5	3	2	3	5	13	4	20	19	27	21
	UCA	1	2	0	0	7	9	6	16	23	40	34
	UCG	4	2	0	0	1	3	0	22	25	21	4
Pro	CCU	0	1	2	3	6	12	21	17	46	48	36
	CCC	2	5	2	0	1	8	8	10	11	19	43
	CCA	1	1	1	1	7	10	15	9	9	21	47
	CCG	1	0	2	0	0	2	0	13	25	16	12
Thr	ACU	1	1	0	2	3	7	25	19	66	59	33
	ACC	5	7	2	2	4	15	15	21	27	46	34
	ACA	3	0	1	0	7	8	16	13	21	38	53
	ACG	2	0	0	0	0	0	0	14	35	19	6
Ala	GCU	0	7	3	7	10	27	45	26	123	97	44
	GCC	2	8	3	6	9	26	6	21	32	39	33
	GCA	1	1	1	1	16	19	12	21	20	34	43
	GCG	0	2	0	1	0	3	0	23	26	16	9
Tyr	UAU	4	4	0	1	3	8	16	9	60	49	39
	UAC	4	3	4	2	7	16	9	32	21	36	30
His	CAU	2	0	0	4	4	8	6	6	30	34	34
	CAC	2	3	2	5	3	13	5	9	11	23	20
Gln	CAA	2	1	3	0	8	12	26	17	48	63	40
	CAG	7	12	5	4	7	28	16	22	62	35	51
Asn	AAU	1	2	0	4	10	16	19	17	73	63	50
	AAC	6	3	3	4	7	16	11	28	43	80	26
Lys	AAA	1	3	1	3	11	18	28	19	91	110	47
	AAG	8	8	3	9	9	29	8	26	56	44	44

Table 4.2 *(Continued)*

		Eukaryotic					Total (excluding HCS)	Viral				
		HCS	RGH	RI	Rab β	COV		SV40	MS2	ϕX174	G4	Py
Asp	GAU	1	1	2	1	8	12	18	28	79	48	60
	GAC	13	9	1	3	6	19	17	22	62	81	46
Glu	GAA	5	4	2	4	18	28	22	16	36	74	65
	GAG	8	10	7	6	15	38	19	28	65	31	64
Cys	UGU	1	2	2	1	5	10	3	6	15	9	27
	UGC	3	2	4	0	1	7	4	6	15	9	16
Trp	UGG	1	1	1	2	3	7	10	23	31	33	32
Arg	CGU	0	2	4	0	0	6	2	21	56	46	4
	CGC	4	6	0	0	2	8	0	20	45	47	10
	CGA	0	1	0	0	0	1	0	10	12	12	8
	CGG	2	1	1	0	0	2	1	11	10	8	16
Ser	AGU	1	1	0	4	6	11	16	8	16	12	24
	AGC	4	6	1	0	9	16	6	6	10	6	36
Arg	AGA	0	1	0	0	10	11	15	7	16	6	56
	AGG	3	1	0	3	3	7	17	6	3	4	27
Gly	GGU	0	1	3	4	3	11	12	37	75	63	30
	GGC	4	3	2	6	8	19	7	16	45	48	31
	GGA	0	1	1	0	6	8	19	12	23	23	52
	GGG	3	3	2	1	2	8	15	16	5	5	36

total base composition of the organism, and χ^2 values were calculated from the 18 amino acids specified by more than a single codon. The χ^2 values and their significances are given in Table 4.3.

Table 4.3 shows some remarkable patterns. ϕX174 and G4 show similar patterns of codon use, although their overall base compositions are quite different and only 67% of their bases are identical. Polyoma and SV40, with 48% identity, show very different patterns of nonrandom codon use. While the first shows nine significant deviations and the second eight, only five of these are in common.

Although ϕX174, G4, and MS2 are all bacteriophages of *E. coli,* the degree of nonrandom codon utilization between the first two and the last is strikingly different. Only 3 out of 18 χ^2 values are significant for MS2, while 11 out of 18 are significant for ϕX174 and 13 out of 18 for G4. These differing amounts of selection are reflected even in the total base composition; all bases occur with approximately equal frequency in MS2, while there is a pronounced excess of T (equivalent to U) in ϕX174 and a pronounced deficiency of G in G4 (Table 4.4). The overt results of selection are far more obvious in ϕX174 and G4.

The pattern of significant values is almost the same in ϕX174, G4 and the mammalian SV40 virus. In most, but not all, cases this is due to an excess of codons ending in U, but a deficiency of those containing CG, apparent in SV40, is not noticeable in ϕX174. The pattern of utilization of arginine codons, significant in both organisms, is very different in each. It is also remarkable that

Table 4.3. Significant deviation from random codon utilization.

df	χ^2 MS2	SV40	Mam	ϕX174	G4	Polyoma
Phe (1)	1.66	9.39†	0.895	2.03	9.45†	0.15
Leu (5)	10.46	10.95	124.2†	38.0†	31.0†	45.2†
Ile (2)	4.0	12.1	23.15†	46.1†	38.7†	6.1*
Val (3)	0.549	7.24*	33.3†	3.54	8.2*	12.73†
Ser (5)	1.28	9.34	6.4	25.6†	95.0†	26.4†
Pro (3)	3.11	13.58†	4.2	23.6†	19.4†	19.1†
Thr (3)	2.20	12.84†	17.0†	15.64†	12.0†	30.6†
Ala (3)	0.821	30.52†	17.3†	87.8†	62.8†	19.9†
Tyr (1)	11.9†	3.58	5.0*	7.6†	1.48	0.28
His (1)	0.475	0.69	2.73	3.4	1.69	2.01
Gln (1)	0.318	0.66	15.0†	2.4	1.81	3.3
Asn (1)	2.23	0.27	0.467	0.75	2.74	4.7*
Lys (1)	0.623	0.7	9.22†	6.97†	11.42†	0.11
Asp (1)	1.01	0.9	3.76	0.52	9.83†	0.44
Glu (1)	2.43	0.94	8.49†	9.52†	6.71†	0.47
Cys (1)	0.006	5.62*	0.06	1.00	0.0089	1.5
Arg (5)	14.1*	25.7†	9.14	88.9†	88.4†	68.5†
Gly (3)	19.2†	2.65	10.66*	52.3†	45.6†	6.6

*Significant at 0.05 level
†Significant at 0.01 level

the nonrandom use of codons is most apparent when there are four or more codons available.

A large number of significant deviations (10 out of 18) occurs in the pooled mammalian data. Pooling of the data was necessary to make the number of codons comparable to those from viral genomes. It will be seen from Table 4.2 that the pattern of codon utilization is similar for leucine and valine among the mammalian messages. Differences seen elsewhere among the various mammalian genes may be significant but concealed by the paucity of data. The significance pattern is different from that of any of the viral genomes, and least like that of the SV40 genome—even though SV40 is a mammalian virus. While again many of the significant values can be traced to an excess of codons ending in U, and to a deficiency of CG (CG is found infrequently in eukaryotic DNA), four significant values are found among the nine two-codon amino acids. It is possible to trace three of these to an excess of codons ending in G over those ending in A, and one to an excess of the codon ending in C over that ending in U.

The data can also be arranged to show the deviations from equal numbers of bases exhibited at the first, second and third positions of the codons in each genome. The total base composition and the deviations from equality at each position are given in Table 4.4. It will be seen that the χ^2 values are highly significant in each case. While the data are not directly comparable except in the case of the ϕX174 and G4 genomes, it is striking that the third position

Table 4.4. Deviations from equal base use. Total base composition was calculated from the coding strand for single-strand viruses and from both strands for the other viral genomes. The mammalian composition is an average of several mammalian genomes. The signs of the deviations from equal base use are shown.

	Position									Total
	1			2			3			composition
	Obs.	Exp.	Sign	Obs.	Exp.	Sign	Obs.	Exp.	Sign	(fraction)
Total mammalian										
U	140	205.25	−	253	205.25	+	198	205.25	−	0.29
C	188	205.25	−	175	205.25	−	250	205.25	+	0.21
A	214	205.25	+	261	205.25	+	144	205.25	−	0.29
G	279	205.25	+	132	205.25	−	229	205.25	+	0.21
$\chi^2_{3\text{d.f.}}$	49.07			56.85			31.04			
MS2										
U	225	267	−	294	267	+	276	267	+	0.248
C	230	267	−	280	267	+	322	267	+	0.259
A	270	267	+	279	267	+	230	267	−	0.237
G	343	267	+	215	267	−	240	267	−	0.256
$\chi^2_{3\text{d.f.}}$	33.4			14.03			19.49			
Polyoma										
U	330	512.5	−	552	512.5	+	538	512.5	+	0.267
C	518	512.5	+	477	512.5	−	452	512.5	−	0.233
A	564	512.5	+	616	512.5	+	593	512.5	+	0.267
G	638	512.5	+	405	512.5	−	467	512.5	−	0.233
$\chi^2_{3\text{d.f.}}$	100.95			48.95			25.09			
SV40										
U	184	199	−	221	199	+	289	199	+	0.311
C	124	199	−	137	199	−	106	199	−	0.19
A	254	199	+	301	199	+	229	199	+	0.31
G	234	199	+	137	199	−	172	199	−	0.19
$\chi^2_{3\text{d.f.}}$	50.75			93.35			92.35			
G4										
U	461	592.25	−	640	592.25	+	802	592.25	+	0.271
C	529	592.25	−	626	592.25	+	659	592.25	+	0.259
A	677	592.25	+	772	592.25	+	511	592.25	−	0.272
G	702	592.25	+	331	592.25	−	397	592.25	−	0.198
$\chi^2_{3\text{d.f.}}$	68.31			173.7			157.32			
ϕX174										
U	464	581.75	−	640	581.75	+	1000	581.75	+	0.313
C	500	581.75	−	573	581.75	−	439	581.75	−	0.215
A	637	581.75	+	737	581.75	+	364	581.75	−	0.24
G	726	581.75	+	377	581.75	−	524	581.75	−	0.233
$\chi^2_{3\text{d.f.}}$	76.34			119.46			422.97			

shows a highly nonrandom distribution of bases. The genomes with the least nonrandomness, MS2 and polyoma, are also those with the least nonrandom use of individual codons. Those with the greatest amount of nonrandomness, ϕX174 and G4, are also those with the most nonrandom use of codons.

Some regularities are apparent from the table. Almost without exception, there is an excess of U and C and a deficiency of A and G in the first position, an excess of U and A and a deficiency of C and G in the second. This reflects the preferential utilization of certain amino acids. The only consistency in the third position is an excess of U, however, and even this is not seen in the pooled mammalian data.

It has been supposed for some time (Fiers et al. 1976) that the third bases of messenger RNA's might be subject to selection if they contribute to secondary structure of the molecule. This may in turn be important in regulation. If this is correct, then it would appear that different secondary structures are being selected for in messages of these different organisms, and that this selection is so strong that it can markedly disturb both the third base and the overall base composition. Such selection might be the source of the very rapid evolution of third bases in sea urchin histone DNA.

The closely related rat growth hormone and human somatomammatropin hormone genes provide another opportunity for investigating the effects of selection. It is possible to align the two genes by the introduction of two deletions in the growth hormone gene between 43 and 48, 96 and 99, and an insertion between 183 and 186. When this is done, 100 out of 186 residues are identical between the genes, as are the majority of the codons for these genes. However, 33 of the 100 codons coding for identical residues differ, reflecting (presumably coincidentally) the proportion of amino acid interchanges between the genes. These codon differences involve 13 of the 18 amino acids specified by more than one codon. We can ask the questions: do the codons that differ tend to be those for particular amino acids? In other words, is the conservatism of codons greatest for certain amino acids? And are the amino acids which show the most conservatism in codon use between the two genes the ones that exhibit the greatest or the least bias in codon usage?

The data are rather sparse, but there is a suggestive trend. Fig. 4.4. shows

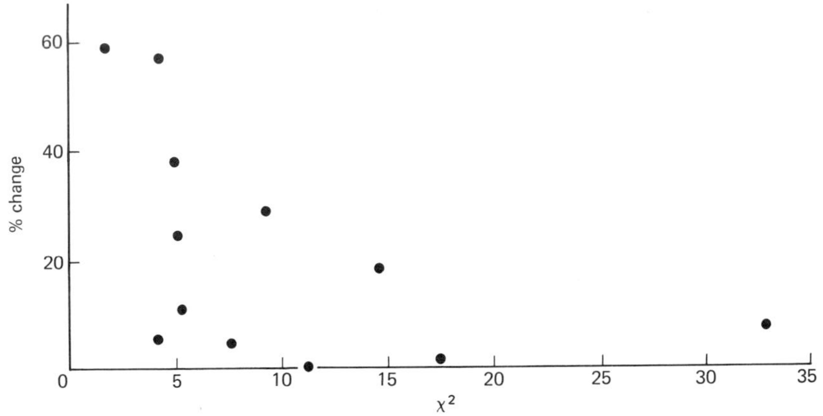

Fig. 4.4. The number of codon differences between the matched regions of human chorionic somatomammatropin and rat growth hormone, compared with the deviation from random codon use as measured by χ^2.

the deviation from randomness of codon utilization, represented by χ^2, plotted against the percentage of codons that have changed between the two proteins. Various statistical compromises have had to be made. Amino acids represented fewer than four times in each protein were not used. The χ^2 values are not directly comparable to each other because of different sample sizes. The expectations for χ^2 were derived from the ratio of U and/or A to C and/or G to be expected from average mammalian base composition, so that each value has one degree of freedom. Finally, the amino acids coded by four or six codons are assumed to have two or three times the opportunity for change that two-codon amino acids have, so the percentage of change was divided by two or three accordingly.

It can be seen from the figure that there is a tendency for sets of codons with high relative rates of change to be associated with low χ^2 values and *vice versa*. From the limited amount of data, it would appear that the amino acids showing the least bias in codon use are also those in which the codons are most freely evolving. This pattern is consistent with strong purifying selection acting on certain codons and much weaker selection or drift acting on others.

Thus, evidence for strong purifying selection acting at every point of translated sequences is now overwhelming. Further, dramatic changes in codon utilization between closely related viral genomes provides *prima facie* evidence for many selection-driven base substitutions.

But what about regions of the DNA which do not code for proteins? There are several points of experimental attack which, in addition to the fascinating light they shed on the function of these DNA sequences, also illustrate that selection has been active here as well.

The nonrandom distribution of satellite DNAs among chromosomes

DNA satellites can be distinguished from the majority of the nuclear DNA by density gradient centrifugation in the presence of antibiotics, such as actinomycin and netropin sulfate, which interact with DNA. The satellites show remarkably uniform short-range periodicity of base sequence, with the repeated sequences ranging from a few bases to a few hundred. Southern (1975) showed periodicities in mouse satellites of about 2000 bases using restriction endonucleases, so that there is a more complex structure to these apparently simple nucleotide sequences as well.

Pardue and Gall (1970) were the first to anneal labelled satellite DNA to chromosomes in which the DNA had been denatured *in situ,* and showed that the labelling concentrated in the centromeric heterochromatin. The satellites can adjoin each other; 'junction' molecules made up of two kinds of satellite DNA were found by Britten et al. (1974). Peacock and his coworkers (summarized in Peacock et al. 1977) have performed the most detailed mapping of these centromeric regions in *Drosophila melanogaster* (Fig. 4.5) and find a specific pattern of annealing characteristic of each centromeric region. Preliminary work with restriction endonucleases as a tool for further characterizing

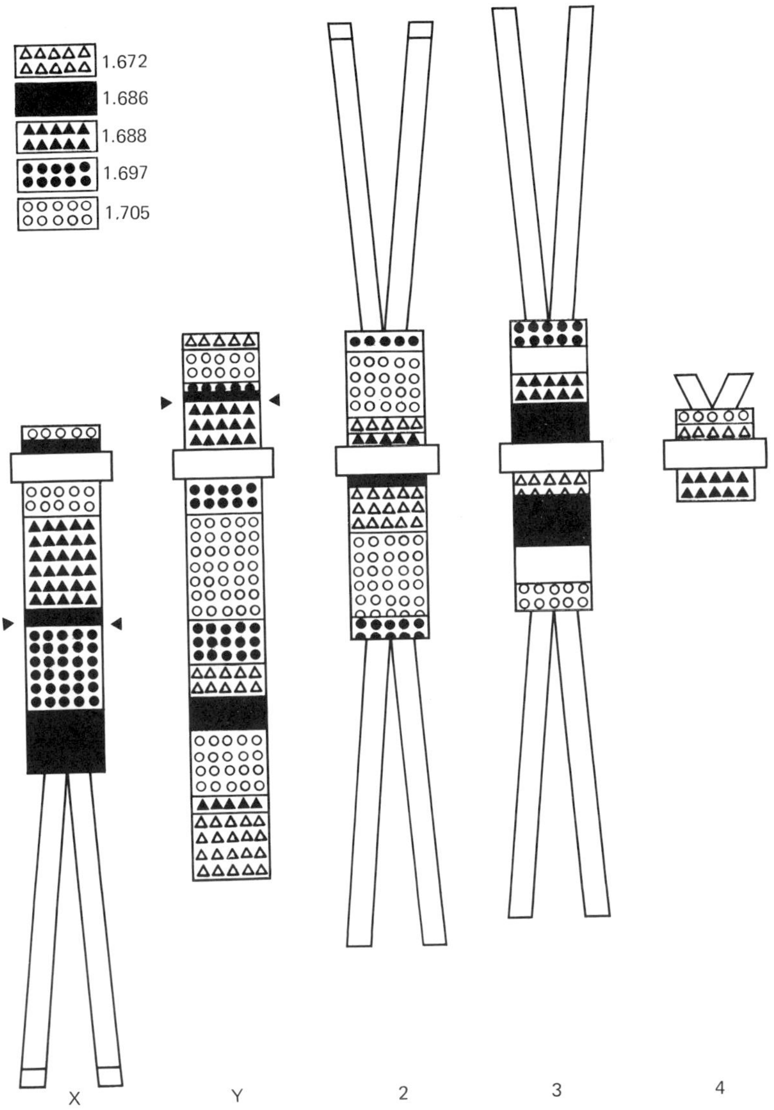

Fig. 4.5. The pattern exhibited by various classes of satellite DNA annealed to *Drosophila melanogaster* centromeric regions. All the classes presently distinguishable by density gradient centrifugation are found associated with more than one centromere, though in widely different amounts. From Peacock et al. (1977, p. 500).

heterogeneities within the satellites leads these workers to expect that even more detailed maps of the distribution of satellites can be constructed in the future. It is striking that apparently identical satellites are found associated with different centromeric regions.

The function of the satellites remains obscure, but because of the importance of centromeric regions in chromosomal recognition during meiosis it was early suggested (Walker 1971; Yunis and Yasmineh 1971) that the satellites have a function in this recognition. This may or may not be because of the precise nucleotide sequence, though it seems likely that specific sequence recognition mechanisms do exist.

The sequence itself may affect the physicochemical properties of the DNA. For example, Renz and Day (1976) showed that for high salt concentrations the retention of DNA on filters by histone H1 depended not only on the size of the DNA and the composition of the reactants, but also on the A + T composition of the DNA. The higher the proportion of A + T, the more binding occurred. As many as ten calf lymphocyte DNA molecules (A + T = 57%) could be bound for each *E. coli* molecule (A + T = 50%) under suitable conditions.

Not only might the satellites play a role in chromosome recognition, but the fact that the satellites are species specific in their distribution may have something to do with the difficulties often encountered in meiosis in interspecific hybrids. The evolution of the satellites of closely related species is very different from that encountered in the lineages of genes that code for proteins. The best-studied case is again that of *D. melanogaster* and *D. simulans,* though similar observations have been made for a number of other pairs of sibling species of *Drosophila.* The two species appear to share five satellites in common—at least, they have identical properties on centrifugation and those satellites which have been compared show identical repeated sequences. *D. melanogaster* has, however, one unique satellite and *D. simulans* two. Further, the 1.695 satellite (the number refers to the mobility relative to marker DNA in CsCl centrifugation in the presence of neotropin sulfate) is common in *D. simulans* and rare in *D. melanogaster,* while the reverse is true of the abundances of the 1.705 satellite.

Smith (1973) has constructed a computer model that simulates the increase and decrease of categories of highly repeated DNA based on the mechanism of unequal crossing-over and subsequent drift. The success of his model suggested that this was a reasonable mechanism by which rare short segments of many different satellites, present in the centromeric regions, could be amplified by chance. The distribution of satellites on the chromosomes may therefore be a fortuitous one. This does not explain, however, the presence of many copies of unique *D. melanogaster* satellites on several chromosomes simultaneously.

Perhaps the strongest argument for selection, at least of the purifying type, comes from studies on the melting of hybrid DNA carried out by Gall et al. (1973) on Satellite I of *D. virilis* and *D. americana,* and by Peacock et al. (1977) on two common satellites of *D. melanogaster* and *D. simulans.* The temperature of melting of hybrid DNA is a measure of the relatedness of the hybrid molecules; each degree Celsius by which the melting temperature is decreased relative to native DNA represents roughly a 1% difference in sequence. In none of these cases could a significant difference in melting between hybrid and native DNA be demonstrated. This can be contrasted with the finding of Laird and McCarthy (1968), who showed an average 3% decrease in melting temperature in hybrid single-copy DNA of *D. melanogaster* and *D. simulans* compared with reannealed intraspecific DNA. The satellites, each of which represents about 4% of the total DNA, have been strongly conserved in sequence, perhaps because they are much more recent or perhaps because they have continually 'edited' by unequal crossing-over. Were such editing to be random, increasing the numbers of mutant sequences as well as the original satellite sequences, one would expect the satellites to have become at least as different from each other as the rest of the DNA of the two species. Indeed, it would be most informative to extend Smith's simulation methods to ask whether, given a reasonable mutation rate and a reasonable amount of unequal crossing-over, highly reiterated satellite sequences would be expected to diverge more rapidly than unique sequences. Because of the 'multiplication effect' of unequal crossing-over, by which some random mutants might be increased manyfold, I suspect that they would show a much greater divergence rate in the absence of purifying selection.

Very long tracts of pyrimidines in cryptic satellites of Drosophila

Sederoff and coworkers (Birnboim et al. 1975; Birnboim and Sederoff 1975; Sederoff and Lowenstein 1975) have described a method using extensive acid hydrolysis for isolating long tracts of pyrimidines from 'cryptic satellites' of the *Drosophila* genome. These are subcategories of DNA that cannot be separated by the usual physicochemical methods. The tracts are not, from the nature of the technique, interspersed with purines, but of course there are equally long purine tracts on the other strand of the DNA which are destroyed during the extraction. The tracts vary from about 700 nucleotides in average length in *D. melanogaster* up to 1300 in some other species, and constitute from less than 1% to 10% of the total DNA depending on the species. Some species have no detectable polypyrimidines. These long tracts have very simple repeating sequences, e.g. TCTCT in *D. melanogaster,* and as with other components of the satellites the pattern seems to be that with increasing evolutionary divergence the predominant sequence in one species is replaced by a different predominant sequence in another species. There may, however, be appreciable amounts of a particular sequence present in closely-related species so that the replacement of one sequence with another is not a sharp species-specific event (Sederoff, personal communication).

Of course, the function of this type of satellite is fully as problematical as that of the other satellites, so that precisely what is shaping these sequences remains to be discovered. It is striking, however, that these sequences, which form part of the germ line, have a low probability of occurrence by chance. Assuming that a purine–pyrimidine and a pyrimidine–purine pair have equal probabilities of occurrence along the DNA backbone, then the chance that a string of 700 pyrimidines occur on one strand by chance would be 0.5^{700}, or one in 10^{211}. There is no evidence that these particular satellite sequences, or indeed any satellite sequences, are synthesized by gene amplification, so if they are passed through the germ line they must be subject to mutation pressure. It may be that an editing mechanism keeps these very long tracts free of mutational alterations, or at least sufficiently free that they are broken at intervals of only about 700 bases. An alternative possibility is that they arose recently enough in the life of the species, as a result of some process of amplification, that mutational noise has only had a chance to accumulate to this very low level. Several species in the *D. hydei* subgroup have been investigated (Sederoff, personal communication) and all have varying quantities of a common sequence. One might expect that older species would have accumulated more mutational noise than younger ones, but there are no great differences in the lengths of this sequence among these species, although the amounts differ.

These tantalizing glimpses into the untranslated genome are remarkable for the degree of organization which can already be perceived. It must be remembered that we have no firm idea of the function of these vast tracts of DNA, amounting in higher eukaryotes to 97% of the total. One of the great tasks of biology in the last part of the twentieth century is to elucidate these functions, and I cannot help but think that these functions will be as varied and fascinating as those we can already attribute to that small portion of the translated 3% that we have begun to understand.

Other evidence for the action of selection on untranslated regions

Min Jou and Fiers (1976) have compared known sequences of the RNA coliphages MS2, R17, and f2, all of which show a high degree of homology. A far smaller number of nucleotide differences was observed in comparisons of pairs of the phage genomes in untranslated regions than in translated ones, suggesting either that the untranslated regions were subject to stronger selective constraints or that there was stronger selection for substitutions in the translated regions. Among all three sequences, there were 257 bases that could be compared in untranslated regions, and 612 in translated regions. There were 31 base substitutions in the translated and only two in the untranslated regions, a difference that is significant at the 0.05 level. Further, only seven out of the 31 substitutions in the translated regions resulted in an amino acid substitution, when about 20 would be expected. This is a highly significant difference, with a probability of occurrence by chance of less than 0.001. Finally, as these authors show, there is a significant excess of transitions (purine to purine or pyrimidine to pyrimidine) over transversions (again significant at the 0.001 level).

The excess of substitutions in translated over untranslated regions is apparently not a general rule. Kafatos et al. (1977) compared the translated and untranslated regions of rabbit and human β-hemoglobin message, and found no significant difference in the numbers of substitutions in the two regions. Further, in the translated regions they found a significant clustering of both silent and nonsilent mutations—that is, both those that had no effect and those that had an effect on the amino acid sequence. As in the RNA coliphages, there is a highly significant excess of transitions over transversions.

It is striking that the data analysed by Min Jou and Fiers should show such apparently contradictory tendencies. In the translated regions, the pattern was the expected one of a higher rate of silent mutations than mutations affecting amino acids (the amino acid differences that were found all were of the type that conserved chemical properties). But in the untranslated regions there was a significantly lower rate of substitution, even though these regions might also be considered 'silent'. As in the case of nonrandom codon usage considered earlier, the selective pressures operating appear to be complex and to vary from one part of the genome to another.

The data presented in this chapter show that every part of the genome that has been investigated is subject to strong purifying selection, and also suggest that base substitutions may be under selective control. As was shown in Chapter 3, and will be developed more fully in Chapter 6, only truncation selection provides the very large per-organism fitness differences needed to accomplish this much selection without an overwhelming genetic load.

5. Evidence for balancing selection

A great variety of selective mechanisms have been found or inferred to be responsible for the maintenance of human polymorphisms. One important mechanism for which there is no evidence is unconditional heterozygote advantage

In this chapter I will turn to a consideration of the cases for which we know, or can infer, the existence of balancing selection. This is selection in which alleles are held at appreciable frequency in the population by opposing selective forces, rather than by an equilibrium between mutation to a harmful allele and its removal by selection. A detailed consideration of the mechanisms of balancing selection will be deferred to Chapter 7.

Most of the examples are drawn from man, for the reason that we know more about our own physiology than about that of any other organism. Further, we know more about the reactions of humans to extreme or unusual environments than we do about the reactions of other organisms, even domesticated animals or plants. It is striking that most of the polymorphisms we have begun to understand are associated with resistance or susceptibility to diseases. Diseases, since they are discontinuous in their distribution, result in a marked environmental heterogeneity. Such spatially or temporally limited heterogeneities provide the best opportunity for observing selective differences. I must emphasize that we have no comparable body of knowledge about other organisms. We are in most cases forced to infer the existence of selection by the association of clines of allele frequency with some environmental factor such as temperature. Associations of that type can be no more than suggestive.

Cases of conditional single-gene heterozygote advantage

Two-allele genetic polymorphisms can easily be maintained in a population if both homozygotes are less fit than the heterozygote. Such cases are, as Lewontin (1974*a*) has pointed out, very rare, or at least authenticated instances are. Cases of heterozygote excess or apparent heterozygote advantage at a single locus have been reported in a variety of organisms (e.g. Richmond and Powell 1970; Wills and Nichols 1971, 1972) but may be due to associative overdominance, the term 'overdominance' referring to the fact that the heterozygote expresses the character even more pronouncedly than the dominant homozygote. One way in which this could occur is if the blocks of genes closely linked to the alleles in question include nonallelic subvital genes. A heterozygote for the locus may be no more fit under the experimental conditions than a homozygote, but may appear to be so because of complementation of the linked subvitals. This is hard to rule out, since little is usually known about the locus or loci being examined. The question of the extent of associative overdominance is considered in more detail in Chapter 8. But in those few cases in which something is known or at least suspected about selection at the locus, overdominance can sometimes be attributed directly to it. It is instructive to examine some of these cases in detail, to look for properties which overdominant or presumed overdominant loci may have in common.

Sickle-cell anemia

Perhaps the best-understood case is that of sickle-cell anemia. Homozygotes for the disease were first diagnosed clinically in 1910 by Herrick, because of the sickle shape taken on by red blood corpuscles when blood was allowed to become depleted of oxygen for 24–48 h under a sealed coverslip. The disease is marked by intermittent but clinically severe symptoms, which may be triggered by stress such as virus infection. The crisis can be *aplastic*—a sudden shutdown of erythropoiesis—or can be marked by *hyperhemolysis,* a rapid lysing of large numbers of red blood cells. This results in turn in severe jaundice as the liver attempts to break down the soluble hemoglobin in the blood and in a strain on the kidneys as urobilinogen excretion increases many times above normal amounts. Or there may be a sudden and often fatal *sequestration* of blood in the spleen and sometimes the liver (Jenkins et al. 1960), as the normal blood-filtering action of these organs is disrupted by sickle-shaped red blood cells. These conditions, drastic as their clinical effects can sometimes be, can be reversed or relieved using the techniques of modern medicine, but they are often fatal under more primitive circumstances.

The heterozygote is little affected and can live a perfectly normal life, but may experience difficulty under conditions of severe oxygen deprivation, such as high-altitude flying with reduced oxygen supply. Tests were carried out on volunteers, chiefly from the military of several countries, who were heterozygous for the condition (carriers of the sickle-cell trait rather than persons with sickle-cell disease, who are homozygous). These volunteers showed no particular problems, so A. C. Allison (1954*a*) was able to state that 'the current view is that the sickle cell trait is devoid of selective value.' In that same paper, however, Allison summarized the fragmentary evidence that the heterozygote is more resistant to malignant falciparum malaria.

Sickle cell anemia holds a great deal of interest at the molecular level. It was the first example of a disease in man determined to be caused by the alteration of a gene product, rather than by its absence (Pauling et al. 1949). Later, Ingram (1957) showed that the alteration could be traced to a single amino acid substitution. This is the sixth residue from the amino terminal end of the β chain. The four chains of normal adult hemoglobin consist of two α and two β chains. The substitution, which in turn can be traced to a single base change in the DNA, is of a valine for a glutamic acid, reducing the overall negative charge of the molecule so that it migrates more slowly toward the anode in an electrophoretic field. This alteration results in a ring structure at the amino terminus of the molecule, in which the N-terminal valine is juxtaposed to the valine at β-6. This structure, with the aid of a sickling cofactor, can 'lock' into a cleft in the adjacent α chain of a neighboring molecule (see Fig. 5.1) (Murayama and Nilbandian 1973). This in turn results in the formation of long chains or cables of hemoglobin molecules, visualizable under the electron

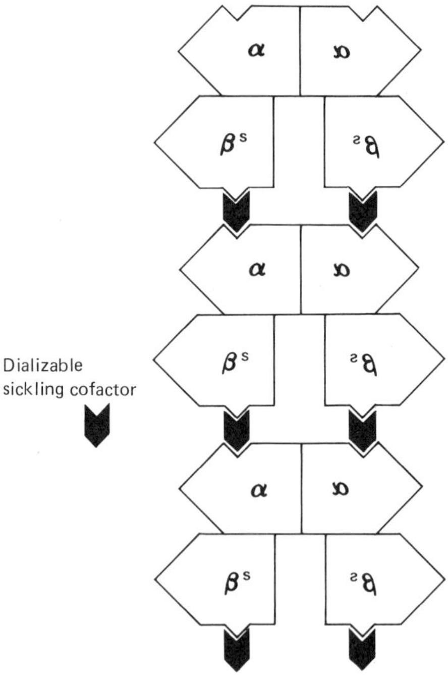

Fig. 5.1. Schematic representation of the formation of chains of sickle cell hemoglobin molecules. Under conditions of low oxygen tension, the β chains move apart. If a ring is present at an amino terminus, they can lock into a cleft of the α chain of an adjacent molecule. The diagram shows a possible role of a sickling cofactor, which has not been fully characterized. From Murayama and Nilbandian (1973).

microscope (Murayama 1966). These form only when oxygen tension drops far enough for the β-chains to move sufficiently far apart. In the homozygote all the molecules are HbS, so that the chains form readily when the partial pressure drops to 45 mbars, distorting the red blood cell into the diagnostic sickle shape. But in the heterozygote each red cell contains a mixture of molecules, so that the sickling does not normally occur in the body. The sickling is almost instantly reversible when the oxygen tension rises, and the RBC's resume their original shape. Some damage occurs to the cells during this process, as is evidenced by the appearance of small blebs of membrane-enclosed hemoglobin in the plasma. Repeated reductions and increases in oxygen tension will eventually result in the cells being permanently sickled.

Even when a homozygote for the sickle cell gene undergoes a crisis, the sickling will usually occur only in the capillary beds. It is this which causes the sudden sequestration of blood in the spleen or the liver. A crisis in the homozygote for sickle cell anemia can be produced by a number of external factors, including virus infections, so that the disease itself is rarely the immediate cause of death.

Allison's was not the first suggestion that resistance to malaria might be a selective factor in the maintenance of a genetic polymorphism. Haldane (1949) had remarked on the association of malaria and the thalassemias, a group of diseases that will be considered in detail later. Beet (1949) had claimed to find a significant deficit of malaria infection among sicklers in Northern Rhodesia, but upon reexamination his data are not significantly different from chance (among sicklers, 10 out of 85 showed parasitemia; among nonsicklers 92 out of 508 were infected; $\chi^2 = 1.64$ after correction for continuity).

Allison (1954*a*) made the connection more explicit, and proposed that the increased resistance of the heterozygote was responsible for the high frequency of the gene (otherwise, he calculated, one would need 3000 times the usual mutation rate to maintain the gene in such high frequencies in some African populations). He performed an experiment (1954*b*) on 15 Luo tribesman volunteers who were sicklers, using as a control 15 who were not. Parasitized blood was injected into all 30, in some cases several times. All 15 of the nonsicklers rapidly developed infections, some so severe that they had to be treated by chemotherapy before the end of the experiment. Despite repeated infection, only two out of the 15 sicklers developed significant numbers of parasites in their blood.

Allison's findings caused considerable controversy. The infection experiment was repeated using black American volunteers, with negative results (Beutler et al. 1955). In retrospect, this could be because resistance to infection had enabled the African sicklers to develop immunological defenses to a greater extent than the nonsicklers, while of course none of the American volunteers had any immunity to the parasite.

Plasmodium falciparum causes the severest type of human malaria, primarily because of its ability to multiply in red blood cells of all ages (Bruce-

Chwatt 1948). This leads to very high infective titers, in contrast to the other common human malarias such as *P. vivax* and *P. malariae* which are only able to grow in young RBCs and in which the number of infected cells only reaches 1–2%. Perhaps as a consequence of this very high level of infection, central nervous system involvement is very common in falciparum malaria. One of the clearest indications that heterozygote advantage is operating comes from the study of Gilles et al. (1967), who showed that fewer children with the sickle cell trait were hospitalized with severe malaria, particularly with central nervous system involvement, than the frequency of the trait in unhospitalized children of the same age would predict. Of 100 children with severe malaria, only four were AS, while 36 out of 200 matched control children were AS. It took these workers 3 years to collect this properly matched set of data.

Other evidence has been found. The frequency of the trait increases with increasing age in parts of West Africa where malaria infestation is heavy (summarized by Rucknagel and Neel 1961). In those regions of the Caribbean that have been malaria-free, the frequency of the gene is lower than it is in malarious regions. The overwhelming weight of evidence now is that heterozygote advantage has led to the high frequency of the gene in malarious regions.

But what is the mechanism of the advantage?

Until recently, it has not been possible to follow the course of infections for more than a few hours *in vitro*. An early study by Raper (1959) showed that the development of the parasite seemed not to be hindered by the presence of sickle cell hemoglobin—once the cell had become infected.

This work was carried out at normal oxygen tension. A likely mode of resistance was demonstrated by Luzzatto et al. (1970), who used the available *in vitro* methods to follow sickling of RBCs over time. The cells were sealed under anaerobic conditions enhanced by the addition of metabisulfite, and after various periods of incubation were fixed rapidly with formalin to prevent any reversal of the sickling. Sickling was found to be a linear function of time with AA and AS cells, but SS and SC cells (the latter heterozygous for hemoglobin S and hemoglobin C, another variant with a lysine substituted for the glutamic acid at β-6) showed nonlinear sickling and were not investigated further.

In mixed cultures of cells with and without parasites, very clearcut differences in the amount of sickling were observed in ten experiments in which the incubation was terminated after periods ranging from 10 to 50 min. The parasitized RBCs always tended to sickle first, and after incubation very few parasitized cells remained unsickled. The same effect was seen in five experiments in which sickling was followed over time. Thus, the parasites led to earlier sickling *in vitro*. The mechanism is presumably that the parasites exhaust oxygen at a greater rate inside the RBC. Because of the sickling, the infected cells are much more likely to be caught in the spleen, which acts as a filter for abnormal RBCs while normal cells can pass easily through small openings in the spleen's stroma. This interrupts the life cycle of the parasites, and keeps the rate of parasitism sufficiently low that central nervous system involvement and other side effects are not as likely to occur. This result was not unequivocal, however. Some earlier experiments involving incubation for longer periods (Miller et al. 1956) and gametocyte counts (Ednington and Laing 1957) had given negative results, indicating that the outcome may be strongly dependent on the experimental conditions.

The development of methods for following the growth and development of malarial parasites in red blood cells for long periods (Trager and Jensen 1978) have made possible far more detailed experiments on the long-term effects of infection.

Friedman (1978) used these methods to follow infected cells of the AA, AS and SS genotypes for 48 hours at high and low oxygen tension, and found that at high tension (18%) there was no significant difference in the course of the infection among the genotypes. After 48 hours at low tension (3%), all the parasites had died in the SS cells and most of the parasites had died in the AS cells. In the AA cells, infection proceeded normally. Most of the lethality occurred at early stages in the parasite development; parasites surviving 48 hours in the AS cells were able to develop normally and infect new cells. The precise mechanism of the killing action remains to be determined. Friedman suggests that it may be mechanical disruption of the parasites by the formation of hemoglobin crystals, inhibition of the plasmodium feeding process, or perhaps a combination of these and other factors. Pasvol, Weatherall, and Wilson (1978) obtained slightly different results in a similar experiment, demonstrating that both young and old parasites were killed in AS and SS cells. They also showed a slight reduction in the numbers of AS and SS cells invaded by the parasite under low oxygen tension. Coupled with the mechanism suggested by Luzzatto, this complex of effects seems to provide protection against the severer consequences of infection. That this is not the whole story is apparent from the experiment of Allison, suggesting that immunological response may also play a role.

The case of sickle cell anemia is a classic story of *conditional* overdominance for a single gene. Once the selective factor is removed, it becomes a gene with a high mutational genetic load, and is rapidly reduced in frequency. Certain interesting aspects of the disease should be emphasized in the present context.

1. The selective pressures involved are very large, sufficiently large to bring about something rather rare in human populations, a clearcut disturbance of Hardy–Weinberg equilibrium in some West African populations (Cavalli-Sforza and Bodmer 1971). In spite of this, there has been disagreement about how selection operates at this locus. Different surveys and experiments have given different results. The situation is a complex one, and as we have seen there is apparently no single mechanism of resistance.

2. The heterozygote advantage at this locus is only apparent under extreme environmental circumstances. Genotype–environment interaction is small over a range of nondeleterious environments, but becomes extremely large and nonlinear when malaria is present. Were a gene similar in effect to the sickling

gene to be found in an experimental organism taken from nature and put into the laboratory, the chances are excellent that it would be considered to be part of the mutational load, and to bear witness to a simple population structure. The disease to which it confers resistance almost certainly would not be part of the laboratory environment. The high frequency of the gene would presumably be attributed to the founder effect.

Hemoglobin C

It seems likely that only one heterotic allele is being accommodated at the β locus in the affected populations. Hemoglobin C is found at high frequencies in West African populations, and has also been implicated, though only inferentially, in resistance to malaria (Allison 1964). The distribution of this gene seems very different, however, radiating out from a central focus as if it had arisen at one point in time and had been spreading since. This impression is borne out by a consideration of the fitnesses of the six genotypes at this locus (Cavalli-Sforza and Bodmer 1971; Crozier et al. 1972). The fitnesses calculated from differences in the observed frequencies among newborn and adults are not such as to lead to a stable three-allele polymorphism. AC heterozygotes show no obvious decrease in fitness but SC heterozygotes are unfit, suffering from the effect of an anemia milder than that of SS homozygotes. The fitness estimates for CC homozygotes vary widely from 0.5 to 1.4, at least in part due to the very small numbers involved. On the basis of a particular set of fitness estimates, which they assume to be unchanging, and which include the very high estimate for the CC homozygotes, Cavalli-Sforza and Bodmer calculate that the C gene should drive out S and then eventually A. This is because with the removal of S, a balanced polymorphism will not be possible. The conditions for stability appear not to be met if this set of fitnesses is taken at face value, and in view of the decided restrictions on stability for a three-allele system (pp. 174–76), this is not surprising. It is also fairly obvious that C will not replace A except locally, if the high fitness of the CC homozygote depends on the presence of malaria.

The mechanism of malaria resistance of this gene, which does not cause sickling, is at the moment quite unknown, but Livingstone (1967) suggests that C might be the older of the two polymorphisms, conferring resistance to *P. malariae*. This parasite is common in regions of high C frequency, and has been in the region longer than *P. falciparum,* a relatively late arrival. The survey by Gilles et al. (1967) mentioned earlier found no difference in the frequency of hemoglobin C in children with severe malaria and those without symptoms (6% and 7% respectively). It may even possibly be that this condition does not confer resistance to malaria at all, but to some other environmental factor. The argument rests entirely on its geographical association with malaria. But whatever the factor may be to which it confers resistance, it would appear to be a local one. Further, despite their apparent high fitness from some population studies, CC homozygotes do show mild but clinically significant anemia. The gene would not replace A if the environmental factor were removed, and if it does prove to be heterotic it would certainly be another example of conditional heterosis. Now that modern medicine is arriving in West Africa, it seems unlikely that we will ever achieve a full understanding of the dynamics of this polymorphism.

The difficulty of establishing unequivocal evidence for conditional heterozygote advantage in the sickle-cell and hemoglobin-C cases bodes ill for anyone trying to do it with any assurance in a less well-studied organism than man. Even in man, the task has been formidable.

The thalassemias

The thalassemias were the first diseases to be implicated in malaria resistance (Haldane 1949). This heterogeneous (Weatherall and Clegg 1972, 1976) collection of diseases is found distributed in Africa, around the Mediterranean basin, and in the Far East. The name was derived by Whipple and Bradford (1936) from the Greek 'θαλασσα,' 'the sea.' These polymorphisms are also highly correlated geographically with the distribution of falciparum malaria as it existed until the advent of public health measures. There are several types of thalassemia, affecting the synthesis of α, β, or both β and δ chains, each with a number of subcategories, and the molecular biology of these diseases is beginning to be understood through a brilliant series of studies involving the most sophisticated tools.

The β-thalassemias. The common thalassemias in the Mediterranean are marked by a depression or absence of the β chain of adult hemoglobin. In the homozygote, this can lead to a serious anemia, Cooley's anemia, or thalassemia major, marked by abnormal erythropoiesis. A child with Cooley's anemia is unlikely to live for long without intensive medical care, and is in any case unlikely to live to reproductive age. In some thalassemias, notably the 'Negro' type and the 'high F' type of β^+ thalassemia, there is an elevation of hemoglobins A_2 ($\alpha_2\delta_2$) and F ($\alpha_2\gamma_2$ or fetal hemoglobin) sufficient that in the homozygote the anemia is severe but not incompatible with life. This condition is known as Thalassemia intermedia.

In the heterozygote for β-thalassemia, the excess α-chains cannot form normal tetramers. They tend to be found as monomers, dimers and (if they are in high concentration) trimers. These molecules denature easily, and tend to nucleate the formation of Heinz bodies, clumps of precipitated hemoglobin which (because the cells carrying them tend to be caught in the stroma of the spleen and in the liver) shorten the lifespan of the red blood cells. This produces the mild anemia known as Thalassemia minor, usually without clinical complications.

Those β-thalassemias which produce small but detectable amounts of the β

chain in the homozygote are grouped together as β^+-thalassemias, and they can be of different severity. As we have seen, the 'Negro' and 'high F' types are partially compensated for by elevated synthesis of γ and δ chains. Another type of β^+-thalassemia, found in the Mediterranean, does not show such compensation so that the homozygote suffers from Cooley's anemia. The molecular deficiency of all three types is unknown, but the small amounts of β chain produced seem to be normal in every respect (Heywood et al. 1964; Bank and Marks 1966), indicating that the lesion or lesions are not in the structural gene itself. Houseman et al. (1973) and Kacian et al. (1973) showed that the amount of β-mRNA was reduced in some β^+-thalassemias, but Nienhuis et al. (1977) showed that normal amounts of premessage were present in nuclear RNA. The processing of the message therefore appears to be defective.

Considerably more is known about the molecular biology of the other main class of β-thalassemias, however, those that produce no detectable amounts of β chains in the homozygote. These are collectively known as β^0-thalassemias, and again they have been shown to be heterogeneous in phenotype. All result in Cooley's anemia in the homozygote. In $(\delta\beta)^0$-thalassemia from Greece and Sicily, no detectable amounts of either the δ or the β chain can be found, and hybridization of purified δ- and β-mRNA (obtained from reticulocytes of infants with hydrops fetalis, in which no α chains are made) to the DNA of patients with this condition indicates that the genes have been deleted (Ottolenghi et al. 1976). But this is not the case in other types of β^0-thalassemia. Old et al. (1977) have classified them into three categories, and they find that more than one category is present in a particular population. The first has no detectable β-mRNA, the second shows incomplete hybridization to β-copy DNA (cDNA) probes, and the third appears to have normal though untranslatable β-mRNA. The different types were found in patients of Greek, English, Saudi Arabian, Chinese, and Pakistani origin.

Some homozygotes of what appears to be this last class have been intensively investigated. Patients in the Ferrara region of northern Italy (at the mouth of the Po river) produce elevated amounts of hemoglobin A_2, so the δ gene is intact. But when cDNA is made from the message isolated from reticulocytes of infants with hydrops fetalis, this in turn hybridizes to mRNA isolated from reticulocytes of homozygotes for β^0-Ferrara in amounts indicating that the mRNA for the β chain is present but is not being translated. Whether this is due to a point mutation somewhere along the message or to a deletion of part of it is unclear, since the cDNA is not a complete copy in these experiments. Synthesis of cDNA sequences starting from both the 3′ and the 5′ ends in the recent experiments of Old et al. (1977) allowed those workers to be certain that the two ends of the message were intact, but β^0-Ferrara patients were not investigated.

There has been one report (Conconi et al. 1972) that addition of reticulocyte factors other than mRNA from a normal donor will induce the β^0-Ferrara mRNA to translate. These data seem particularly convincing since the donor was a sickle cell homozygote and normal β chain was made in the cell-free system. This indicated that the effect was not due to accidental contamination by mRNA from the donor. But other β^0-Ferrara patients checked in another laboratory (Ramirez et al. 1976) appear to have abnormal β-mRNA, suggesting that even β^0-Ferrara may be heterogeneous.

A heterogeneous collection of other β-thalassemias has been found (Clegg and Weatherall 1976), affecting the level of β-chain, but very little is yet known about them.

The suspected association of the β-thalassemias with malaria was reinforced by a careful survey of Sardinian populations in which the genes reach frequencies as high as 15%. The survey was carried out by Carcassi et al. (1957). The frequency of thalassemia was negatively correlated with altitude, as was the incidence of malaria. Carcassi et al. could not find any significant difference in the frequencies of blood groups between the lowland and highland populations, which suggested that stratification of the samples was not producing the effect.

The thalassemias are obviously a heterogeneous group of blood diseases. It is difficult to imagine, in view of the severity of their homozygous effects, any mechanism other than heterozygote advantage which could be responsible for their high frequencies. But the situation is strikingly different from that of sickle cell anemia. Different and even multiple forms of thalassemias appear to be present in different areas—and indeed there may be considerable heterogeneity in the selective pressures involved. *P. vivax* has historically been the prevalent malarial parasite in northern Italy, while *P. falciparum* is found in the south.

Because of the great technical difficulties involved in determining the molecular nature of the disease exhibited by each patient, there is little information on the distribution of thalassemias of the different types. The 'Negro' type is not common in the Mediterranean basin, and the β^0 types studied from Ferrara and elsewhere are obviously different from each other. But are there subtypes yet to be discovered which make the story even more complicated? Is it possible, leaving migration aside, for more than one type of thalassemia to be common in a particular area, or do they tend to exclude each other? As we will see, theory based on constant selection coefficients would suggest that it is extremely difficult to have more than one allele exhibiting heterozygote advantage in a single population without one eventually winning out, and it is not yet clear whether the thalassemias provide a confirmation of or a series of exceptions to this rule.

To make the story even more complex, there is as yet no evidence concerning the nature of the resistance of these heterozygotes to malaria. Does infection by the parasite accelerate the production of Heinz bodies, so that the cells are selectively removed from the circulation? Do different thalassemias have different effects? What about the effect of different malaria parasites? The recent reduction in the endemicity of malaria in the Mediterranean basin makes data difficult to collect, and conclusions based on the rate of infection of volunteers may be complicated, as we have seen in the case of sickle-cell anemia, by the previous history of malaria infection of the volunteer.

About all we are certain about is that the β-thalassemias provide another undoubted case of conditional heterozygote advantage. Since the theoretical constraints on multiple-allele polymorphisms maintained by heterozygote advantage are so severe, the mechanism may be as follows:

The heterozygote advantage (which, as noted above, has not yet been demonstrated *in vitro* and for which a mechanism is not certain) may only be expressed under extremely severe environmental conditions, for example at seasons and in regions in which malarial infections are very severe. Any of a great many thalassemia genes, arising *de novo* or existing in the population at low levels, would be raised in frequency under these conditions, but would tend not to spread outside the local patch both because of low migration rates and because the heterozygote advantage disappears beyond the boundaries of the small patch. In this model, the distributions of the alleles would also tend to be highly patchy, with no more than one or two in each patch. Unfortunately for the population geneticist, but fortunately for the population, any evidence for such a distribution is undoubtedly disappearing rapidly.

The heterozygote advantage for the β-thalassemia genes seems to be even more highly conditional than that for the sickle cell allele. This leads to the prediction that it will be much more difficult to demonstrate such an advantage in *in vitro* experiments involving heterozygotes for the thalassemias than it has been in heterozygotes for the sickle cell allele. The results of experiments carried out in different laboratories will be even more likely to differ from each other.

The α-thalassemias. The α-thalassemias are, in many ways, even more difficult to understand. Although less is known about their molecular biology, they are again distributed in a way strongly suggestive of their association with falciparum malaria. But because of the duplication of the α genes in most human populations, it is quite possible to be homozygous for a condition which represses α-chain synthesis and still to suffer little effect from it.

The duplication of the α genes, long suspected because of the fact that α-chain mutants tend to affect only about one quarter of the α chains in heterozygotes while half the chains are affected in β-chain heterozygotes, was difficult to demonstrate for certain. Because few polymorphic α-chain structural gene mutants are known, families segregating for two different variants are rare and the few data available are ambiguous. Kan et al. (1975) showed, using cDNA made from the α-mRNA of a patient with Cooley's anemia, that the cDNA hybridized to a level consistent with the presence of four structural genes in the normal genome. It had already been shown that DNA from infants with the invariably fatal condition hydrops fetalis, marked by an absence of α chains, would not hybridize with α-mRNA. This suggests that some of the structural genes must be absent in the cases of α-thalassemia that have been investigated (Ottolenghi et al. 1974).

The α-thalassemias vary in severity, and this fact is assumed to be related to the number of α genes available for transcription. Absence of one or two α genes has little clinical effect, while absence of three will produce a pronounced effect with higher levels of hemoglobin Bart's (γ_4) and hemoglobin H (β_4). Absence of all four is invariably fatal at around the time of birth, leading to hydrops fetalis. In this condition virtually all the circulating hemoglobin is γ_4.

The situation is complicated, particularly in Southeast Asia, by the presence of significant frequencies of some α read-through mutants producing depressed quantities of an abnormal α chain in which a mutation has occurred in the termination triplet, allowing the rest of the mRNA to be translated. In some cases, e.g. Hb Constant Spring, the evidence is that this read-through may continue to the very end of the message (Bunn and Forget 1975). When heterozygous with an α-thalassemia gene, this condition may lead to clinically severe anemia. About 45% of the cases of hemoglobin H disease in Thailand seem to be due to heterozygosity of α-thalassemia genes with the gene for hemoglobin Constant Spring (Wasi et al. 1972). These read-through mutations are at a high frequency, particularly when compared with other α-chain point mutants which are almost invariably familial. This certainly suggests that some sort of balancing selection has been acting on them.

The connection of the α-thalassemias with malaria resistance is even more tenuous than that of the β-thalassemias. Rigas and Koler (1961) have shown that α-thalassemia RBCs have a shorter lifespan than normal RBCs and that large numbers of red blood cells taken from the spleens of patients with hemoglobin H disease carry a single large Heinz body. But, as with the β-thalassemias, there is no indication of the mechanism of their interaction with the malaria parasite. The double homozygote is lethal, so that there is selection against the thalassemia genes. But because the homozygotes are so rare, it would appear that most of the selection acting to maintain these genes acts on either the heterozygote or on what is effectively a heterozygote because of gene duplication. It is difficult to imagine what if any advantage is conferred on the possessor of one or two doses of the gene, with at the most 5% of Hb Bart's. But several scenarios are possible when it comes to imagining what forces have raised these alleles to appreciable frequencies in some human populations.

It may be, for example, that the greatest benefit is conferred when the carrier has one or two doses of defective α genes. This would be straightforward heterozygote advantage, and the advantage of gene duplication in this situation would lie in its tendency to reduce the genetic load associated with homozygosity. Indeed, if only one out of the two α genes carried mutant alleles, the heterozygote advantage would not be coupled with a homozygote disadvantage. The fact that mutations at *both* loci are present in some populations in which there is a significant frequency of α-thalassemia could simply be a historical accident, or it could mean that an even greater advantage is conferred on the carrier of three doses of the gene, despite his clinically significant anemia. If such a carrier had the highest fitness, then heterozygote advantage would still be maintaining the genes, since the carrier of three doses could be heterozygous at either locus. Because of the invariable lethality of four doses of the gene, it is difficult to imagine anything but some form of heterozygote advantage maintaining it. Again, however, the advantage would appear to be conditional on the presence of malaria or something correlated with it.

It has recently been found (Dozy et al. 1979) that only one of the α genes has been deleted in black populations, in contrast with the high frequency of double deletions in Asia. Such a disjunct distribution might be expected as a result of strong selection against the double homozygote. This is particularly likely since, as Haldane (1962) pointed out, such selection will allow gene frequencies at each locus to vary widely and reciprocally, provided that the frequency of the double homozygote remains constant.

The high frequency of hemoglobin Constant Spring in Thai populations introduces another complication. We will eventually be able to ask whether Constant Spring and other read-through mutations are confined to only one of the α genes and whether the α-thalassemias found in association with them tend to be those affecting the other structural locus.

The α-thalassemias may turn out to be less heterogeneous than the β-thalassemias. This would suggest a less highly conditional heterozygote advantage. This in turn would lead to the expectation that in-vitro experimentation will be more productive in the α-thalassemias than in the β-thalassemias, since less severe experimental conditions might be required to reveal a selective advantage.

Polymorphisms in which heterozygote advantage is not certain

It has been assumed in a number of cases that heterozygote advantage *must* be maintaining certain human polymorphisms, if only because theory dealing with other types of balancing selection has not tended to be incorporated into the literature of human population genetics. But the evidence for this is uncertain at best. Two polymorphisms which I consider to fall into this category are again associated with malaria, so that something can be determined or inferred about selection at these loci. In neither case is there strong selection against the homozygote. These are the polymorphisms for hemoglobin E and for G6PD-deficiency. It is striking that these two polymorphisms have very different kinds of distribution in human populations.

Hemoglobin E

Hemoglobin E, like hemoglobins S and C and unlike the thalassemias, is due to a single amino acid substitution in the β chain. In this case it is in a different position, a substitution of a lysine for a glutamic acid at position 26. Far less is known about it than about the African structural gene polymorphisms, but it is quite widespread in distribution, extending through Southeast Asia from India to Indonesia, and found at low frequency in Taiwan and the nearby coastal areas of China. In Eastern Thailand, the frequency of the gene can be over 50% in some tribes, higher than the most extreme frequencies of Hb S in Africa.

The homozygote for hemoglobin E shows a slight but not clinically significant anemia, but the heterozygote for E and β-thalassemia shows symptoms intermediate between thalassemia minor and thalassemia major (Flatz et al. 1965*a*). Homozygotes can and do live perfectly normal lives. Flatz et al. (1965*b*) investigated a sample of AA, AE and EE persons in eastern Thailand and found, first, no disturbance of Hardy–Weinberg equilibrium in any of the three populations surveyed. (This is hardly surprising, in view of the difficulty in detecting a disturbance in this measure.) But there was also no significant difference in fertility. Though the EE homozygous women did have a smaller number of children on average (4.6 ± 1.8 *v*. 6 ± 2.8 for AA homozygotes and 6.5 ± 2.1 for AE heterozygotes), this was not significant because of the sizes of the samples—9, 118, and 53 respectively. It is suggestive that the heterozygotes had the largest numbers of offspring.

Is this a case of heterozygote advantage? These authors found a significant negative correlation between the distributions of Hb E and thalassemia, which could not simply be a mechanical exclusion of one or the other allele because the majority of thalassemia in Thailand is α-thalassemia. It is not clear whether this is a case of tribal admixture or of one polymorphism driving out the other. If the latter, and if it could be shown that HbE was the 'winning' polymorphism, then this might constitute evidence that HbE is being maintained in the population by heterozygote advantage. Yet, because of the high fitness of the homozygote, there is another interpretation. It may be that the homozygote is actually, in the presence of the selective agent or agents, the fittest genotype. The gene would therefore be spreading in certain populations, and it is only because of a constant immigration of A genes from outside the regions subject to the unusual selection that the gene for HbE has not taken over these populations completely.

Militating against this latter interpretation is the observation that the gene for HbE has not taken over in *any* population, but has remained polymorphic everywhere, even in isolated villages. Of course, this could simply be a combination of weak selection for the E gene and a significant immigration rate.

Das et al. (1975) investigated HbE in two tribes of Mongoloid ancestry in Assam on the northeastern border of India, and again found that the maximum frequency was about 50%. They postulated as two alternatives a balanced polymorphism and the takeover of A by E. As in Thailand, this region is malarious. There is as yet no explanation why frequencies of above 50% are not observed.

In the case of a heterotic polymorphism, the internal frequency reached is a result of a balance between the advantageous effects of the alleles when heterozygous and their deleterious effect when homozygous. A high frequency of an allele can also be achieved as a result of mutation pressure if selection against it in the homozygote and heterozygote is very weak. A combination of recurrent mutation pressure and very slight selection against the homozygote will produce quite high frequencies, particularly if, as shown recently by Hartl (1977), the selection coefficients involved fluctuate appreciably. So it might be argued that HbE is held in the population at a high frequency by mutation pressure because the advantage of the homozygote is more than cancelled by the disadvantage due to the mild anemia which the homozygote exhibits. On the face of it, this seems highly unlikely because it would require repeated mutation to the specific HbE allele. This is not impossible, but there is no reason why other mutations conferring an advantage against malaria should not also have become frequent—HbS or C mutations, for example. Further, the interaction between β-thalassemia and HbE shows that selection against the HbE allele is quite strong. Thus, selection for the heterozygote may also be quite strong. It is striking to compare the polymorphism of the single allele HbE with that of the β-thalassemias, in which the pattern seems to be one of repeated mutations each with a strong but highly conditional heterozygote advantage, none becoming very frequent in the affected population because of homozygote lethality but each apparently confined geographically as if there had been multiple centers of origin. HbE seems much more like HbS, but without the strong homozygote disadvantage.

G6PD deficiency

The distribution of the polymorphism. Another polymorphism without drastic effects in the homozygote (or in this case chiefly the hemizygote) is that found at the sex-linked gene for glucose 6-phosphate dehydrogenase. High frequencies of alleles leading to a deficiency in activity of this enzyme are again positively correlated with the incidence of falciparum malaria. The majority of the alleles appear to affect the structural gene itself, so that the distribution of the numerous alleles at this locus can be determined more easily than in the case of the thalassemias. It is remarkable, however, that just as for the *Drosophila* xanthine dehydrogenase locus mentioned on p. 59, more and more alleles are continually being found. In a recent survey (Beutler 1978), 106 distinctly different variants are listed, most of them familial. Many other variants have been described, but because nonstandard methods were used for their assay it is not possible to tell how many of them differ from these 106. Among the criteria used in distinguishing them are electrophoretic mobility, heat stability, K_m and K_i, and rate of utilization of analogues. At least 16 of these variants are presumed to be polymorphic (Luzzatto 1975). That is, these sixteen alleles are found in at least one population at a frequency greater than 1%, so that presumably it is not simply mutation pressure that is keeping them there. There is no guarantee, however, that careful examination of these alleles will not reveal each of them to be a heterogeneous collection of alleles in turn.

Between one quarter and one third of patients with congenital nonspherocytic hemolytic anemia are G6PD-deficient (Kirkman 1971). Luzzatto (1975) has summarized approximately 50 cases of such anemia (that is, anemia not associated with the production of abnormal RBCs) due to G6PD deficiency, and found them to fall into 23 family groups. Only two unrelated families showed the same characteristics and therefore shared the same allele. Many different variants, each at low frequency, are thus found in populations in which this character is found at high frequency. Hemizygotes or heterozygotes for 'rare' alleles make up 1–2% of the *total* sample in Mozambique and up to 40% in the samples from Greece.

This is a very high frequency of rare variants. By way of contrast, the frequency of detectable non-polymorphic hemoglobin variants is only 0.125% and the probable total proportion of variants is only about 0.3% (Harris et al. 1974). Their calculation assumes that about two thirds of the variants are not detectable in electrophoretic surveys.

The mechanism of protection against malaria. G6PD deficiency is remarkable for several reasons. First, because of its sex-linkage, it probably affects more human beings than any other genetic variant (200 million, chiefly male, according to one estimate). Second, at least in the Mediterranean basin, it is frequent in precisely those areas where one would not expect it to be so, because of the effect of local diets on hemizygotes and homozygotes, so that selection both to increase and decrease the frequency is obviously operating. Third, it has been demonstrated quite unequivocally that RBCs are more likely to be parasitized by *P. falciparum* if they are normal than if they are G6PD-deficient. This is because in the circulation of a female heterozygous for G6PD there will, because of the Lyon effect, be two populations of circulating RBCs, one deficient and one normal. The relative proportions of these two populations that are parasitized can be determined.

Luzzatto et al. (1969) distinguished the normal and deficient red cells by the method of methemoglobin elution, which selectively removes denatured hemoglobin and thus distinguishes not only between normal and deficient cells but also between young and old ones. The degree to which these populations are confounded is unclear, but in patients homozygous or hemizygous for the condition between 0.5 and 17% of the cells resisted elution and were presumably young (a reflection, in all likelihood, of the severity of anemia in these patients), while in heterozygous females the percentage ranged from 25 to 75% and represented young plus enzymatically normal cells.

In the 20 heterozygous females examined, the ratio of parasitized to unparasitized eluted cells ranged from 1.9 to 79, strong evidence for preferential infection of the normal cells. Unfortunately, the ratio was also pronouncedly skewed for homozygous and hemizygous patients, indicating a preferential infection of younger cells. In view of the confounding of these effects, it would seem desirable to repeat these experiments using an enzymatic assay to distinguish between normal and deficient red cells.

The mechanism of protection against the parasite is also somewhat clearer for G6PD deficiency than it is for—for example—HbC or HbE. Approximately 10% of the glucose utilized by the RBC passes, not through the usual glycolytic pathway, but through an alternative pathway, the pentose shunt. Because the red cell loses not only its nucleus but also its mitochondria as the cytoplasm fills up with hemoglobin, this shunt is the only source of reduced NADP, provided by G6PD. This cofactor is in turn utilized by the enzyme glutathione reductase to produce reduced glutathione. Among its other functions, this molecule acts as a 'policeman' in reducing disulfide bonds that form within or between hemoglobin molecules or between the molecules and the cell membrane. Allen and Jandle (1961) found that under oxidative conditions the sulfhydryl (SH) at β-position 93 is oxidized first, which in turn exposes other interior sulfhydryl groups to oxidation. The consequent breakdown in the tertiary structure of the molecule leads to the formation of Heinz bodies, which may be associated with the cell membrane and will, as in the thalassemias, prevent the cell from passing through the stroma of the spleen.

If the level of G6PD in the cell is low at the time the cell loses its nucleus, the cell will have a reduced lifespan. Luzzatto and Afolayan (1971) showed that a rare variant which can cause chronic nonspherocytic hemolytic disease causes more rapid destruction of RBCs than does the more common A^- African variant which shows a greater G6PD activity. In both cases the lifetime of the deficient RBCs was significantly shorter than normal.

It may be postulated that infection of G6PD-deficient cells by the parasite may accelerate the production of Heinz bodies, perhaps in part because the parasite itself requires reduced glutathione for metabolism (Kirkman 1971). In the absence of malaria, however, the condition is deleterious. The biochemistry of the condition was not worked out until the 1950s, but it had been known since ancient times that the eating of broad or fava beans, a staple in some parts of the Mediterranean basin, could bring on a sudden and often fatal anemia. This anemia is characterized by the lysis of large numbers of RBCs and the release of hemoglobin into the circulation. Conditions short of death may include severe permanent eye damage. The fava bean contains strong oxidizing agents, the biochemistry of which is not fully understood. Other compounds which are themselves strong oxidizing agents such as quinacrine, or are capable of being converted into them such as naphthalene, can bring on a hemolytic crisis. Traditionally, Greek army recruits were issued blankets that had been stored in mothballs only after they had been thoroughly aired!

Fava beans are not a part of the diet in tropical Africa, although the deficiency is common. Here the commonest form is apparently a single allele A^-, and there is also a polymorphic allele A segregating in the same populations (most thoroughly studied in Nigeria) which results in normal amounts of an enzyme with altered electrophoretic mobility and kinetics but no apparent clinical manifestations. Luzzatto (1975) noted from his observation that hemizygotes or homozygotes for the A^- variant seemed to be involved in fewer clinical complications, including neonatal jaundice, than were carriers of some of the rarer variants.

Neonatal jaundice is a severe complication in the Mediterranean, where Russo et al. (1972) found it to occur commonly in males and much more rarely in females in a Sicilian population. In Greek populations, particularly in certain villages, the jaundice can be so severe as to result in brain damage. The variation between villages and even within families for this condition can be considerable, as can that for favism. Luisada (1941) commented that it may take a small or a large amount of fava beans to bring on an attack, and that there was considerable variation for this within families.

It seems certain that jaundice brought on by the birth trauma is a contributing element to selection against the allele, particularly in the absence of sophisticated medical care. But it is not clear how large a contribution to mortality is made by it, or by favism. Livingstone (1967) used some figures given by Crosby (1956) on the incidence of favism in Sardinia to calculate that selection against the homozygote or hemizygote might involve selection coefficients greater than 0.1. This should be detectable as differential survival to adulthood, but it has not. Selection both for and against these alleles is undoubted. Allison (1960) and Motulsky (1960) independently suggested that the condition might protect against malaria, and many workers starting with Siniscalco et al. (1961) have shown a strong correlation of this condition with endemic malaria. But the details of selection, including the magnitude of selection coefficients and their fluctuations, continue to escape us.

We have seen that at least some of the 'rare' variants, which taken as a class are not necessarily rare, are involved in greater clinical complications than the more common variants. But the mode of selection on the common alleles is difficult to determine. Bienzle et al. (1972) determined the G6PD genotype of 700 children who appeared at a Nigerian hospital because of acute illness. Allele frequencies compared were those of GdA^-, GdA and GdB, where GdB is the normal allelic variant of GdA. The genotypes did not differ significantly in frequency from those of a random sample of 186 healthy adults. They first compared the genotype frequencies of patients who were and who were not affected with malaria. The data lent themselves to analysis by a 2×2 contingency table using χ^2. I reanalysed their data using a continuity correction, and found no significant differences between the affected numbers of different genotypes. They then determined the numbers of parasites in the affected subsample (their Fig. 1). The enormous variation within the data is striking. The numbers varied from $\sim$1000 to 400 000 parasites/μl of blood. In order to find a significant comparison, they were forced to use a median test, in which those above the median number of parasite counts were compared with those below—including those with no malaria at all. When the data are considered as a whole, the χ^2 values are not significant. Only one individual comparison out of nine, that of Gd^{A-}/Gd^{B} females against all other females, is significant at the 0.05 level. This could easily have occurred by chance, but it is suggestive that it is in the direction expected for heterozygote advantage.

Interestingly, the data for males show no significant differences among the three genotypes when the numbers of individuals below and above the median infection level are compared. Again the problem of small numbers plagues this study as it does all others investigating selection, but it is striking that there are actually more G6PD-deficient (A^-) patients with infection above the median level than below (43 compared with 35).

The dynamics of the polymorphism. The authors of this study interpret their data as indicating that selection for heterozygotes is operating, while there is no noticeable selection for hemizygotes for the condition. If amount of parasitism is a good indicator of selection, which is an arguable point in view of other complications such as the unknown use of antimalaria drugs among the patients in this study, then selection acting in favor of the heterozygote cannot be very strong. Selection against the hemizygote and the homozygote, however, through the agency of neonatal jaundice and other complications, may be much stronger. Various assumptions can be made about the equilibrium frequencies likely to be achieved with various selection coefficients, but all lead to the conclusion that the frequency of the G6PD allele is unlikely to become high unless the selection on the heterozygote is large compared with selection against the hemizygote.

It is possible to make certain statements about the selection acting on common and rare alleles. Both heterosis and selection in a fluctuating environment could be responsible for maintenance of the common alleles, since the hemizygotes and homozygotes do not normally show gross clinical symptoms. Mutation–selection balance maintaining the rare alleles at high frequency, however, can be ruled out by the following argument.

The precise frequency of particular rare variants in any human population is difficult to determine, because it is not possible to distinguish unambiguously between different alleles with similar effect as it is in the case of some of the hemoglobin variants. Further, large surveys of populations free of the allele have not been carried out. But their frequency is certainly low. Mourant et al. (1976) report on studies finding only five deficients out of 2350 in a Spanish population and none out of 1336 in a Chilean survey. The frequency in the Spanish population is about 0.1%, so that the 'allele' does not qualify as polymorphic by the usual criteria. The frequency in northern European populations is probably very low. Because juvenile jaundice and nonspherocytic hemolytic anemia are associated with many of the rare variants, it is not surprising that they do not reach appreciable frequency in regions where there is no falciparum malaria. For a sex-linked gene, the criterion for mutation–selection balance is that $\hat{p} \cong 3\,\mu/s$, where $\hat{p}$ is the equilibrium frequency, μ is the mutation rate to the deleterious allele and s is the selection coefficient against the deleterious hemizygote—the assumption being that there is no large difference in fitness between normal homozygotes and heterozygotes in the females, so that the allele is unlikely to reach a high enough frequency to appear often as a homozygote in the female. If we assume a selection coefficient of 0.1 against the hemizygote or homozygote deficient and a mutation rate of 10^{-5}, then the equilibrium frequency will be approximately 3×10^{-4} and the allele will be unlikely to appear more than once or twice even in a very large sample. This frequency may be contrasted with the equilibrium frequency of 0.01 expected for an *autosomal* recessive, selected against with the same intensity and introduced by mutation at the same rate.

The frequency of rare G6PD-deficiency alleles in populations subject to malaria is very much higher that 3×10^{-4}, however, as noted by Luzzatto (1975). In that paper Luzzatto summarized 50 cases of nonspherocytic hemolytic anemia associated with deficiency, and found that they fell into 23 allelic classes with only one class occurring in more than one family. Nonetheless, the total frequency of rare variants was often greater than 1%. For a frequency of 1%, the selection coefficient needed to maintain a sex-linked allele in mutation–selection balance must be quite small, only 3×10^{-3} for a mutation rate of 10^{-5} per gene per generation. This would not be detectable and is certain to be smaller than the actual selection against these alleles, many of which can lead to clinically serious consequences as we have seen.

We have already noted that there is no obvious difference in amounts of parasitism among the genotypes. Luzzatto et al. (1969) used this fact to suggest that the heterozygous females should be at an advantage, since they are asymptomatic and have a population of parasite-resistant red cells. Can heterosis maintain these alleles at their observed frequency? The answer would appear to be yes, but only if selection against the mutant hemizygote and homozygote is weak. The conditions for heterotic maintenance of a sex-linked polymorphism are such that small selection coefficients can easily result in an internal equilibrium, but the conditions are more restrictive than for an autosomal polymorphism because selection in the hemizygous sex may work counter to that in the dizygous sex.

Crow and Kimura (1970) give the conditions for maintenance of a sex-linked polymorphism, in which W_1 and W_2 are the fitnesses of the wild type and mutant hemizygotes, and W_{11}, W_{12} and W_{22} are the fitnesses of the female genotypes. Then if $k = W_1/W_2$, the ratio of the wild type and mutant equilibrium frequencies for the two alleles in the females, $\hat{y}$, is given by

$$\hat{y} = \frac{(k+1)W_{12} - 2W_{22}}{(k+1)W_{12} - 2W_{11}}$$

and the ratio of the frequencies in the males, $\hat{x}$, is given by $\hat{x} = k\hat{y}$. Both numerator and denominator of the equation for $\hat{y}$ must be of the same sign for the equilibrium to be stable, so that many fitness sets lead to instability in a sex-linked situation, even though at an autosomal locus the same set of fitnesses in both sexes would lead to stability. Table 5.1(a) shows a variety of fitnesses leading to stable and unstable polymorphism when the selection coefficients are small. Too great a difference in the fitnesses of the hemizygotes counterbalances the effect of heterosis and leads to instability. Further, it is remarkable that the mutant allele in a sex-linked heterotic polymorphism cannot be too rare or the polymorphism will move to fixation for the wild type allele. If $W_1 = W_{11}$ and $W_2 = W_{22}$, then it can easily be seen that for a wide range of values of W_2 the equilibrium frequency of the mutant allele in the females cannot be less than about 30% without instability of the polymorphism. Some representative figures are given in Table 5.1(b). The mutant allele must be held at a very high frequency if heterozygote advantage in the females is the balancing mechanism.

An alternative, and perhaps more likely, explanation than heterosis is that there may be a localized advantage to the hemizygote which has so far escaped detection, leading to a localized buildup in allele frequency.

The great variety of alleles at this locus argues that neither heterozygote nor hemizygote advantage occurs over a widespread geographic area, or else the 'best' allele or alleles would be common in the population. As with the thalassemias, the advantage of these alleles appears to be highly conditional. It may be, however, that the African A^- variant exerts a far less conditional advantage

Table 5.1. Representative sets of fitnesses leading to stable or unstable sex-linked polymorphisms.

(a) Examples showing how too great a difference in the fitnesses of the hemizygotes (W_1 and W_2) may lead to instability even with strong heterotic selection in the dizygous sex.

W_1	W_2	W_{11}	W_{12}	W_{22}	$\hat{p}$(♀)	$\hat{p}$(♂)
0.8	0.9	0.8	1.0	0.9	0.235	0.215
0.8	0.93	0.8	1.0	0.93	0.0008	0.0007
0.8	0.95	0.8	1.0	0.95	Unstable	Unstable
0.7	0.9	0.7	1.0	0.9		

(b) Examples showing how small shifts in the equilibrium allele frequency can also lead to instability.

$W_2 = W_{22}$	$\hat{y}$	$\hat{p}$(♀)	$W_1 = W_{11}$	Stability
0.995	2	0.667	0.998	Stable
0.995	3	0.75	1.00001	Unstable
0.99	2	0.667	0.9967	Stable
0.99	3	0.75	1.00005	Unstable
0.95	2	0.667	0.9839	Stable
0.95	3	0.75	1.00135	Unstable
0.8	2	0.667	0.9454	Stable
0.8	2.7	0.73	1.0076	Unstable

than the others, which would explain its relatively high frequency in some populations. This variant seems to be a prime candidate for heterotic maintenance for two reasons: first because it has attained high frequency in certain populations, and second because it appears to be only one allele.

It seems likely that resistance to malaria infection can indeed cancel out or even reverse the harmful effects of these alleles, but only for a limited time or over a limited area. Such selection, localized in space and time, would be enough to permit the survival of appreciable numbers of these alleles, but as soon as the selection due to malaria diminished they would be strongly selected against. Whether selection is heterotic, is acting directionally on the hemizygotes, or is some unique mixture of these for each allele and set of environmental circumstances cannot be determined at the present time.

Polymorphisms for which balancing selection can be inferred, although a clearcut mechanism cannot be implicated

The ABO system

The genetics and biochemistry of ABO. The clinical effects of the ABO blood group polymorphism were investigated by Landsteiner (1900), although the genetics of the system were not worked out for another quarter century (Bernstein 1924). Von Dungern and Hirszfeld (1910) demonstrated that the factors were inherited as Mendelian characters, and Bernstein showed that they were in fact multiple alleles at the same locus. Hirszfeld and Hirszfeld (1919) were the first to show that the frequencies of the blood groups varied from one population to another. As a consequence of its extreme clinical importance, this blood group system, the first genetic polymorphism to be found in man, has been studied intensively and a tremendous amount has been discovered about it. It is remarkable that in spite of all this study, and despite the repeated demonstration of various kinds of selection acting on this locus, we are not much wiser about why the polymorphism has persisted in the human population or what the primary function of these antigenic molecules might be.

There are four common alleles at this locus, specifying antigens A_1, A_2, B and 0. The first three are co-dominant, and all are dominant to the last. The usual nomenclature for the alleles specifying the antigens is I^{A_1}, I^{A_2}, I^B, and i. Because of the dominance relationships, it is normally impossible to determine directly the proportions of homozygotes and heterozygotes for blood groups A and B, which in turn makes the detection of differential selection for homozygotes and heterozygotes at this locus much more difficult. The I^AI^B heterozygotes are of course easily determined (though not always without error), but their numbers are low in all populations. Normally, the frequencies of the alleles are determined by maximum likelihood methods.

The clinical importance of this polymorphism stems from the fact that, unlike other polymorphic blood groups, antibodies against those ABO antigens that are not made by the individual appear early in life. Their appearance is apparently stimulated by bacteria and foodstuffs that carry antigens very similar to the A, B and 0 antigens. The evidence for this is chiefly based on some experiments by Springer et al. (1959), who raised White Leghorn chicks in a germ-free environment and found they did not develop an anti-B agglutinin for human cells that was developed by chicks raised normally. The effect of this stimulation of the immune system by the environment is that a person of genotype *ii* and therefore blood type O will, from an early age, carry anti-A and anti-B antibodies in his plasma. This leads to problems in blood transfusion, since if for example type A blood is given to a type O person the anti-A of the recipient will cause the donor's red cells to agglutinate, leading in many cases to a fatal embolism. The reciprocal transfusion will not cause clinical symptoms if the amount of blood transfused is relatively small, since the donor's antibody is diluted into the recipient's bloodstream. Repeated transfusion, of course, requires the donor's and recipient's bloods to be matched much more carefully.

The apparent prevalence in the natural world of antigens similar to the A and B antigens is remarkable, and underlines the importance of the ABO system. The antigenic determinants are different combinations of terminal sugar residues on a carbohydrate molecule, which in turn is covalently coupled to a lipid moiety forming part of the cell membrane of the blood cells and many other tissues. Glycolipid chemistry is difficult because these molecules are insoluble in water, so it is lucky for the chemist that another unlinked locus, the secretor locus, is also polymorphic in human populations. Persons homozygous or heterozygous for the dominant Se gene produce, in a variety of bodily secretions including saliva, sweat and semen, soluble glycoproteins with antigenic carbohydrates covalently coupled to protein rather than lipid. The brilliant work of Watkins (1966) has elucidated the structure of the antigenic sites of these carbohydrates, and a diagrammatic scheme of the additions to the backbone specified by different alleles is shown in Fig. 5.2.

It is of particular interest that the O type, like A and B, is an addition to the backbone which in a few rare cases is absent. Persons homozygous for the rare *h* gene lack this addition and exhibit the so-called Bombay phenotype. Because they do not have the O antigen, anti-O antibodies are produced by their immune systems. This makes matching their bloods for transfusion a very serious problem! It also indicates that O-like antigens must be prevalent in the environment.

The fact that different carbohydrate moieties are added by each allele, at different points on the backbone, strongly suggests that the galactosyl transferases involved are specified by different closely linked genes or by pseudoalleles within one gene. The genetic evidence that the ABO locus is a complex one is unfortunately ambiguous (see Race and Sanger 1975, pp. 33–35, for a summary of the evidence). This is partly because of the apparent infrequency

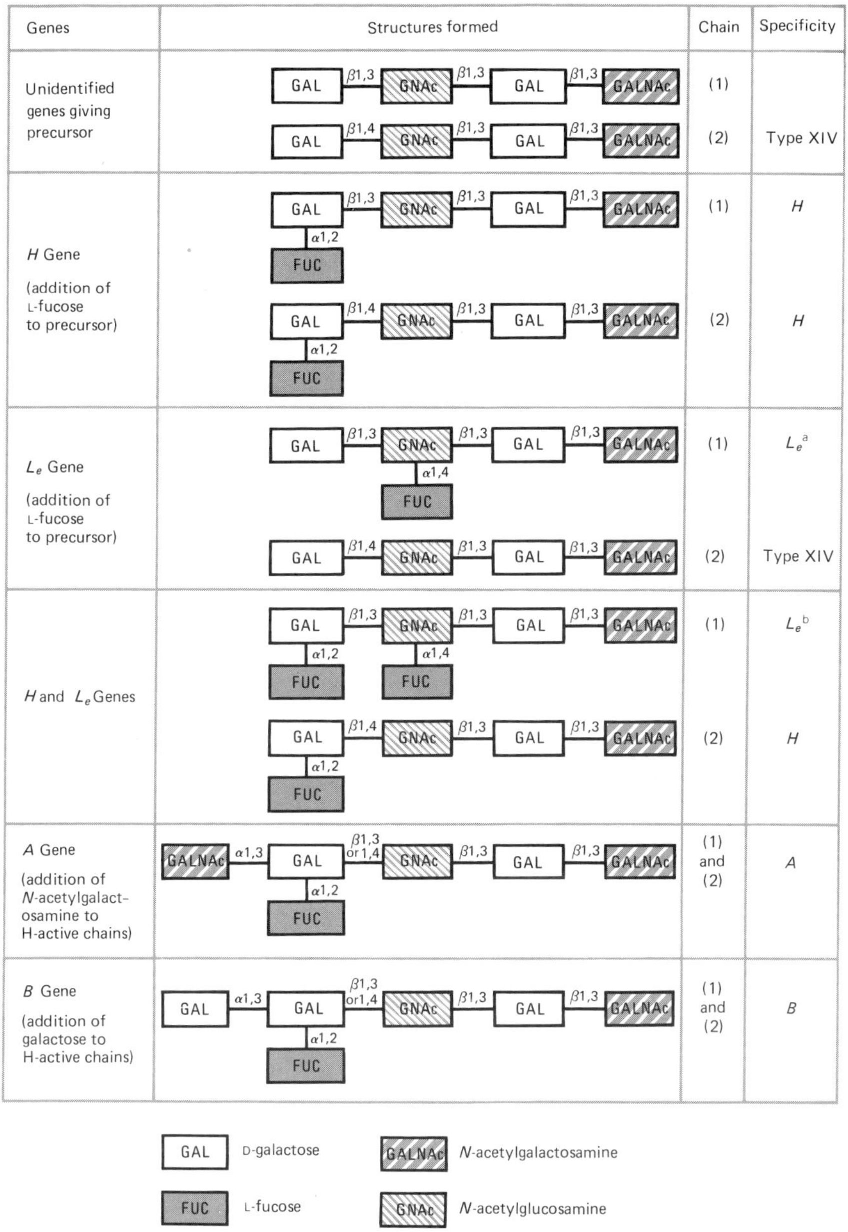

Fig. 5.2. The addition of sugar moieties to a carbohydrate backbone yields the O (H), A and B antigenic determinant. Note how two slightly different precursor chains give rise to the Lewis (*Le*) specificities upon interaction with the H gene. This provides an excellent example of epistatic interaction between two apparently unlinked genes. From Watkins (1966).

of the crossover events and partly because of difficulties in obtaining unambiguous family trees. Because of the differences in function of the different 'alleles', however, it seems likely that the locus is complex.

ABO incompatibility. Many kinds of selection have been demonstrated or postulated to act at the ABO locus. The possibility of ABO incompatibility between mother and offspring was first suspected by Dienst (1905), though he failed to find convincing evidence because he was looking for eclampsia (convulsions caused by toxemia) in the mother rather than effects in the infant. We now know that because of the circulating antibodies in the mother, an A or B child of an O mother, an A child of a B mother or B child of A mother, and an AB child differing in genotype from his mother would all be in danger of exhibiting adverse reactions because of the mother's antibodies. The reverse effects do not occur because the fetus' immune system is not yet functioning.

Unequivocal evidence of incompatibility was not seen until some time after the discovery of another, and clinically more striking, maternal–fetal incompatibility, due to differences in the Rh genotype.

As we will see, Rh maternal–fetal incompatibility does not usually become a clinical problem until the second incompatible pregnancy, since an Rh-negative mother does not normally have circulating anti-Rh antibodies. During birth, however, because of the disruption of the closely interdigitated maternal and fetal tissues in the placenta, a certain amount of fetal blood escapes into the maternal circulation; if the mother and child are incompatible this may result in the formation of maternal antibody. But ABO incompatibility should put the child at risk from the very beginning, and it is remarkable that studies designed to investigate the effects of such incompatibility have revealed so little impact on the population.

Fujiki (1975) summarized five studies in which the numbers of children of parents of incompatible and compatible marriages had been obtained. In only one out of the five was there a significant effect of ABO incompatibility on number of children per family. This argues either that there has been compensation on the part of families at risk—an overproduction of children to make up for an increased number of stillbirths caused by incompatibility—or that the clinical effects of ABO incompatibility are not sufficiently severe to show up in population data.

Hiraizumi (1964) examined data collected by Matsunaga and Itoh (1958) in the city of Ohdate. Here there was no sign of the effects of incompatibility, but in another study conducted in a coal-mining community on Hokkaido, an effect could be seen. This, Hiraizumi suggested, could be due to the severity of the environment of the people examined in the latter study. In this, as in so many other cases, the effects of selection seem to be enhanced by a stringent environment.

Since so many children are at risk in a normally polymorphic population, why are the effects of ABO incompatibility not obvious? There are apparently a number of reasons. If large quantities of soluble antigens or antigens on the surface of other tissues are present in the fetus, this may tend to lower maternal antibody titre to levels below those needed to agglutinate the fetal red cells. Further, there is some evidence that fetal antigens may be more weakly expressed than the maternal ones (Tovey 1945). Or the placental barrier may be sufficiently protective to keep the maternal antibodies from equilibrating with the fetal system.

This latter situation appears not to be the case, since the small maternal IgG antibodies carrying anti-A or anti-B determinants pass freely across the placental barrier (Kochwa et al. 1961). Voak (1969) examined in some detail a number of cases that showed increased bilirubin levels immediately after birth. When this could be attributed to ABO incompatibility, there was a good but not perfect correlation between maternal IgG and hemolytic disease. It had been known for some time (Wiener and Unger 1955) that much more IgG antibody crosses the placenta from type O mothers than from those of type A or type B, and incompatible children of type O mothers are at correspondingly greater risk of hemolytic disease.

It seems likely, then, that the protection against ABO maternal–fetal incompatibility is due to a complex of factors, not least of which is the genotype of the mother. It is worth noting that Voak (1969) and others found a significant number of ABO-incompatible cases with 'mild' symptoms of between 5 and 10 mg % of bilirubin. There appears to be a continuum of effects, dependent on maternal and fetal genotypes and on circumstance.

A factor which must also be taken into account is the increased frequency of early abortion due to ABO incompatibility. This is another way in which ABO incompatibility differs from that due to the Rh locus, in which clinical complications tend to arise near the end of term. In a large study, Lauritsen, Grunnet, and Jensen (1975) examined 288 spontaneous abortions occurring during the first four months of pregnancy. It had been possible to type 555 of the parents. On the assumption that the differences in numbers of chromosomally-normal stillbirths between compatible and incompatible pregnancies were due to ABO incompatibility, they calculated that a maximum of 18% of the abortions were due to the incompatibility. The numbers of chromosomally-abnormal stillbirths were not significantly different between the two groups.

A striking qualitative observation made by these workers was that many of the incompatible fetuses exhibited disturbed organogenesis, which may indicate that the effects of incompatibility involve other tissues.

This increase in spontaneous abortion relatively early in pregnancy, presuming that incompatibility is in fact the cause, might not under unstressed conditions make much of a difference to family size or to fitness of the different genotypes. But repeated abortions under more primitive conditions and in the absence of modern medical care could have made a much more pronounced difference. Further, while only about 0.1% of newborns exhibit serious complications of ABO incompatibility, milder symptoms of the disease including intermediate bilirubin levels are more common and might affect neonatal survival under primitive conditions.

Incompatibility between mother and fetus results in selection against the heterozygote. This results, in turn, in an unstable equilibrium in which the rarer alleles will eventually be lost. In most populations these are the I^A and I^B alleles, and it remains a mystery why these have been retained.

The mystery is compounded by the observation that in studies in which parents and offspring have been typed, there is regularly a deficiency of AB offspring. This is often statistically highly significant. Fujiki (1975) tabulated the results of ten studies in which parents and children were typed for ABO, and found a striking deficit of AB children in nine of them. The tenth, a study carried out in Japan in 1973, showed a significant excess of this genotype in the children. There is no correlation with the time at which the studies had been done, which tends to make technical problems less likely. Wiener (1943) had noted the apparent deficit of AB in population surveys, and attributed it to the occasional misclassification of A_2B as B since A_2 is a weaker antigen than A_1. But such an argument presumably cannot be applied to studies in which a difference is found between parents and offspring, and in Japanese populations at least, in which two out of three studies showed a deficit, A_2 is at very low frequency (Chung et al. 1960).

If this deficiency is real, then the persistence of I^A and I^B becomes even more of a problem. There must be some counterbalancing selection for these alleles, sufficient to maintain the polymorphism.

Associations with disease. An obvious direction of investigation of these clinically highly important alleles is to enquire whether they are associated with particular diseases. Hirszfeld (1928) was the first to look for such associations, but was unsuccessful, presumably because his sample sizes were limited. When Aird et al. (1953) published the first data showing an association between A and carcinoma of the stomach, an immense literature was spawned and a great deal of controversy resulted.

The mere fact that there is an association between a particular genotype or phenotype and a disease does not mean that one is the cause or a contributive cause of the other. The association could arise by chance, although if it occurs in many studies in different parts of the world, this likelihood is greatly diminished. The sample being studied may not be representative of the population—for instance, the people coming to a particular hospital may not be a random sample of the population surrounding it, because of economic or social stratification. If the association turned up in one or two studies, and not in others, the stratification argument or the argument that the association is due to chance would carry a great deal of weight. But as we will see, in at least some cases the associations have been detected repeatedly in many parts of the world.

A third possibility is that the association is due not to the ABO alleles themselves, but to alleles at loci closely linked to them. Given sufficiently close linkage, this could certainly happen, provided also that the hypothetical selected alleles were in the coupling configuration and at the same frequencies as the alleles at the ABO locus. At least for some associations, this linkage relationship would have to be maintained throughout a diversity of populations from many parts of the world.

Associations with blood groups have now been found for a great variety of diseases, and have been ably summarized by Mourant et al. (1978). Their monumental volume also includes associations of diseases with many other human polymorphisms, and their review and the recent book edited by Dausset and Svejgaard (1977) summarize the rapidly growing information about association of human leukocyte antigen (HLA) genotypes with diseases.

In Table 5.2 I have summarized the enormous number of associations given in the survey of the literature by Mourant et al. (1978). The table gives the number of associations that are significant, at least at the 0.05 level, for the polymorphisms (excluding HLA) for which extensive data are available. It is hoped that by pooling the data in this way, individual studies that might be erroneous for a variety of reasons will not contribute much cumulative error to the totals.

A very large amount of data is available for ABO, far more than for all the other polymorphisms combined. The number of significant comparisons is roughly three times what would be expected from chance, and this excess is found in virtually all the disease categories.

A striking pattern, when the observations are considered in more detail, is the tremendous variation from study to study within a particular disease. Even for well-known associations, the heterogeneity is striking.

One strong association is that of type A with cancer of the stomach. Of 161 comparisons of affected A and O, the A/O ratio (corrected for frequencies in the controls) was greater than one in 133 cases, and less than one in only 28. But it was significantly greater in only 54 studies, with some very large studies showing no significant excess. The same pattern is seen with the association of O with duodenal ulcers. Of 102 comparisons, 95 showed an excess of O over A, but only 53 of these were significant. Again, some very large studies showed no significant association.

Some suggestive patterns in the associations of ABO with diseases can be seen. While gastric cancers are associated with type A, peptic ulcers, both gastric and duodenal, are associated with type O. The duodenal ulcer association is particularly strong. Further, Langman and Doll (1965) showed that bleeding ulcers of all types were most closely associated with type O. While cancers tend to occur later in life and have presumably little effect on reproductive ability, ulcers might easily have an effect. It is impossible at the present time to assess the effect of a bleeding ulcer on the reproductive success of a primitive man or woman, but we may assume that it was not negligible.

Another association, particularly apparent when large amounts of data are

Table 5.2. Summary of studies on associations of human polymorphisms with disease. Data summarized from Mourant et al. (1978). Each comparison represents an individual study, often involving a very large sample.

	ABO			MN		Rh(D^+/D^-)		ABH		Plasma haptoglobin		PTC tasting	
Disease group	No. of comparisons	No. significant (<0.05 level)	% significant	No. of comparisons	No. significant	No. of comparisons	No. significant	No. of comparisons	No. significant	No. of comparisons	No. significant	No. of comparisons	No. significant
Neoplasms	1636	183	11.2	19	0	167	14	37	3	25	2	10	5
Infective and parasitic diseases	871	133	15.3	40	2	72	10	10	3	20	4	22	3
Endocrine and metabolic disease	312	41	13.1	8	0	19	0	12	1	4	0	71	13
Disease of blood	131	16	12.2	3	1	15	0	4	0	—	—	1	0
Mental disorders	273	21	7.6	18	1	25	5	4	1	6	0	3	2
Diseases of nervous system and sense organs	132	17	12.9	6	0	9	3	5	2	6	0	19	7
Disease of circulatory system	297	56	18.9	17	0	32	3	12	3	5	1	—	—
Disease of respiratory system	44	4	9.1	—	—	3	0	1	0	—	—	—	—
Disease of digestive system	662	190	28.7	4	0	44	6	36	25	3	0	—	—
Disease of genitourinary system	80	7	8.7	3	0	13	2	2	1	1	0	2	1
Perinatal disorders	57	16	28.1	—	—	—	—	—	—	—	—		
Diseases of skin	82	16	19.5	3	1	7	2	2	1	6	1	1	0
Diseases of musculoskeletal and connective tissue	42	10	23.8	2	0	—	—	—	—	2	0	—	—
Congenital anomalies	135	9	6.7	9	1	28	1	6	0	3	0	—	—
Allergies	54	9	16.7	5	0	16	5	3	0	—	—	—	—
Total	4808	728	15.1	137	7	453	51	134	40	81	8	129	31
				% significant = 5.1		11.3		29.9		9.9		24	

gathered together, is of heart disease with type A and B. Jick et al. (1969) pointed out that associations of A and O with thrombo-embolism and bleeding, respectively, could be at least partially explained by the higher levels of anti-hemophilic globulin observed in type A.

Cavalli-Sforza and Bodmer (1971, pp. 303–4) used the association of O with ulcers to calculate possible selection coefficients acting on the ABO genotypes, and found them to be very low. Their assumption was that the majority of the diseases which show an association occur late in life, when their effect on reproductive ability is minimal, or (as in the case of ulcers) have relatively minor effect. But if there are real differences in the blood-clotting ability of A and O, manifested at the present time in different incidences of these diseases, it is possible that differential selection on these alleles was much greater under primitive conditions.

Associations with infectious diseases are equally complex. Type O shows a significant association with scarlet fever when compared with A, though not when compared with B. Scarlet fever is often accompanied by infection with Group A β-hemolytic streptococci, and type O has been shown to have a greater immunological reaction against streptococcal infection. For example, Haverkorn and Goslings (1969) showed that there were fewer streptococci in the saliva of persons secreting large quantities of H substance (the type O antigen) than in type A secretors. On the other hand, rheumatic fever and rheumatic heart disease show a strong association with A and B, based on very large samples. Rheumatic fever is due to sensitization to *Streptococcus pyogenes,* and an intense immunological response increases the probability of contracting the disease. It would appear that type O, with its already high titres of antibody to A-like and B-like antigens, is not as easily sensitized to the rheumatic-fever organism. This lack of sensitization may explain why type O persons appear more susceptible to the streptococcal infections accompanying the symptoms of scarlet fever, and less susceptibility to the consequences of rheumatic fever.

An early report (Vogel et al. 1960) on the discovery of A-like antigen on the surface of *Pasteurella pestis* led to the intriguing possibility that plague has been a selective agent acting on the ABO locus. The finding has been called into question by Springer et al. (1961), who found that only one out of three strains of *P. pestis* tested showed even weak O and B specificity. Pettenkoffer et al. (1962) replied that a strong antigenic reaction could be obtained only with the smooth variant of *P. pestis,* which presumably had its polysaccharide coat intact. A similar controversy over the nature of antigenic determinants has been fought over the vaccinia virus, which in the hands of Vogel's coworkers (Pettenkoffer and Bickerrich 1960) showed A antigenic specificity. Moreover, Vogel and Chakravartti (1966) investigated an unvaccinated, primitive population in West Bengal and Bihar State of India, and found that the course of smallpox was markedly less severe in B and O than in A patients. Downie et al. (1965) and Sukamaran et al. (1966) found no such association in hospitalized patients from Indian populations.

Wiener (1960) suggested that the high rate of mutation of surface antigens on bacteria and viruses makes it unlikely that a specific constitutive antibody would confer much protective effect. It is quite obvious, however, that not enough is understood about the interaction of blood groups with diseases. The well-established interaction of ABO with diseases caused or triggered by varieties of streptococcus certainly suggests that it can play a role in other diseases.

The interactions of ABO and other loci are also complex and controversial. We have already seen that there is a large effect of the secretor locus on the way the ABO polymorphism is expressed, and an additional complication has been discovered. Arfors et al. (1963) investigated the electrophoretic patterns of alkaline phosphatases in the plasma, and found that two isozymic forms secreted from bone and liver tissue were always present. A third band showed variable intensity, but it was not a true polymorphism. This third isozyme is always present in the intestinal mucosa, but its secretion into the plasma is dependent on the presence of the Se allele. Further, the secretion can be repressed by the I^A allele. This repression led Beckman (1968) to postulate that the blood groups, as well as the secretor locus, may be instrumental in regulating the secretion of glycoproteins. He pointed out that this is the first non-immunological differential function of the blood groups that has been reported. Rendel and Stormont (1964) have reported an association of the presence of serum alkaline phosphatase with blood groups in cattle and sheep. To make the situation even more intriguing, Bottini, Lucarelli, and Pigram (1972) showed that the distribution of polymorphic alleles of placental alkaline phosphatase is significantly different in ABO-compatible and incompatible fetuses. A correlation found between placental alkaline phosphatase alleles with ABO in the population at large was suggested by Bottini, Lucarelli, and Gloria (1972) to be compatible with the observations on fetal survival.

This last set of studies pinpoints the great weakness in the above discussion of selection and the ABO blood groups. Despite their having been intensively studied for over three quarters of a century, we still have no firm idea of the function in the cell membrane of the molecules to which these antigenic determinants happen to be attached. At the risk of sounding sententious, is it any wonder that we cannot ascribe a function to most of the polymorphisms that have been described in other organisms?

Other selective forces. Much that is known or suspected about the interaction of ABO with other loci will not be considered at length here. I have tried to concentrate on situations which have been thoroughly described, and in which the evidence for selection or differential activity is firm. But complex interactions will continue to emerge. For example, there is suggestive evidence concerning an interaction of ABO with the α-haptoglobin locus. Haptoglobins are globular proteins found in the plasma in a great variety of multimeric states. They have a high affinity for free hemoglobin and have been shown to aid its breakdown in the liver with the subsequent recovery of iron. The polymorphism at the α-haptoglobin locus is complex, with one allele being the result of a partial gene duplication (Smithies 1964). In population studies alleles at the locus have been divided into two major categories, Hp-1 and the partial duplication Hp-2. There are a number of Hp-1 alleles, but the molecular weight difference between Hp-1 and Hp-2, along with the tendency of the latter to associate more readily into multimers, has made distinguishing these two categories on starch gels very easy.

Ritter and Hinkelmann (1966) were the first to describe an excess of Hp-1 among ABO-incompatible offspring, and this observation was confirmed by Kirk et al. (1970) and Kirk (1971). Kirk initially supposed that some protection was conferred by Hp-1, which has a higher affinity for hemoglobin than Hp-2, acting as a more efficient scavenger of iron. There are problems with this interpretation, however, not the least of which is that in many infants synthesis of haptoglobins is delayed until after birth. This delay is not correlated with haptoglobin genotype (Beckman and Grivea 1964). Further, Vana and Steinberg (1975) have questioned the original observation of association with ABO incompatibility, for they have found no evidence of such association in the Hutterite populations which they have studied. Rather, they find that the Hp-1 allele is non-randomly associated with ABO genotypes in the population itself, leading to a spurious correlation. Because of the association, the incompatible offspring will tend to have a higher frequency of Hp-1 than the compatible ones. Whether this is due to selection or to linkage disequilibrium in these small populations remains to be discovered.

Selection pressures which should lead to a disappearance of the ABO polymorphism have been amply demonstrated, primarily as a result of ABO incompatibility but also because of a lower rate of survival among children with the I^AI^B genotype. Such selection was undoubtedly stronger in the past than it is at the present time. Selection which should lead to a maintenance of the I^A and I^B alleles has yet to be demonstrated, however, and the mechanism of this maintenance remains a mystery. There is no indication of heterozygote advantage from population data presently available except for the statistical study of Yamazaki and Maruyama (1974) mentioned on page 57. Another very real possibility is that there is now, or has been in the past, geographically localized homozygote advantage for I^AI^A and I^BI^B sufficient to maintain the polymorphism.

With all the selection that has been demonstrated at this locus over years of investigation, it is not surprising, perhaps, that the worldwide distribution of the alleles at this locus is decidedly nonrandom. This was pointed out by Brues (1954), whose Fig. 1b is reproduced as Fig. 5.3. She plotted 215 representative human populations, and found that the ABO frequencies were confined to only about a fifth of the possible range. There is a pronounced negative correlation of I^A with I^B frequency. Brues suggests that this could have been due to historical circumstance—any sort of distribution could be generated by the appropriate series of accidental differentiations followed by the formation of clines—

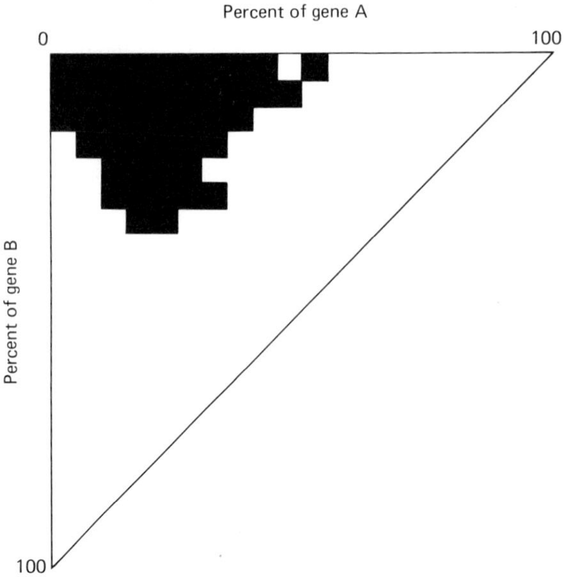

Fig. 5.3. The possible and actual range of frequencies at the ABO locus. The actual worldwide range is much smaller than the possible range. From Brues (1954).

but considers this to be unlikely because there is so much evidence for diversity in allele frequencies in small primitive populations such as the Eskimos, the Australian aborigines and the Ainu. She anticipates the argument made in connection with the Yanomama (p. 63) as follows (p. 564):

> The evidence for the vigor of genetic drift under more primitive conditions is so convincing as to be disturbing. If it has acted as strongly even after the formation of certain races and subraces as we now know them, why has it not produced, during the whole period of the differentiation of the species, often under conditions of isolation and smallness of numbers, more marked than have existed subsequently, a range of variation much greater than we now see?

The answer is, of course, that the alleles at this locus are not neutral. It is remarkable that very few human populations are not segregating for the I^A and I^B alleles; only the Indian tribes of eastern North America, and Central and South America are monomorphic for the *i* allele. If drift is as strong a generator of genotypic diversity as the data from the Yanomama indicate, why are there not more monomorphic populations?

We have already seen some possible answers to these questions. In addition to selection, the role of continued migration in diminishing the differences between groups has not been fully explored, and may be powerful enough to reverse the differentiation seen in localized groups such as the Yanomama.

The Rh system

Selection is also acting on this blood group polymorphism, which was discovered by Landsteiner and Wiener (1940, 1941). Fetal–maternal incompatibility at the Rh locus, even mitigated as it can be by the effects of ABO incompatibility (Race and Sanger 1975), was until recently a significant cause of infant mortality in the developed countries. It continues to be important in many parts of the world where both D^+ and D^- alleles are segregating in the same population. Because the Rh-negative mother does not normally carry antibodies to Rh-positive in her bloodstream, she must be immunized by her first Rh-positive child. The placental barrier is normally very effective against the passage of fetal red-blood cells carrying the Rh^+ antigen, but during disruption of the placental tissue at birth some fetal red cells are released into the maternal circulation. They may or may not stimulate the production of maternal antibody, in part depending on whether the mother carries antibody against the child's ABO blood type. In the latter case, the fetal RBCs are destroyed before they stimulate the maternal immune system. But the mother may not produce antibody even though conditions appear to be met for such production, for reasons that are not understood. At the other extreme, because of a fault in the placenta, the mother may begin producing antibodies during the very first incompatible pregnancy.

The hemolytic jaundice which results from the incompatibility reaction becomes progressively more severe as the pregnancy advances, sometimes necessitating an early delivery or a transfusion *in utero*. But, perhaps because the Rh antigens are confined to the surface of the red cells, there does not appear to be a high incidence of early stillbirths in this condition as there is with ABO incompatibility. Since the mortality occurs at term, this constitutes a much more severe selection against the heterozygote.

Because selection against the heterozygote leads to the loss of the rarer allele, one would expect some human populations to be 'monomorphic' for Rh^+ and Rh^-. I emphasize the term monomorphic because Rh^+ is itself a heterogeneous collection of alleles and pseudoalleles that constitutes one of the more complex human polymorphisms. Of C, D, and E, the three pseudoallelic loci postulated by R. A. Fisher (1947), D is primarily responsible for the incompatibility reaction. The question of the probable structure of the Rh locus, and its bearing on the nature of closely-linked polymorphisms in general, will be considered in more detail in Chapter 8.

It is notable that the frequency of the alternative allele *d* is very high among the peoples of Western Europe, the Basques and Celtic peoples such as the Welsh and Irish, who form in part relict populations pushed to the extremes of the continent by succeeding waves of migration (Chalmers et al. 1949). At the other end of Eurasia, the Oriental peoples are almost completely *D*. The possibility exists that by chance, the *d* allele was fixed or nearly so in Western

Europe (though it is now found at a frequency greater than 50% only among the Basques), while the *D* alleles were fixed in the east, so that the intermediate frequencies now seen through much of Europe and the Middle East could be due primarily to migration. Similar patterns of migration may be postulated to have occurred in Africa, though less convincingly. This interpretation would thus not require a counterbalancing positive selection to maintain the polymorphism, but only that the early history of the human species was such that different founding populations were on both sides of the unstable equilibrium. Such an explanation, while possible for Rh, is perhaps less likely for ABO because all human populations are clearly on one side of the unstable equilibrium and should become monomorphic for O. In short, the need to postulate some sort of balancing selection preventing fixation is greater in the ABO than in the Rh system.

ABO, Rh, and MN, or at least systems which cross-react immunologically with them, are present in the other primates (Wiener et al. 1964). But the chemistry of these blood groups has not been examined in detail, and it is possible that the alleles arose independently in these species. Further, as one moves away from man, the immunological cross-reaction becomes progressively weaker (Wiener et al. 1968). So far as Rh is concerned, our closest relatives the chimpanzees are (from a limited study) monomorphic for Rh^+, while the gorillas are polymorphic. The 'great age' of the polymorphisms has been used (Wills 1973) as an argument for their selective maintenance, especially since the polymorphisms appear to have passed through the speciation process. But there is, in fact, a perfectly simple way that the ABO polymorphism, for example, could be generated anew in each species. If I^A and I^B were to pass through the bottleneck, the null allele *i* could arise subsequently by mutation. Should this have been the origin of the human ABO system, the current distribution of the ABO alleles cannot be explained without the invocation of selective advantage for the null alleles. A less likely scenario is that the alleles which passed through the speciation process were null alleles, which then became 'reactivated' by back mutation. For them to have achieved appreciable frequency would require that the reactivated alleles confer an advantage sufficient to overcome selection against the heterozygote. Either way, even if the polymorphism is newly-arisen in our species, the current ABO distribution cannot be explained without the intervention of balancing selection.

If the same events occurred during speciation in the Rh system, we are required to explain how different human founding populations had different frequencies of Rh alleles.

Disease associations with the Rh system are not as marked as with ABO (Table 5.2). It may be that incompatibility is the primary source of selection on the Rh alleles at the present time, and there are no obvious current sources for the balancing selection which would be needed to maintain the polymorphism.

A survey of the data compiled by Mourant et al. (1978) strongly suggests that there may be some interaction with the organisms responsible for typhoid and paratyphoid, mumps and infectious mononucleosis, with D^+ having the advantage in each case. All these association were found in the same study (Paciorkeiwicz 1970); they are very much in need of confirmation. The advantage of D^+, if it proves to be real, would simply reinforce the directional selection already known to be occurring at this locus.

Perhaps the most plausible explanation for the current distribution of the Rh polymorphism is that the positive and negative alleles were strongly selected for in different parts of the species distribution, by some factor or factors unknown, and that the distribution we see is due to subsequent migration.

The MN system

The MN system is remarkable for two facts. The first is that no selective advantage or disadvantage has been demonstrated for any of the genotypes at the MN locus itself. The second is that not only the MN antigenic determinant, but the entire molecule on which it is found, glycophorin A, may be absent without noticeable clinical effects. Glycophorin A, a sialoglycoprotein, makes up about one quarter of the total membrane protein in the red cell. Tanner and Anstee (1976) report on members of three families in which there appears to be a complete absence of glycophorin A, and a greatly lowered amount of the MN antigens, with no apparent clinical symptoms. There seems to be no obvious need, not only for the antigenic determinants, but for the whole molecule on which they are found!

Mourant et al. (1978) list 137 different studies of disease association with MN, for 69 different conditions. No more than the number of significant associations to be expected by chance in such a large body of data are found. As will be seen from Table 5.2, this is in vivid contrast to the situation at the ABO, secretor and PTC loci, and suggests that the large numbers of significant disease associations with alleles at the latter loci are indeed measuring *something*. At the present time, the MN system seems to be the best human candidate for a neutral locus.

Yet there are certain puzzling aspects of its distribution. The frequency of the M allele ranges from over 90% in the Indian tribes of the Pacific Northwest to less than 2% in the highlands of New Guinea. In 1288 studies reported by Mourant et al. (1976), many involving small sample sizes, there is only one instance of monomorphism for M and none of monomorphism for N. This suggests that there may *possibly* be balancing or frequency-dependent selection here, although it would appear to be weak.

Harrison and his colleagues (Harrison et al. 1976) have used the interesting approach that a population in a primitive state, certainly subject to stronger selection pressures than are most human populations at the present time, might be revealing in its associations of polymorphisms with disease. Two New Guinea populations were studied. These workers examined the association of various clinical symptoms—proteinuria, which is symptomatic of kidney disease, splenomegaly and hepatomegaly, infection with malarial parasites, goitre—with a variety of polymorphisms. The expected association of ABO with goitre was found, as were a number of other associations of variable significance. It is difficult to know how much weight to give the data, since the sample sizes were small. The strongest associations were with haptoglobin, a locus which we have seen may be involved in incompatibility reactions although the evidence is unclear at present. Hp-1 homozygotes were found to be at significantly greater risk of proteinuria, splenomegaly and hepatomegaly, the first indication of such an association with clinical symptoms and one which would certainly not be apparent in a less primitive population. There was also a significant association of MN with splenomegaly and hepatomegaly, and of the closely-linked locus *Ss* with parotid enlargement, but these associations were found only in males and only in some of the groups surveyed.

Our ignorance of MN is still vast. There is no indication of its function, and it appears to be virtually invisible to our own immune system. There is also no appreciable body of information on the association of diseases with the genotypes at the nearby S blood group locus, also antigenically invisible and sufficiently closely linked to the MN locus that recombination is a rare event (Gedde-Dahl et al. 1967). It is notable that the allele distributions at these two linked loci are quite different. While M is found at virtually every frequency somewhere in the world, S is essentially never found at a frequency higher than 0.5. The two loci are in linkage disequilibrium, but the pattern of disequilibrium is consistent with the effects of sampling (see pp. 243–44).

The HLA system

In contrast to the MN system, in which selection is quite undetectable by any means currently at our disposal (except perhaps by the approach taken with the New Guinea populations studied by Harrison and his coworkers), the HLA system shows strong signs of selection due to association with disease. HLA stands for human leukocyte antigens, and the system encompasses a very large number of different antigens, normally studied on the surface of white blood cells, which can be detected by cytotoxicity assays. Four closely-linked loci, linked in the order A, C, B, and D, are currently understood to specify these antigens, and there is a series of multiple alleles at each locus. The total length of the HLA complex is about a map unit, which provides for constant generation of new combinations of alleles, but which also ensures that the complex is usually inherited as a unit.

The antigens undoubtedly play a role in determining whether a donor tissue graft will be accepted or rejected, but attempts to increase the success rate of grafts from unrelated donors by HLA typing have had only mixed success. It is clear from these studies of tissue rejection that the full complexity of the HLA system has yet to be understood. Tissues from a sib with the same HLA type have a much higher chance of acceptance by the recipient than do tissues from an unrelated donor, no matter how carefully the HLA type of the latter is matched to that of the recipient. This indicates that other subgroups within each locus or perhaps other closely-linked loci influencing the probability of rejection remain to be discovered.

The polymorphic genes specifying the second and fourth components of complement have been located in this complex as well (Teisberg et al. 1976; Alper 1976). The complement system, a necessary component in the destruction of cells by the immune system, is thus intimately involved with HLA. Because of linkage disequilibrium, certain complement and HLA genotypes will tend to be inherited as a unit.

The association of HLA types with diseases is a rapidly-developing field of investigation (Dupont et al. 1976; Dausset and Svejgaard 1977). The most remarkable set of associations have been those of antigenic type B27 with a variety of arthritic conditions, particularly ankylosing spondylitis, or 'bamboo spine', in which the vertebrae gradually fuse together. Over 90% of those affected by this disease are B27, though only 9% of the general population has this type. A person with type B27 has 87.8 times as great a chance of suffering from this disease as does a person from the general population. And, significantly, this association with B27 has been found everywhere it has been looked for, providing *prima facie* evidence that the antigenic type is directly responsible for the disease—though, of course, because of disequilibrium it may not be the currently detectable alleles that are responsible.

Of course, only a small fraction of those with B27 develop the disease, and not everybody with ankylosing spondylitis carries B27. There are complicating influences of genetic background, as were seen with ABO and disease association, which makes a direct causal relationship less certain.

The association of other antigenic types with diseases is much less striking, but some remarkable associations have been discovered. Type B8 is very strongly associated with coeliac disease, resulting in a nine times greater risk for the B8 carrier. Slight associations have also been found between B8 and autoimmune diseases, both lupus erythematosis and the more generalized group of symptoms lumped together under the term Sjögren's disease. B8 and B18 are also associated weakly with juvenile diabetes.

If a disease is associated with more than one determinant, there is a good possibility that the disease is really more strongly associated with, and perhaps causally related to, another determinant at another locus in linkage disequilibrium with these (Mourant et al. 1978). Further, different ethnic groups and geographical regions often carry different gametic types in excess, again suggesting that only a small portion of the genetic complexity of the major histocompatibility system has been unravelled.

There have therefore been intensive efforts to examine the relationship between the histocompatibility determinants themselves and the diseases with which they are associated. The investigations have centred on the role of the histocompatibility antigens as cell-surface receptors. Rubinstein (cited in Dupont et al. 1976) has shown that HLA-identical sibs of children with juvenile-onset diabetes mellitus have an unusually high binding capacity for insulin on their monocytes—a condition which in extreme form can lead to diabetes. Arniaz-Villena and Festenstein (1975) have shown that specific histocompatibility antigens are acting as complement receptors. This may, in a way not fully understood, be a direct link with autoimmune diseases in which tissues are attacked by a person's own immune and complement systems.

Among the diseases associated with human histocompatibility antigens, malignancies are notably absent. This is quite the opposite of the situation at the mouse H-2 loci, which appear to be highly homologous with the human HLA system. Certain combinations of alleles at these loci have been shown to influence strongly the development of viral leukemia and mammary-tumor virus susceptibility. It may be that in man similar genotypes for viral susceptibility are not linked to HLA. This would suggest, in turn, that mouse H-2 alleles may not be responsible for the susceptibility to cancer viruses, but rather that it is due to alleles at closely linked loci in disequilibrium with them in the mouse but not in man. Few malignancies in man appear to be directly related to viruses, however.

While association of the histocompatibility antigens with particular disease states is undoubted, the causal relationships if any remain to be elucidated in detail. It is striking that so many associations are found with the autoimmune diseases, including in addition to the ones mentioned above, myasthenia gravis, psoriasis and multiple sclerosis (Ryder and Svejgaard 1976), and it seems certain that the antigens themselves will be found to play an important role. But so far as the human studies are concerned, these associations once again do not suggest in the least why the enormous multiplicity of alleles is found at these loci, or why they have been maintained and their analogues have been maintained in mouse and rabbit.

Because of the large number of alleles, over 90% of the human population is heterozygous at each of the major genes that have been thoroughly investigated. Several suggestions have been made to account for the maintenance of this variability. One, which would tie in very strongly with the differences in viral susceptibility observed in mice of different antigenic types, has to do with the fact that many mammalian viruses carry an outer coating derived from the cell membrane of the cell in which they are grown. This viral envelope, though modified, can be shown to contain host antigens (Doherty et al. 1976). When a virus from an antigenically different organism invades a new host, the antigens it brings with it may 'disguise' the newly-invaded cells. In order for the immune system to mediate cytotoxicity, the target cell must carry antigens allied to those of the host immune system in addition to the different antigens. Heterozygotes for HLA (or H-2 in mice) would have a greater range of host antigens. Their immune system would be more effective in destroying virus-modified host cells, since the chance would be greater that the modified cells would still carry a recognizable antigen. Evidence for this was found by Doherty and Zinkernagel (1975) in mice; their experiments showed that virus infected host cells were more readily destroyed in H-2 heterozygotes than in homozygotes.

If this proves to be the selective raison d'etre for the histocompatibility loci, it would appear that disease association with particular antigenic types is not strong enough to remove them from the population. Thus the disease association may be an unfortunate side-effect of the diversity of types, and reflect accidental combinations of alleles that for some reason trigger the immune response inappropriately. This might be enough to change the relative frequencies of alleles in the population. It seems not to be enough to lower the B27 antigenic type, for example, to less than the 5–10% in which it is normally found in Caucasian populations. In turn, this could be because of the rarity and generally late onset of the associated diseases.

Parallels may be drawn with the pollen incompatibility loci that have been found in a number of plants and have been thoroughly investigated theoretically (Fisher 1930; Wright 1939). In these systems, pollen will not grow in a style with which it shares an allele, discouraging self-fertilization. Selection has therefore been for a multiplicity of alleles, more than 200, for example, in clover, with an associated load which is apparently easily compensated for because of the excess of pollen involved in fertilization. In both situations, pollen incompatibility and HLA histocompatibility, selection appears to have produced enough alleles to assure virtually 100% heterozygosity. A practical requirement for the stability of such an extremely multiallelic system, as we will see on pp. 174–76, is that there be no great difference in fitness between any of the heterozygotes and between any of the homozygotes, and that the homozygotes all be less fit than the heterozygotes. This is apparently fulfilled in these two cases, and has led to extensive and stable polymorphism. Indeed, it is striking that the variation in frequency of the HLA alleles among different racial groups of the human population is quite small, much less on the average than the Rh, Duffy, MN or Gm polymorphisms (Bodmer et al. 1972). This is suggestive of, though it does not prove, stabilizing selection in these systems.

Because of its potential clinical importance, the HLA polymorphism is under the most intensive investigation. The HLA antigens are glycosylated polypeptides. Preliminary sequencing of the N-terminal amino acids of the HLA-A, B, and C antigens indicates that all arose from a common ancestor, with C more closely related to A than B (Crumpton et al. 1978). Further, the allelic differences between different A and different B alleles are due to more than one amino acid substitution (Bridgen et al. 1976). One possibility, which awaits more sequence information for confirmation, is that the multiple alleles at each locus may have arisen in part by intragenic recombination, so that the HLA system has the capacity to regenerate alleles by recombination, even though they may be lost through drift. This in turn might help to explain the remarkable stability of this multiallelic system, even in the face of the weak directional selection due to association with autoimmune and other diseases.

It should be noted parenthetically that intragenic recombination may be a significant evolutionary force. The haptoglobin-2 allele is a partial duplication of two slightly different alleles of haptoglobin-1, and it may have occurred only once since all investigated Hp-2 alleles are spliced at the same point. Homozygotes for 2 may mispair and recombine, giving rise to a triplication and a Hp-1 allele. Triplications are found rarely, indicating they are probably selected against. Unequal crossovers will thus tend to remove Hp-2 alleles from the population. Nance (1963) has pointed out the possible importance of such processes during evolution. The high frequency of Hp-2 in the population seems very likely to be the result of balancing selection, working against the genetic instability of the Hp-2 allele.

The Duffy system

One of the most exciting new avenues of investigation of the influence of polymorphism on disease susceptibility has come with the investigation of the interaction of cell-surface receptors and malaria parasites. It has been known for many years (Bruce-Chwatt 1948) that different parasites differ greatly in their ability to invade RBCs—*P. vivax, P. malariae,* and *P. ovale* attack only younger cells, so that parasitemia rarely rises about 1%, while *P. falciparum,* though it prefers younger cells, can invade cells of all ages, greatly increasing the average clinical severity of the disease. But more detailed investigation of the mechanism of penetration, aided by the use of similar systems in other primates, has revealed some unexpected findings.

Innate susceptibility has a great deal to do with the success of the infection. For instance, *P. knowlesii* infects rhesus monkeys more easily than man, and this is known to be correlated with the relative susceptibilities of monkey and human RBCs to infection *in vitro.* Miller et al. (1973) investigated this difference using selective digestion with various proteases. They found that chymotrypsin and pronase-treated human erythrocytes were resistant to infection *in vitro,* but that trypsin and neuraminidase had no effect. This indicated that the receptors were not a component of the major sialoglycoprotein in the membrane, or neuraminidase treatment would have destroyed them. They also observed that chymotrypsin did not block infection of rhesus erythrocytes.

Further investigation (Miller et al. 1975) showed that chymotrypsin destroyed the Duffy (Fy) blood-group determinants on human erythrocytes, but left the Duffy-like determinants present on the surface of rhesus RBC's intact. This appears to be because the determinants are chemically (though not antigenically) distinct, consisting of a glycolipid on the rhesus cells and a glycoprotein on the human cells.

These observations led to a fascinating conjecture. There are three common alleles at the Duffy locus, Fy^a, Fy^b, and *Fy,* the last of which seems to be analogous to the i allele of the ABO system, in that it is recessive. It is also not antigenic except under special circumstances. The frequency of Fy is very high ($>$ 99%) in central and western Africa, and it has been shown (Butcher, Mitchell, and Cohen 1973) that West Africans are resistant to *P. vivax,* even when inoculated with infected blood. Might not the Fy antigens be associated with the site of entry of this parasite, as seems to be the case with *P. knowlesii?* With recent advances in the *in vitro* culturing of human malarial parasites, discussed on pp. 96–97, the possibility of further experimentation is now very real.

P. vivax appears to have been common in Africa before the widespread appearance of *P. falciparum,* and if so, this might explain the very high frequency of the *Fy* allele. What is not explained is why the *Fy* allele is not more common in other regions of *P. vivax* infection, particularly the Mediterranean basin. But it is tempting to speculate that the spread of the *Fy* allele has something to do with the present disjunct distribution of *P. vivax* in Africa, and perhaps, ironically, with the subsequent success of *P. falciparum.*

The reader may be reminded at this point of Van Valen's Red Queen hypothesis (p. 56). The adaptation of Central African populations to one species of malarial parasite has apparently led to its replacement by another and even more dangerous species. This deterioration of the environment has forced further genetic changes in the human population.

Summary

The foregoing is by no means a complete survey of the polymorphisms which are known or suspected in man to be subject to selection. At this juncture, however, certain points can be made.

Each of the polymorphisms we have considered in this chapter appears to be subject to a different type of selection:

Conditional and geographically localized heterosis accounts for the high frequency of the *Hb-S* allele.

The most likely explanation out of several is that under conditions of strong malarial infection the homozygote for *Hb-C* is at an advantage, leading to a localized high frequency. In surrounding areas carriers of the gene are at a disadvantage.

Possibly, weak heterosis maintains the *Hb-E* allele, though there is certainly no evidence for it. If not, then the high frequency can most easily be explained by a mechanism similar to that for *Hb-C,* in which the spread of the gene is limited by migration of the wild type allele into regions where the selective pressures are strongest.

The G6PD deficiencies, caused by a heterogeneous collection of alleles, seem most likely to be held in the population by a variety of forces. The more common alleles may be maintained by heterosis, though again there is no evidence for it (and a great paucity of evidence althogether). But it seems possible, because of their severe hemizygous effects and restrictions on the heterotic maintenance of alleles at a sex-linked locus, that many of the rarer alleles have been built up in frequency only because of very strong, but geographically very localized, selection for survival of the hemizygote.

So far as the thalassemias are concerned, we are hampered by lack of information on the true extent of the heterogeneity of the thalassemia polymorphisms. Because they are autosomal, there is a stronger argument for heterotic selection than at the G6PD locus, and this seems to be the most likely cause of the high frequency of the α and β thalassemias. The β-thalassemias in particular, however, are presumably heterotic over a more restricted range of environments than is hemoglobin S, so that many independent mutations have achieved low frequencies in the affected populations.

Turning to the blood groups, we are hampered by a lack of knowledge of the function of these antigenic molecules, so that the associations seen with disease might be due to the direct action of the molecules themselves or to the interaction of the environment with a closely-linked gene or genes in disequilibrium with the blood groups. It is striking, especially with ABO, that many classes of disease seem to be associated with the polymorphism. There is not indication as yet of heterotic selection at the ABO, Rh or MN loci, and indeed what selection can be detected is in the opposite direction, involving selection against the heterozygote or against one or another allele. Yet, at least for the ABO blood group, there seems to be no doubt that balancing selection is acting or has acted in the past. There are therefore two general possibilities: heterotic selection which is too weak to be detectable, but which might have been stronger in the past, or selection in favor of different alleles in different environments. It is impossible at the moment to distinguish between these. As far as the MNS system is concerned, there is no indication why it should be polymorphic, but the closely related simian V–A–B system is polymorphic in a number of primates (Wiener et al. 1972) and the impression gained from large numbers of surveys of human populations is that the MN polymorphism is remarkably tenacious even in small isolates in which the frequency of one or the other allele is low. As with the other blood-group loci, distinguishing between heterotic and multiple-niche selection is impossible, if indeed selection does occur.

The HLA system is quite different from all of these. If it is selectively advantageous for each member of a population to have a high probability of being genetically different from his neighbors, then the HLA system would appear to be ideally suited for this purpose. Continual generation of alleles by intragenic recombination would, in conjunction with the differential rate of loss of these alleles through selection and drift, result in a steady-state distribution which might be quite constant throughout the human species. This is in fact what is observed. Because the heterozygotes are all approximately equally fit, and the regeneration of alleles through recombination is formally equivalent to a very high and frequency-dependent mutation rate, such a situation could explain a stable multi-allele polymorphism in which the conditions might otherwise be too restrictive for stability. The selective advantage of such a system should only be apparent under certain environmental circumstances, for example infection by viruses which carry surface antigens of their previous host. But when the selection is manifest it should be heterotic.

Whether this or some allied mechanism is responsible for the HLA polymorphism is, of course, not possible to determine at the present time. But it is difficult to imagine another mechanism generating the situation at this locus, in view of the undoubted selection against certain alleles through increased disease susceptibility. The inclusion in this small region of chromosome 6 of immune response genes and at least two of the complement loci, and the high probability that other polymorphic loci remain to be discovered, suggests that this complex is extremely important in the protection of the organism against disease.

Strong disease associations with the secretor and PTC-tasting loci are an indication that much remains to be understood about the maintenance of these polymorphisms by selection as well.

What generalizations emerge from this multiplicity of patterns? One is that a suprisingly high proportion of the loci considered—hemoglobin β polymorphisms, perhaps some G6PD alleles, the thalassemias and the HLA polymorphism—appear to be subject to heterotic selection under some unusual circumstances. The circumstances, however, all appear to be connected with disease, and to confer resistance, therefore, to particular environmental agents which by their very nature are distributed discontinuously. It has also been suggested that the ABO polymorphism is maintained by its interaction with various diseases, such as smallpox and streptococcal infections, though with less evidence. It is, of course, impossible to tell whether this selection, if it occurred, was heterotic, though a more complete understanding of the undoubtedly very complex way in which these glycoproteins and glycolipids function on the surface of cells may cast some light on the matter. In none of the cases we have considered here is the heterosis unconditional. When considered in detail, each polymorphism is unique, and contributes to the genetic variability of the population in its own way.

A second generalization is that even when the effects of selection are clear cut, as in the sickle cell and G6PD polymorphisms, the results of different experiments and different surveys of populations are highly inconsistent. The slightest changes in the experimental conditions, or in the circumstances in which the populations are living, produce large differences in the outcome. This inconsistency bespeaks a high degree of interaction of these polymorphisms with their environment. This point is central to the subsequent arguments, particularly those of Chapter 7.

I would like to suggest that this sample of human polymorphisms differs from the polymorphisms observed in other organisms only in that we know so much more about them.

Let us imagine a race of creatures from another planet who are so large that they are forced to observe us through a low-power microscope, just as we are forced to observe *Drosophila*. Suppose a sample population of humans were abducted from Hawaii, or some other portion of the planet where most of the polymorphisms discussed in this chapter are likely to be present, stripped of their culture and forced to live and reproduce under what the experimenters imagine to be ideal conditions. The small amount of mortality due to maternal–fetal incompatibility might easily pass unnoticed—autopsies on such insignificant objects as human infants would after all be impractical. Further, the effects of disease would certainly be minimized under these idealized 'culture' conditions. The only alleles subject to detectable selection would be those of sickle cell, G6PD and the thalassemias, and they would appear to be unconditionally deleterious in the absence of malaria. Further, if the sample were taken from most parts of the Earth, the frequencies of these alleles would be low enough that they would appear to make up part of the mutational load.

Under these circumstances, then, the differential contribution to fitness of these polymorphisms would either be undetectable, or would appear as partially dominant and unconditionally deleterious alleles. Assuming that these were the only polymorphisms they were capable of measuring, these creatures from another planet could be forgiven for presuming that we harbored a large number of apparently neutral polymorphisms, and that our genetic structure was a simple one.

The human polymorphisms provide the best-studied examples of polymorphisms subject to selection. It is dangerous to generalize from them, since almost all of the selection studied has been the result of response to disease. But, in this sample, the majority of the balancing selection which can be measured or inferred appears to be heterotic, though only conditionally so. In the absence of balancing selection, some polymorphisms will revert to apparent neutrality and some will become partially dominant. The way in which these polymorphisms behave might give some clues to the nature and function of polymorphisms in other organisms such as Drosophila which *we* are forced to observe through a low-power microscope, and which we habitually raise under what we fondly imagine to be ideal conditions—except when there is a mite infestation in the laboratory.

In the next chapter I will deal with the ways in which polymorphisms such as those considered here might be acted on by selection, bearing in mind their highly condition-dependent properties. In Chapter 7, I will examine further theoretical constraints on their behavior, and the rather sparse experimental evidence concerning how they actually behave.

6. Matching modes of selection to classes of variability

Alleles with pronounced selective effects are likely to show rather limited interaction with the environment, and to be acted on by multiplicative or approximately multiplicative selection. Those with much smaller effect will both be more numerous and more likely to be acted on by truncation or quasi-truncation

A classification of the types of genetic variability subject to selection

In Chapter 3 we have seen the great difference between the numbers of variants that can be acted upon simultaneously by truncation selection, both balancing and directional, and by multiplicative selection producing the same decrease in total fitness. In this chapter, I wish to follow up those initial observations by exploring the following question. What classes of selected variants in populations are likely to be acted upon by these two general types of selection? Having established the probable properties of these classes, we can turn to the matter of the relative numbers to be expected in each class, in a population that has accumulated genetic variants for long periods of time.

Harmful variants

Unconditionally harmful variants. Suppose, as is commonly done, that the majority of new alleles arising as a result of mutation are deleterious over a wide range of environments, so that they never achieve high frequency. A subclass of these will be unconditionally deleterious, in that they kill the organism or otherwise prevent it from reproducing when homozygous (if recessive) or heterozygous (if dominant). Others will have more or less deleterious effects which again are not strongly dependent on environmental circumstances. For these classes of genes, which include lethals, severe subvitals and possibly some genes of lesser effect, one is justified in making the assumption that selection coefficients associated with these genotypes can be treated as constants. Such mutants form part of the mutational load (Muller 1950), in that their frequency in the population is a function of the mutation rate.

One is certainly justified in using a multiplicative model to determine the average fitness of an organism in a population segregating for many lethal genes, and it may be that this model can be extended to many subvital genes as well—with the caveat that as soon as a gene does not kill its carrier or render it sterile with certainty, the environment inevitably plays a role. In man, medical intervention is one obvious source of varying selection coefficients. The fitness of a homozygote for the sickle-cell gene will vary greatly depending on the medical care available. Under natural conditions as well, changed circumstances may alter the selection pressures on a gene of severe subvital effect. For instance, there is strong selection against albinism in both animals and plants. It is difficult to image circumstances under which this selection would be reduced in plants depending on photosynthesis, but unpigmented animals and fungi are commonly found in cave environments. In these cases, strong selection against a character has been changed to selection in favor of it. Perhaps this is because the energy saved by an organism that does not have to produce pigment is substantial, perhaps because their transparency protects these organisms against the occasional sighted predator on the periphery of their environment.

These and other exceptions aside, it seems likely that for the majority of severe-subvital genes as well as for lethals, selection coefficients can be treated as constant and multiplicative, at least over most environments. Their frequencies can be calculated directly from the mutation rate and the selection coefficients. Given an appreciable mutation rate, the number of loci giving rise to such mutants is severely limited, in all probability to a few thousand (pp. 43–45).

Conditionally harmful variants This leaves a very large class of deleterious mutants of much smaller effect. These alleles do not kill the organism even when homozygous, and as a consequence might reach rather high frequency levels through mutation pressure alone. Further, it is to be expected that they will be greatly influenced in their effect by the environment, and by epistatic interaction with other loci. Many such mutants are the result of small changes in gene products that do not destroy their function, but render that function slightly less efficient. Others might be alterations in untranslated parts of the genome which have a slight harmful effect on the rate of transcription or translation elsewhere, changes in regions of the genome which affect the fidelity of chromosome disjunction during cell division, and so on. We have seen many examples of such mutations in Chapter 4. None of these mutations would have an unconditionally deleterious effect—some might under certain circumstances have no detectable effect at all. But the expression of all will be very much under the control of the environment.

What factors in the environment will influence the strength of selection against such alleles? Let us consider a model population of organisms of size N which reproduces synchronously. The number of young produced during this period of reproduction is GN, where G is a fertility factor, and is greater than one. In many marine organisms of this type, for example, G may be very much greater than one. Assuming a constant set of environmental conditions from one generation to the next, which is certainly the norm for most populations most of the time, then to maintain a stable population size, this new population must be reduced by the factor $1/G$. Let us call this the total truncation factor. This reduction may be brought about by predation, limited food supply, limited space, and so on, and will occur regardless of the genotypic composition of the population. If all organisms were homozygous for the same allele at all loci, this reduction would still perforce take place. If the population were segregating for a moderate number of unconditionally deleterious recessive and dominant alleles, then part of the reduction would be due to the deaths of homozygotes for these alleles. So long as the average fitness calculated from the effect of these deleterious genes, $\overline{W}$, is greater than $1/G$, the final population size after selection might not be greatly changed. Of course, fewer organisms would be available for predation, for example, since some are dying from the effects of their own genotypes, so predation might act to depress the average value of N.

If the population were also segregating for a large number of conditionally deleterious mutations, the ultimate population size would again be GN/G, but part of the reduction would be due to the effects of these mutations. Organisms heterozygous or homozygous for such alleles might be more susceptible to predation or starvation or otherwise less able to compete, and thus be the first to be destroyed. But it should be emphasized that if such a population were to become homozygous for one or more of these conditionally deleterious mutations, the final population size would still be approximately N. These organisms might compete very poorly against organisms carrying the wild-type allele at these loci. But if the latter organisms were not present in the population, those organisms homozygous for the mutant alleles would simply fill up the spaces available for them in the environment. This is the point made cogently by Wallace (1968, 1975), who distinguished between 'hard' and 'soft' selection. The former is multiplicative, the latter truncation with a moveable threshold.

Given repeatability of environmental conditions from one generation to the next, the total truncation factor $1/G$ should remain approximately constant. Further, while the selective pressure acting on unconditionally-deleterious mutations will remain constant, the pressure acting on conditionally-deleterious mutations will vary as a function of the distribution of genotypes in the populations. An organism homozygous for a certain number of conditionally deleterious mutants may be at an advantage when competing against organisms with one set of genotypes, but at a disadvantage when competing against organisms with another set. Thus, for repeatable environmental conditions resulting in stability of the population size, the primary environmental factor acting to determine the selection coefficients associated with conditionally deleterious mutations will be the distribution of phenotypes of the other organisms in the population. These will in turn be at least partly determined by the distribution of the genotypes of the other organisms in the population. For this set of conditions, which is certainly the common one, we are therefore justified in thinking of truncation as removing a certain proportion of the organisms each generation, as Milkman (1967) and Maynard Smith (1968) suggest, rather than removing all organisms below a certain genotypic value, as in the model of King (1967).

Beneficial variants Presumably, a smaller proportion of newly arising mutations are at least conditionally advantageous, though how much smaller this proportion may be cannot even be guessed at. These alleles would be expected, unless they were lost in the first few generations, to replace the less advantageous alleles at the same loci in the population. The process may be initiated or accelerated by an environmental change, so that there may be a wave of fixation of alleles that were previously disadvantageous or neutral. But the advantage may disappear before complete fixation as a result of a reversal or a further shift of environmental conditions. Fixations or near-fixations may occur locally, as in the alleles for melanism in many moths or the Duffy-negative allele in man (see pp. 132–33). Many other alleles may be selected, either as a result of heterozygote advantage or as a result of different selective pressures at different times or in different parts of the environment, so that they move to some internal equilibrium in the population. As we have seen in Chapter 5, however, all the known examples of this phenomenon are strongly environment-dependent.

I will make an extrapolation from the very considerable body of data now known, and suggest that virtually all the advantageous variants which arise are conditionally so. Those that reach internal equilibria, by whatever selection mechanism (and a number will be considered in detail in Chapter 7), are mostly what I have called 'selected functional polymorphisms' (Wills 1973). All the alleles making up these polymorphisms are functional, so that organisms homozygous for them are quite capable of survival. They, and conditionally advantageous mutations destined to replace older alleles at a locus, should be subject to truncation selection in a way quite analogous to that operating on conditionally-deleterious alleles. For these classes of alleles as well, under stable conditions the factor in the environment having the most to do with determining fitness will be the distribution of other phenotypes in the population. Assuming that the ABO and Rh polymorphisms are being or have been maintained by balancing selection, they would be excellent examples of selected functional polymorphisms. Populations monomorphic for type O or Rh positive are known and perfectly capable of survival. Unknown selective conditions have often, however, given the advantage to polymorphic populations in spite of the introduction of an incompatibility load.

Not all balanced polymorphisms are selected functional polymorphisms. Sickle cell anemia and the thalassemias are examples of polymorphisms in which one homozygote is deleterious, and in which the selection against that homozygote is sufficiently constant over a range of environmental conditions that the effects on fitness of these polymorphisms can be treated as approximately multiplicative. This may not be true in detail, just as it was not for the severe subvital genes considered earlier. But even if the fitnesses are only approximately multiplicative, the number of balanced polymorphisms the population can support in which one or both homozygotes have severe effect is severely limited, as is the number of unconditionally harmful alleles. Again, the limitation is that $\overline{W}$ cannot fall below $1/G$. Although their number must be kept very low in man by selection, they have been relatively easy to discover because of their large effect on fitness.

On the other hand, homozygosity for alleles of selected functional polymorphisms poses no problem of survival *per se,* but only of survival in the presence of organisms with adaptively superior phenotypes. Thus, the upper limit to the number of such polymorphisms that can be maintained in the population is not imposed by any reduction in average fitness, since none will occur. Rather, it will be imposed by the effective size of the population, which will govern their loss through drift, and the mutation rate to such polymorphisms, which will govern the rate at which they are introduced.

Obviously, there must be a continuum of effect, and very few selected functional polymorphisms can have no absolute effect on fitness. Yet, if selected polymorphisms are permitted to accumulate in a population, those with the least absolute effect on fitness should be the most numerous.

Fitness potential and rank-order selection

Definitions and consequences of the model

Milkman (1973) introduced the concept of 'fitness potential,' which is somewhat analogous to King's 'survival factor parameter.' In King's concept, the genotype of an organism at a locus contributes a certain amount to a survival-factor parameter. When the contributions of all loci are summed, if this parameter is greater than a certain size, it ensures the organism's survival. In Milkman's concept, the fitness potential is determined by summation in the same way, but the survival of an organism with a given total fitness potential is then contingent on the severity of the selection, and on the fitness potentials of the other organisms in the population. Thus, the *relative* rather than the *absolute* values of the fitness potentials are important in determing survival. The Milkman concept is thus applicable to that subset of selected functional polymorphisms that have no *absolute* but only a *relative* effect on survival.

I extended this concept a little further (Wills 1978) by suggesting that selection might act on fitness potentials in the following way. It would result in the death or inability to reproduce of a certain proportion of the organisms with the lowest fitness potentials. The size of this proportion would be dependent on the severity of the environment, but the rank-order rather than the absolute values of the fitness potentials would determine which organisms died and which survived. For the reasons detailed in the two previous sections, I considered that this would be the predominant mode of selection acting on alleles with slightly harmful or beneficial effects that are conditional on the environment. It would also be the predominant mode acting on selected functional polymorphisms. Alleles with pronounced deleterious effect, whether they be entirely harmful or result in conditional heterosis like the allele for hemoglobin S, should be acted upon by some approximation of multiplicative selection. It would seem, from the above arguments, that the two major types of selection which can be most productively compared for their effects on a population are multiplicative and rank-order truncation, since they represent the extremes of a continuum of effects.

As with any attempt to classify something as complex as different modes of selection, exceptions and complicating factors immediately become apparent. There are undoubtedly alleles that fall into one category under some conditions and into the other category under others. Further, pure rank-order selection, in which fitness-potential rank is the sole underlying variable on which selection acts, is unlikely to operate in real situations. This is because, just as a statistical test based solely on rank discards much quantitative information present in the data, selection based on the rank-of-fitness-potentials does not employ other information present in the fitness potential array.

The model of pure rank-order selection can easily be modified to accommodate this objection without altering the major premise on which it is based, that relative rather than absolute values of fitness potential form the underlying variable on which selection acts. Modified rank-order selection need merely take into account the magnitudes of the *differences* between the fitness potentials of the organisms in the population. In the example given in Fig. 6.1, the spread of fitness potentials in the population influences the shape of the fitness curve which the environment imposes on the population. But, as with pure rank-order selection, all the organisms in the population would be quite capable of survival in the absence of competition from other organisms with higher fitness potential. Fitness is still entirely relative, and the concept of absolute fitness inherent in the multiplicative model and in truncation models with a specified survival cutoff has not been reintroduced. Neither pure nor modified rank-order selection leads to a decrease of average fitness upon inbreeding.

While an experimenter may impose a sharp truncation on a population, nature is unlikely to do so. The shape of the fitness curve imposed on a population is more likely to be something like that shown in Fig. 6.1. How might this affect the great difference between truncation and non-truncation models, which is that the spread of fitnesses can be much greater in the former than the latter?

Milkman (1978), Kimura and Crow (1978), and Crow and Kimura (1979)

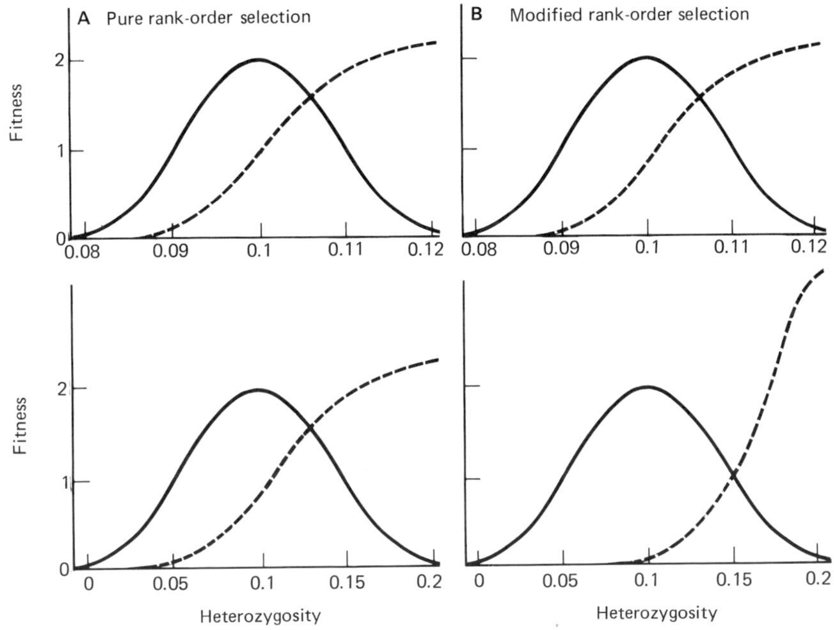

Fig. 6.1. Diagrams contrasting pure and modified rank-order selection. Solid lines show the distribution of heterozygosities in the selected populations, ranked from lowest to highest heterozygosity (which we take to be equivalent to 'fitness potential'). The upper diagram in each set shows a population with a small variance of heterozygosity, and the lower diagram shows a population with a larger variance. The ordinate, in units of fitness, refers to the selection curve imposed on each population, which is shown by a dotted line. The shape of the curve is dependent on the severity of the environment. In 'pure' rank-order selection, represented in A, the shape of the fitness curve imposed by a particular environment does not alter as the variance of heterozygosity (fitness potential) changes. In modified rank-order selection, shown in B, as the variance of heterozygosity increases the intensity of selection also increases, giving a higher fitness to organisms with an extremely high fitness potential. Thus, the differences between organisms on the fitness potential scale play a role in modified rank-order selection. From Wills (1978).

have investigated a variety of simple truncation and quasi-truncation models that lead them, through somewhat different paths, to the conclusion that apparently quite different models still yield similar results. Kimura and Crow point out that a sharp truncation of the underlying fitness-potential distribution is the most efficient mode of selection, but that a variety of other selective curves imposed on the population do not greatly affect the efficiency of the selective process.

Drawing from quantitative genetics, Kimura and Crow show that for a sharp truncation, the selective advantage of an individual allele may be very great. If s_i is the selective advantage (measured from the mean, and not from some ideal genotype as in the multiplicative model), A_i is the average amount by which individuals carry the allele exceed the population average, $F(c)$ is the ordinate of the fitness-potential distribution at the point of truncation, and $\overline{w}$ is the proportion of the population saved, then

$$s_i = A_i \frac{F(c)}{\overline{w}} \tag{6.1}$$

Imposing a selective curve based on the cumulative normal distribution rather than a sharp truncation reduces the efficiency of selection only slightly, *provided* that the variance of this cumulative normal distribution is about the same as the fitness potential distribution. If it is, we have the situation illustrated in Fig. 6.2 (a). It will be seen that appreciable numbers of organisms in the population have very low and very high fitnesses—that is, the wide spread of fitnesses that is the source of the great effectiveness of truncation selection is present in this population even though the truncation is not sharp. If the variance of the fitness curve is greater than that of the fitness-potential curve, as in Fig. 6.2 (b), the effectiveness of truncation selection is sharply reduced. Most of the organisms fall within a narrow fitness range. Of course, in selection of this type far fewer organisms are being removed, so that the average fitness of the population in (b) is much greater than that of the population in (a).

In fact, the requirement that the two variances be comparable is not very

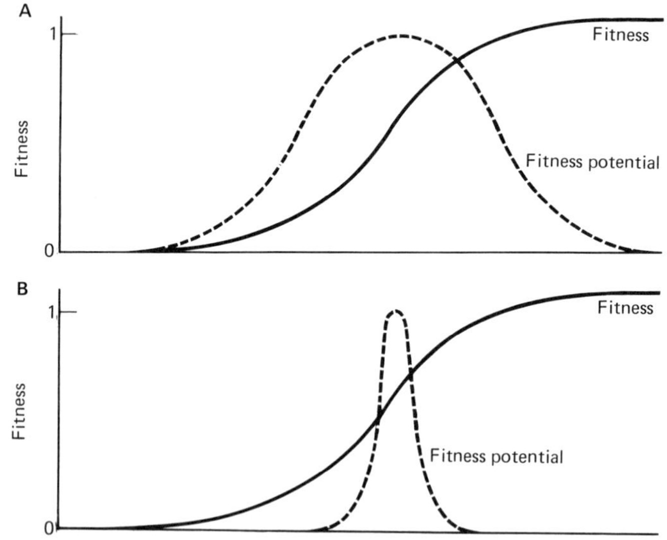

Fig. 6.2. Modified rank-order selection of the Kimura and Crow type. The selective curve is a cumulative normal. **A.** The variances of the selective curve and the underlying fitness-potential curve are comparable in size. **B.** The variance of the selective curve is greater than that of the fitness-potential curve.

stringent. Kimura and Crow were able to show that as the variance of the selection curve imposed on the population is allowed to increase relative to the variance of fitness potential, the efficiency of selection remains high. For a 50% drop in efficiency below that of a sharp truncation, the variance of the selection curve is three times that of the selected character.

In pure rank-order selection based on a cumulative normal-selective curve, the two variances change together. In modified rank-order selection they do not. Thus, factors which tend to change the variance of fitness potential may be very important in changing the effectiveness of rank-order selection.

The concept of modified rank-order selection will become useful later when we consider the effects of linkage (Chapter 8). In the meantime, we will contrast the extreme models of multiplicative and pure rank-order selection with regard to both the removal of harmful alleles from, and the maintenance of balanced polymorphisms in, a population.

Selection against harmful alleles

Morton, Crow, and Muller (1956) presented a powerful argument against the presence in the genome of a large number of genes capable of mutating to lethal alleles. Even though each individual lethal might be rare, the chance that an organism would be homozygous for at least one, if many such loci were segregating, would produce an insupportable genetic load.

Selection against a lethal allele will, at equilibrium, be balanced by the introduction of the lethal by mutation. So if μ is the mutation rate to the lethal, $\hat{q}$ the equilibrium frequency and h the partial dominance, then at equilibrium in an outbreeding population,

$$\mu = \hat{q}^2 + 2h\,\hat{p}\,\hat{q} \qquad (6.2)$$

The number of loci L for which lethal alleles are segregating in a population can therefore be calculated directly from the mutation rate to lethals and the average fitness of the organisms in the population:

$$\overline{W} = (1 - \mu)^L \qquad (6.3)$$

We have seen (pp. 25–26) from a variety of lines of evidence that the number of genes in *Drosophila* with indispensable function, and which therefore can mutate to lethals, is at the most 5000. For a mutation rate of 10^{-5}, $\overline{W} = 0.95$, so that the genetic load generated by this number of loci is most certainly supportable in this case. Further, as we have seen in Chapter 3, the population can easily survive inbreeding. Suppose all lethal alleles showed 5% dominance. Then at equilibrium, $\hat{q} = 2 \times 10^{-4}$. Now imagine that the population were suddenly subject to inbreeding as a result of a size bottleneck. Since the number of loci segregating for lethals would not change, the initial impact of this sudden inbreeding would be considerable but not overwhelming. Now, the fitness of the first generation after inbreeding would be,

$$\overline{W} = [1 - 2h\,\hat{p}\,\hat{q}\,(1 - F) - \hat{q}^2\,(1 - F) - F\hat{q}]^L \qquad (6.4)$$

where F is the coefficient of inbreeding. For $F = 1/16$, equivalent to first cousin matings, $\overline{W}$ would be reduced to 0.85 in the first generation. For $F = 1/4$, $\overline{W}$ would become 0.72. Even this reduction should pose no great difficulty for an organism such as *Drosophila.*

But the decline in fitness upon inbreeding becomes precipitate as the number of loci segregating for lethal genes is allowed to increase. Table 3.3 shows the decline on the assumption of various degrees of partial dominance. It could be seen that a population with many more than 20 000 loci segregating for partially dominant lethals would be in imminent danger of extinction if it passed through a severe bottleneck.

Therefore, loci capable of mutating to lethals and to other alleles subject to multiplicative selection should be severely limited in their numbers—presumably, if our estimate of the average mutation rate is correct, to no more than a few thousand.

We have already seen, however, in Chapter 4, much evidence from molecular evolution and from recent advances in molecular studies of the genome that there is strong selection against mutational substitutions, not only in parts of the genome that are translated, but also in untranslated regions. The conservation of codon usage in phyletically related genes and the conservation of long tracts of purines and pyrimidines in cryptic satellites both suggest that there has been strong selection against substitutions in these regions. It has also been estimated that there are 35 000 translated messages in a HeLa cell (Bishop et al. 1974); only a fraction of these, as we have seen, can be read from genes coding for some essential function. If all should show tendencies towards conservation of base sequence similar to the tendencies already found in the β-hemoglobin and the chorionic somatomammatropin and growth-hormone genes, then another type of selection must be operating. This selection seems most likely to be rank-order selection or some variation of it.

Essentially, the capacity of a population to accumulate conditionally deleterious variants selected through competition would be unlimited if only rank-order selection were operating. This seems unlikely on the face of it; very low values of fitness potential must have some depressive effect on fitness itself. We can, however, approximate the effects on population structure of small reductions in fitness brought about by truncation selection imposed on a population in which the organisms are ranked according to their fitness potential.

Suppose fitness potential were a direct function of the variance of the number of harmful genes per organism. If the frequency of harmful alleles at each locus is q, then the variance of the proportion of homozygotes for harmful alleles per organism will be

$$\sigma^2_{\mathrm{Hom}} = q^2(1 - q^2)/L \qquad (6.5)$$

where L is the number of loci. The number of homozygotes per organism could be quite high, since homozygosity for this class of alleles does not bring about the death of the organism, but merely lowers its fitness potential. Similarly, the variance of the proportion of heterozygotes per organism will be (assuming that all harmful alleles at all loci are at frequency q):

$$\sigma^2_{\text{Het}} = 2pq(1-2pq)/L \tag{6.6}$$

These distributions will determine, at least in part, the distribution of fitness potentials in the population. I will assume, in the subsequent calculations, that all of the variance in fitness potential is due to variance in the selected genotypes, although of course much fitness potential will be due to environmental factors and the influence of other classes of genetic variability.

We wish to impose some sort of truncation on the population such that the mean fitness potential is increased—in this case by lowering the frequency of the harmful alleles at all the loci in the population. Many types of truncation might be envisaged, though it is important to remember that the efficiency of truncation selection—in the sense of its ability to shift fitness potential—is not greatly altered if the truncation is sharp or if it is less so, so long as the same total number of organisms is removed from the population (Milkman 1978; Kimura and Crow 1978; Crow and Kimura 1979). In these and succeeding examples, I will impose a sharp truncation on the population, but the truncation threshold will be such that only a relatively small proportion of the population is prevented from reproducing. If such a sharp truncation is imposed on the ranked population, so that a fixed proportion of the organisms with the lowest fitness potential is removed by truncation, and the same proportion of the organisms with the highest fitness potential is given a fitness of two, then if H_a is the truncation threshold in standard deviations, the decrease in q due to selection at each locus will be:

$$\Delta q_s = [2/\sqrt{2\pi}]\ \sqrt{q^2(1-q^2)/L}\ e^{(-H_a^2/2)} + [2h/\sqrt{2\pi}]\ \sqrt{2pq(1-2pq)/L}\ e^{(-H_a^2/2)} \tag{6.7}$$

where h is the partial dominance. Since the increase in q due to mutation is

$$\Delta q_m = \mu\,(1-q) \tag{6.8}$$

the equilibrium frequency $\hat{q}$ is that value at which $\Delta q_m = \Delta q_s$.

Table 6.1 shows the size of q for various values of partial dominance, L and truncation. It can be seen that even for very large values of L, large enough for the entire genome, the average value of $\hat{q}$ per locus can be kept very low even if only 5% of the population is removed by truncation each generation and 5% is assigned a fitness of two, with the remaining 90% of the population untouched by selection. This will be so only if h is substantial, and there is evidence from Drosophila (Mukai and Yamaguchi 1974; Wills 1966) that the

Table 6.1. The removal of harmful genes by truncation selection.The first column gives values of q for various values of L, partial dominance (h) and strength of truncation and replacement selection. The mutation rate is assumed to be 10^{-5}/gene/generation. The second column gives the average number of homozygotes for conditionally harmful genes per organism. The third column gives the size of the per-locus selection coefficients acting against the homozygotes.

L	h (Partial dominance)											
	0			0.1			0.25			0.5		
	$\hat{q}$	No. of Homozygotes	s	$\hat{q}$	No. of Homozygotes	s	$\hat{q}$	No. of Homozygotes	s	$\hat{q}$	No. of Homozygotes	s
(a) 20% truncation and replacement												
10^5	0.006	7.2	0.277	0.0013	0.34	0.038	2.5×10^{-4}	0.0125	0.08	6×10^{-5}	7×10^{-4}	0.167
10^6	0.02	800	0.025	0.01	400	0.005	0.002	80	0.01	6×10^{-4}	0.72	0.0167
(b)5% truncation and replacement												
10^5	0.015	45	0.042	0.005	5	0.01	0.0014	0.4	0.0143	0.0005	0.05	0.02
10^6	0.05	5000	0.004	0.025	1250	0.0018	0.01	200	0.002	0.004	32	0.0025

partial dominance of both newly-arisen and equilibrium harmful genes of small effect can be very high, reaching values of 0.4 in some experiments.

The partial dominance of those lethals and severe subvitals that survive in the population is very low (Crow 1979), in contrast to the high partial dominance of harmful mutants of small effect. It can be seen from these calculations that if large numbers of mutants of small effect are arising and they are subject to truncation selection, they will be rapidly eliminated (particularly if their partial dominance is high) without a drastic decrease in the average fitness of the organisms carrying them.

Further, this class of harmful variants, in contrast to those acted upon by multiplicative selection, will not depress $\overline{W}$ if homozygosity is increased by sudden inbreeding.

Table 6.1 also includes two other values of interest for each hypothetical population. The first is the average number of conditionally harmful homozygotes per organism at equilibrium. It will be seen that this number is very low if partial dominance is high—far fewer than one per organism, for example, for 20% truncation and replacement selection, $h = 0.5$ and 10^5 loci being selected simultaneously. The number can become very high if more loci are being selected simultaneously and if partial dominance is low—5000 homozygotes per organism for 5% truncation and replacement selection, $h = 0$ and 10^6 loci being selected simultaneously. This would certainly constitute an impossibly high genetic load if these alleles were all unconditionally deleterious.

The second calculation of interest is the *per locus* selection coefficient, as calculated from the equilibrium frequency $\hat{q}$. At equilibrium,

$$s = \frac{\mu}{\hat{q}^2 + 2h\hat{p}\hat{q}} \tag{6.9}$$

These values can be seen to be very large, and reflect the geat difference in fitness between the truncated and the nontruncated members of the population. The average fitness of a population subject to 5 percent truncation and replacement selection is roughly 0.95. If 10^6 loci were being acted upon by selection, each harmful allele being completely recessive with a selection coefficient of 0.004 and each at a frequency of 0.05 as in the table, then the fitness of such a population under the multiplicative model would be $(0.996)^{5000} = 2 \times 10^{-9}$. By contrast, the same amount of selection can be accomplished by rank-order truncation and replacement selection with only a 5% loss due to selection each generation.

The great advantage of truncation over multiplicative selection is most apparent when the variance of the selected character is large, so that the truncation imposed on the population is effective. If the equilibrium frequency of the harmful alleles is low, so that the variance is also low, there are circumstances under which multiplicative selection is the more effective of the two. In Table 6.2, the number of loci on which selection can act, given $\overline{W}$, s, h, and $\hat{q}$,

Table 6.2. The numbers of loci that can be acted upon by multiplicative and truncation selection to give the specified value $\hat{q}$ for harmful alleles at each locus

				L	
$\overline{W}$	s	h	$\hat{q}$	Multiplicative	Truncation
0.8	1	0	0.00316	22 300	31 300
0.8	1	0.05	0.0001	22 300	2 040
0.8	—	0	0.01	—	980 000
0.8	—	0.05	0.01	—	2 800 000
0.8	—	0.05	0.001	—	102 000
0.8	—	0.2	0.001	—	965 000

has been calculated for multiplicative selection and for truncation selection. In the first two cases, $\hat{q}$ is calculated by solving eqn (6.2). The L for the multiplicative case is determined from equation 6.3. For the truncation case, L is determined by rearranging and solving equation 6.7. For the remaining cases in the table, $\hat{q}$ is specified at the outset, so that the multiplicative model does not apply.

It will be seen (line 2 of the table) that for very low $\hat{q}$, truncation selection is actually less effective than multiplicative selection. But if we arbitrarily set $\hat{q}$ higher, as in the last line of the table, then a million loci can be 'purified' by 20% truncation and replacement selection to the extent that the equilibrium total of all harmful alleles at each of the loci is only 0.1%.

Thus, both multiplicative and truncation selection have their optima, and each is most effective with different classes of harmful alleles. One interesting aspect of these calculations is that conditionally-harmful mutants would be expected to be very rare in a population if they are partially dominant and truncation selection is operating. Is it possible that functional alleles at frequencies even as low as 1% may be acted upon by some form of balancing selection (if they are not neutral) rather than being maintained by mutation–selection balance?

Selection acting on polymorphisms

In addition to its great effect on selection against conditionally harmful alleles, truncation selection acting on organisms ranked according to their fitness potential can be highly effective at maintaining polymorphic loci. In this section I will deal with two-allele polymorphisms maintained by simple heterosis, in which both homozygotes are less fit than the heterozygote. If the magnitude of this decrease in fitness is dependent on the array of genotypes in the population, then these will be selected functional polymorphisms and they will be subject to rank-order selection. If, on the other hand, the lowered fitness of the homozygote would be expressed even in the absence of competition, then the polymorphisms would be maintained by some approximation of multiplicative selection. As in the previous section, the populations are considered to be random-mating with discrete generations.

In contrast to the calculations in the previous section, we must now take into consideration the effects of population size. This is because polymorphisms, no matter how strongly selected, may be lost through drift.

A difficulty with all finite population studies is to decide what might be a reasonable rate of introduction and loss of selected polymorphisms. I have deliberately chosen a very low rate of loss, to ensure that a low mutation rate to new selected functional polymorphisms would be sufficient to keep the number of polymorphisms segregating in the population constant. The rate of loss of polymorphisms is a function of their distribution around their selective equilibrium. This in turn is a function of both the strength of selection moving the alleles towards the equilibrium and the amount of genetic drift. The latter tends to move them, on average, away from equilibrium. A selective criterion C may be chosen as follows: balancing selection must be sufficiently strong that, for a given size of population, the standard deviation of the distribution of polymorphic loci is a small fraction of the distance to the nearest fixation point. In other words, the polymorphisms must be so tightly grouped around their selective equilibrium that, in order for fixation to take place, an allele must drift several standard deviations from its selective equilibrium.

This criterion has two useful features. The first is that only a very low mutation rate need be postulated to maintain the polymorphisms. Let us assume a C of $\hat{p}/5$, where $\hat{p}$ is the equilibrium frequency of the rarer alleles at each locus. Then, if new selected functional alleles arise at a rate of 10^{-9} per generation, which is a low mutation rate, and if there are at any one time 100 loci currently monomorphic, but capable of becoming polymorphic (surely a modest estimate), then we can calculate the population size needed to maintain the balance in terms of the selection coefficient s acting on the alleles of the polymorphisms. At steady state, the rate of loss of polymorphisms will be one in 3×10^7 per generation. Recalling that the probability of retention of an advantageous mutation (which is what a new polymorphism is in the early stages) is approximately $2s$, we can calculate that a population of size $3 \times 10^{-7} / (10^{-9} \times 2s \times 100) = 1.5/s$ will be sufficient to maintain the balance. As will be seen, the selection coefficients obtained from rank-order selection tend, even for very large numbers of segregating loci, to be much larger than Kimura's most pessimistic lower limit of $1/4N_e$, so that population sizes need not be large. Of course, if the pool of monomorphic loci is larger than 100, which is highly likely, then even lower mutation rates can be postulated.

The second advantage of this criterion is that the distribution of frequencies around the selective equilibrium is so small that all the loci can be treated as having the same frequency without introduction of an appreciable error. Thus, it is possible to estimate the average heterozygosity per locus as $\overline{H} = 2\hat{p}\hat{q}$, and the variance of heterozygosity as $\sigma_H^2 = [\overline{H}(1 - \overline{H})]/L$, where L is the number of polymorphic loci.

Truncation selection. A truncation-and-replacement selection model, the same as that in the previous section, was investigated (Wills 1978). The fitness-potential criterion by which organisms were ranked was the proportion of heterozygous loci. A certain proportion of those at the lower end of the ranking scale was assigned a fitness of zero, and the same proportion at the high end of the fitness scale was assigned a fitness of two. The truncation I chose to investigate happens to be very efficient, though more effective ones could easily be devised—for example by truncating more organisms at the low end and increasing the fitness of the group at the high end of the fitness-potential scale. It should be emphasized, however, that for 5% truncation and replacement selection, which as we will see is quite efficient, 90% of the population is untouched by selection and retains a fitness of one.

This type of truncation and replacement selection increases the mean heterozygosity of a population, $\overline{H}$, as follows:

$$\Delta \overline{H} = [2/\sqrt{2\pi}]\ \sqrt{\overline{H}(1-\overline{H})/L}\ \exp(-H_a^2/2) \tag{6.10}$$

where H_a is the truncation threshold (in standard deviations) for no reproduction, since the two truncation thresholds are placed symmetrically. Random mating will subsequently reduce this value $\overline{H} + \Delta\overline{H}$ to a new heterozygosity value H', which will still (provided $\overline{H} > 0$) be greater than $\overline{H}$. At the same time, drift will be reducing $\overline{H}'$ by a factor $1/2N_e$, so that what we seek is the point, for a given population size, number of loci and strength of selection, at which these two factors just balance out.

Since the population size is kept constant in selection of this type, we can treat the selection as simply the replacement of some of each type of homozygote with heterozygotes at each locus. We have further made the simplifying assumption that all loci have the same selective equilibrium, which will be called $\hat{p}$. We now introduce a value p_H, which is the average allele frequency which would give the observed $\overline{H}$ if the rarer alleles at all the polymorphic loci were at this frequency. Of course, as a result of drift they are not all at a particular frequency, but for purposes of our calculation we can treat them as if they all were. Then, provided that the difference between p_H and $\hat{p}$ is not too great, the ratio of the proportions of the two types of homozygotes replaced by heterozygotes is

$$R = \frac{p_H^2\, \hat{q}^2}{q_H^2\, \hat{p}^2} \tag{6.11}$$

where $\hat{q}$ and $\hat{q}_H$ are the corresponding frequencies for the commoner alleles. This ratio is 1 if $\hat{p} = p_H$. If p_H deviates in either direction, selection will tend to move the ratio back to unity. This relationship is unique to balancing truncation-selection, and will be very useful when we consider the effects of linkage disequilibrium in Chapter 9.

The truncation selection will tend to move p_H towards $\hat{p}$, at the same time as drift moves it away. The amount by which p_H^2 is decreased by selection is given by $\Delta\overline{H}$ times $R/(1 + R)$, and of course the heterozygotes are increased by $\Delta\overline{H}$. After selection, the new value of p_H, p'_H, is given by:

$$p'_H = p^2_H - \left[\left(\frac{p^2_H\,\hat{q}^2}{p^2_H\,\hat{q}^2 + \hat{q}^2_H\,\hat{p}^2}\right)\Delta\overline{H}\right] + \left(\frac{\overline{H} + \Delta\overline{H}}{2}\right) \quad (6.12)$$

After random mating $\overline{H}' = p'_H(1 - p'_H)$. Finally, the variance of loci around their equilibrium is given by the well-known relationship

$$\sigma^2_p = \hat{p}\hat{q} - \frac{\overline{H}}{2} \quad (6.13)$$

The calculations proceed as follows. N_e and $\hat{p}$ are specified, and p_H is calculated from eqn (6.13) to fit the given criterion C. Then different values of L are used in repeated iterations of eqns (6.10, 6.11 and 6.12) until the increase in $\overline{H}$ due to selection, $\overline{H}' - \overline{H}$, just equals the decrease due to drift, $\overline{H}/2N_e$.

Values of the numbers of loci that can be maintained for given strengths of truncation and replacement selection are given in Fig. 6.3. Three things can be seen from this figure.

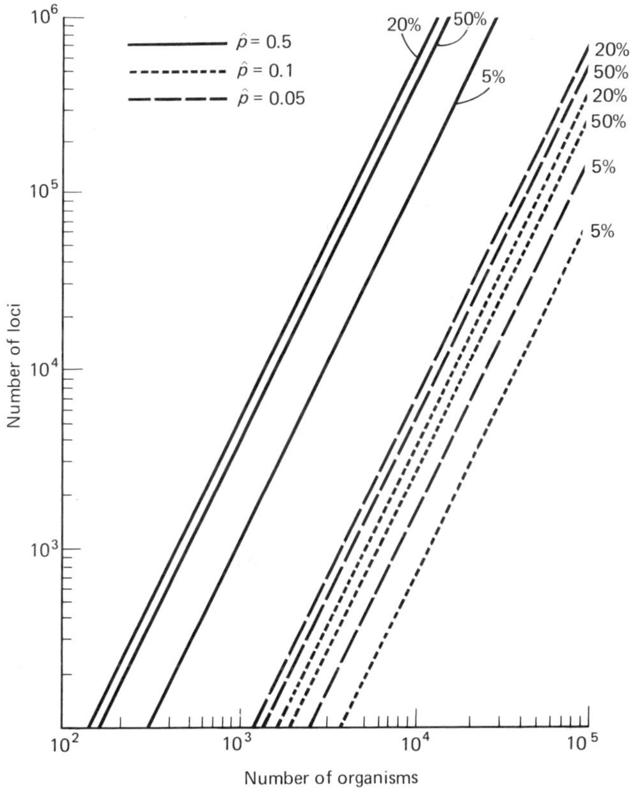

Fig. 6.3. The number of polymorphic loci that can be maintained near a given equilibrium value of $\hat{p}$ for various strengths of truncation and replacement selection. The selective criterion C is $\sigma_p = \hat{p}/5$. From Wills (1978).

1. The number of polymorphisms that can be maintained increases as the square of the population size. We have already seen examples of this advantage which truncation selection has over multiplicative selection: in this case it leads to the maintenance of very large numbers of polymorphisms even though the population size is quite moderate. Of course, it is unlikely that selection will be this intense. Only a small fraction of the organisms that die each generation must do so for genetic reasons, and it may be that many of these do so because of homozygosity for conditionally and unconditionally harmful mutations. The enormous power of truncation selection, however, means that thousands of balanced polymorphisms can be maintained by a small fraction of the selection employed in the present examples.

2. The number of polymorphisms that can be maintained under equilibrium conditions declines dramatically as $\hat{p}$ is moved away from 0.5. The advantage of the approach taken above, is that it is possible to quantify this decline, and it is indeed startling. For the criterion that $\sigma_p = \hat{p}/5$, which is the one used in constructing the graph, it appears that 150 times as many polymorphisms can be maintained at $\hat{p} = 0.5$ as at $\hat{p} = 0.05$.

3. The effect of finite population size can be seen from the figure. N_e is taken to be approximately the number of organisms remaining after truncation. The effect of drift in reducing $\overline{H}$ is far greater for 50% truncation and replacement than for 20% truncation and replacement, so the former is actually less efficient.

We can also examine the size of the selection coefficients measuring the strength of selection acting on each polymorphism. These coefficients are given by $\Delta p_H^2/p_H^2$ and $\Delta q_H^2/q_H^2$, where the numerator in each case is the change in p_H^2 and q_H^2 due to selection over one generation. These values will remain constant for a given selection criterion regardless of the value of N_e and the strength of truncation selection.

For the current selective criterion and $\hat{p} = 0.5$, these values are 18 times Kimura's (1971) limit of $1/4N_e$. It must be remembered that these figures are for a population essentially saturated with polymorphisms—48 million if the population has a size of 10^5! Of course, no population will have this many polymorphisms at 0.5. Consequently, C can be reduced and the selection coefficients per locus will be correspondingly greater.

When $\hat{p} = 0.1$, far fewer polymorphisms can be maintained. Now, at saturation, the coefficients associated with the rarer homozygotes are 2000 times Kimura's limit, and those associated with the more common homozygotes are 45 times the limit.

Selection coefficients of this size would, of course, result in an insupportable segregational load in a population with these large numbers of heterotic polymorphisms and subject to multiplicative selection. For $\hat{p} = 0.5$ and $N_e = 0.8 \times 10^5$ (to make the size approximately comparable to that of a population of 10^5 subjected to 20% truncation and replacement selection), the average fitness would be:

$$\overline{W} = (1 - 4.5 \times 10^{-5})^{2.4\times10^7} < 10^{-99}$$

It should be noted that the precise type of truncation is specified only by eqn 6.10. Any mode of truncation or quasi-truncation loading to an increase in $\overline{H}$ could replace this equation and the rest of the theory is left unchanged.

Multiplicative selection. How many polymorphisms can in fact be maintained by multiplicative selection to a particular criterion C? As we have seen, the number of polymorphisms maintained by this type of selection is simply a function of the average fitness of the population $\overline{W}$ and the average fitness conferred by each locus:

$$\overline{W} = (1 - s_1\hat{p}^2 - s_2\hat{q}^2)^L \tag{6.14}$$

Once again, as in multiplicative selection against harmful alleles, the number of polymorphisms that can be maintained is severely limited by the reduction in fitness, both in an outbred population and in a population with inbreeding.

However, it is possible to determine the *minimum effective population size* needed to maintain polymorphisms subjected to multiplicative selection, given that they are selected strongly enough to fit a selective criterion C. Given $\hat{p}$, we calculate p_H for a given C. Then, since we must specify the selection coefficients associated with the two homozygotes for each polymorphism, we can obtain the change in p_H in one generation from standard population-genetics theory:

$$\Delta p_H = \frac{p_H\, q_H\, (s_2 q_H - s_1\, p_H)}{1 - s_1 p_H^2 - s_2 q_H^2} \tag{6.15}$$

The change in heterozygosity due to selection is:

$$\Delta\overline{H}' = (p_H + \Delta p_H)\,(1 - [p_H + \Delta p_H]) - p_H\,(1 - p_H) \tag{6.16}$$

Now $N_{e_{min}}$, the minimum effective population size, can be calculated by noting that the decrease in heterozygosity due to drift is $\Delta\overline{H}_D = \overline{H}\,/2\,N_{e_{min}}$. When $\Delta\overline{H}_D = \Delta\overline{H}'$,

$$N_{e_{min}} = \overline{H}/2\,\Delta\overline{H}' \tag{6.17}$$

In Fig. 6.4, I have taken C to be $\hat{p}/5$, and plotted $N_{e_{min}}$ for continuously varying values of s_1 and s_2 leading to a range of values of $\hat{p}$. The two sets of branching lines represent values derived when either s_1 and s_2 is kept constant. It must be remembered that $N_{e_{min}}$ is the minimum population size needed to keep even *one* polymorphism sufficiently close to its selective equilibrium that it meets the criterion $C = \hat{p}/5$.

$N_{e_{min}}$ must be very large under some circumstances. For example, to maintain even one polymorphism at a selective equilibrium of 0.05 when $s_1 = 0.19$ and $s_2 = 0.01$, which is quite strong selection, with the additional requirement that $\sigma_p = 0.05/5 = 0.01$, the minimum effective population size must be 48 000.

To determine the *number* of loci that can be maintained for a given average fitness, we can make the slight approximation of assuming that all loci are at

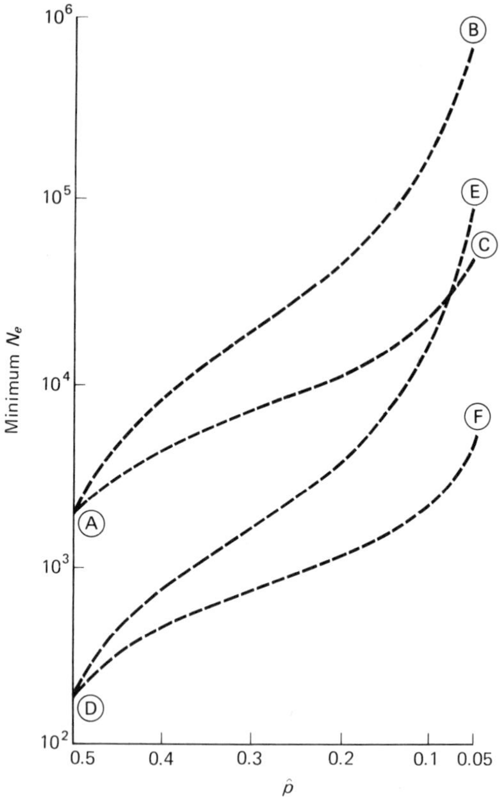

Fig. 6.4. The minimum effective population sizes for multiplicative polymorphisms selected to the criterion $C = \hat{p}/5$. Circled letters indicate points on the graph analyzed further in Table 6.3.

their selective equilibrium (valid if C is small). Then, L can be determined by solving eqn (6.14).

Table 6.3 shows the number of polymorphisms that can be maintained by multiplicative selection which produces a reduction in population size equivalent to that produced by 5%, 20% and 50% truncation and replacement selection. The numbers are for the lettered points of Fig. 6.4. It can be seen that under no circumstances are the numbers large. The intense selection at each locus ensures that few loci can be maintained to the selective criterion used. The numbers in parentheses show the numbers of polymorphisms that can be maintained at the same selective equilibrium with the equivalent intensity of truncation and replacement selection. The disparity between the two types of selection is particularly pronounced for small $\hat{p}$, where the two numbers can easily be different by a factor of 10^4 or more. Of course, more polymorphisms could be maintained by multiplicative selection if the size of the selection coefficients were reduced—but then $N_{e_{min}}$ would, of course, have to increase and

Table 6.3. Numbers of polymorphisms that can be maintained by multiplicative selection for various average fitnesses at the labelled points in Fig. 6.4. Numbers in parentheses are the numbers of polymorphisms that can be maintained by truncation and replacement selection giving the same average fitness.

$\overline{W}$	Ⓐ $s_1 = 0.01$, $s_2 = 0.01$	Ⓑ $s_1 = 0.01$, $s_2 = 0.000526$	Ⓒ $s_1 = 0.19$, $s_2 = 0.01$
0.5	138 (25 000)	1386 (2.7×10^7)	72 (60 000)
0.8	44 (30 000)	446 (3.2×10^7)	23 (70 000)
0.95	10 (7 000)	103 (7×10^6)	5 (16 000)
	Ⓓ $s_1 = 0.1$, $s_2 = 0.1$	Ⓔ $s_1 = 0.1$, $s_2 = 0.00526$	Ⓕ $s_1 = 0.9$, $s_2 = 0.1$
0.5	13 (270)	138 (270 000)	6 (250)
0.8	4 (320)	44 (320 000)	2 (300)
0.95	1 (70)	10 (70 000)	0.5 (60)

the disparity between the two types of selection would be even greater, since the numbers maintained by truncation selection increase as the square of N_e while the numbers maintained by multiplicative selection increase only linearly with N_e.

It can be seen, then, that if alleles subject to these two types of balancing selection arise at perceptible rates, at equilibrium between the introduction by mutation and loss through drift the overwhelming majority of the selected polymorphisms that survive in the population should be those subject to rank-order truncation selection.

We have seen that the problem of genetic load is removed by the adoption of rank-order selection accomplished by some form of truncation. This certainly allows a large number of polymorphisms to be maintained, and I have suggested (Wills 1973, 1978) that if selected functional polymorphisms arise at all, they should accumulate in the population. There appears to me to be no obvious impediment to their accumulation, though the rate at which they accumulate will be governed both by the mutation rate and the number of other selected functional polymorphisms already in the population—since the intensity of selection acting on them is partly a function of the level of heterozygosity already achieved.

There are, however, two very large problems. The first is that the polymorphisms retained by this type of selection will be those with the smallest absolute contribution to fitness, so that selection acting on them will be very difficult to detect. The second is the discrepancy between the observed and expected distributions of these polymorphisms.

If many polymorphisms are maintained by balancing selection, such that they move to some intermediate equilibrium, then the ones which tend to be retained in the population will be those furthest from fixation. The ones nearer fixation will have a far greater chance of being lost after they arise, even though they might be maintained in the population for a while. Thus, particularly in populations which we suspect to have remained for long periods of time under relatively constant conditions, we should see a vast excess of polymorphisms at frequencies around 0.5, regardless of the relative rates at which polymorphisms at different equilibrium frequencies arise. But, as was shown in Chapter 4, we do not.

This relative deficiency of polymorphisms at intermediate frequency is, to my mind, the greatest difficulty remaining in the long battle to knock down the theoretical impediments to a selectionist view of genetic variability. In Chapter 3, we reviewed the status of a number of other difficulties. The importance of this problem is such that it will need a long chapter of its own to be dealt with at all adequately. I do not pretend to come to a resolution of this problem at the end, but I hope to suggest ways in which different types of balancing selection can lead to the observed distributions of allele frequencies. In this attempt I will be aided by our growing theoretical understanding of those varieties of balancing selection which depend only slightly on heterozygote advantage. Further, I will suggest ways in which the very small and highly conditional selection pressures acting on selected functional polymorphisms may be detected.

7. The nature of balanced genetic variability

The distribution of polymorphisms subject to conditional heterosis and marginal overdominance may be very different from that expected from unconditional heterosis, and may mimic that of neutral genes. Such modes of selection will be missed by the design of most population genetics experiments

In Chapter 5 we saw that several human polymorphisms appear to be maintained, or may have been maintained in the past, by conditional heterosis. Such intermittent heterotic selection may help to explain the excess of intermediate allele frequencies, above neutralist expectation, observed in human blood groups by Yamazaki and Maruyama (1974). But if the conclusions of Chapter 6 are correct, and there is nothing to prevent selected functional polymorphisms from accumulating in populations to very high numbers, why does the distribution of most polymorphisms appear to conform to the neutralist expectation, or perhaps to the very similar distribution to be expected from mutation–selection balance? Are most of the variant alleles in fact neutral or selected against? Where are the balanced polymorphisms we would expect if rank-order truncation selection is operating?

In this chapter I will suggest a variety of ways in which balancing selection can act to give a distribution of allele frequencies mimicking that of neutrality. Further, I will suggest that many of the experiments that have been designed to examine the nature of genetic variability tend to miss such classes of genes. I will draw from the findings of quantitative genetics, *Drosophila* viability studies, and studies on selection acting on isozymes. In the course of the chapter I will suggest a variety of ways in which data from such experiments can be reanalysed, or the experiments can be redesigned, to give more information about the classes of variability being examined. The reanalysis and different experimental approaches may indeed reveal that the balanced component of genetic variability is too small to be detectable, but some hints from the data suggest that it may be substantial.

What kinds of balancing selection are possible?

Single-gene overdominance

This first and simplest mechanism for the maintenance of variability was proposed by Fisher (1922), and relies on the heterozygote exhibiting a higher fitness than either homozygote. In the usual notation, which implies multiplicative fitnesses and introduces a normalization step if the population is to be kept at a constant size, we can assign the fitnesses at a two-allele, three-genotype locus as in Table 7.1.

Because heterozygotes have the highest fitness and contribute equal numbers of *A* and *a* gametes to the next generation, a stable polymorphism will result. The allele frequencies, which will be approached from either side of the stable point, will be a function of the relative sizes of s_1 and s_2. This stable equilibrium is reached when as many *A* as *a* alleles are eliminated by selection each generation, so that $s_1\hat{p} = s_2\hat{q}$. The equilibrium frequency $\hat{p} = s_2/(s_1 + s_2)$.

However, the *rate* at which this equilibrium is approached is a function of the absolute sizes of s_1 and s_2, as is the variance of frequencies around the equilibrium once the equilibrium has been reached. The latter is also a function of the effective population size. Thus, s_1 may be very different from s_2, so that the equilibrium is very close to 1 or 0. Or s_1 and s_2 may be small so that random drift plays an appreciable role in shifts of allele frequency. In addition, the population size is small so that even large selection coefficients cannot overcome the effects of sampling. In all these cases a theoretically stable polymorphism subject to overdominance selection can be lost. It can also be lost if it is subject to swamping amounts of immigration of one or the other allele from elsewhere in the population. And of course it can be lost if conditions change such that the heterozygote is no longer the fittest genotype.

Table 7.1. The effects of single-gene overdominance on fitness, assuming the multiplicative model

Genotype	*AA*	*Aa*	*aa*	Total
Frequency	p^2	$2pq$	q^2	1
Fitness	$1 - s_1$	1	$1 - s_2$	
Proportion remaining after selection on the developing zygotes	$p^2(1 - s_1)$	$2pq$	$q^2(1 - s_2)$	$1 - s_1p^2 - s_2q^2 = \overline{W}$
Proportion after normalization	$p^2(1 - s_1)/\overline{W}$	$2pq/\overline{W}$	$q^2(1 - s_2)/\overline{W}$	

If overdominance is widespread in natural populations, then the optimum genotype would be heterozygous at all the loci at which overdominance is acting in that particular environment. If there is a mixed bag of selected polymorphisms in a population, held there by overdominance and by other balancing mechanisms, then polymorphisms which are overdominant in a particular environment should be more likely to be retained if that environment becomes the norm—as it may do under laboratory conditions. Eventually, if the environment remains extremely uniform for long periods of time, the only selected polymorphisms that are left should be held in the population by overdominance. Whether this ever really happens in nature or can be made to happen in the laboratory is, to say the least, moot.

Observational evidence of overdominance at the single gene level is difficult to come by, for a variety of reasons—not the least of which, as will be developed below, is that it might not be very common! First, a sample taken from a natural population is usually limited in numbers. The expectation to which the worker taking the sample fits his data is usually derived from the allele frequencies calculated from the data themselves, unless the data are from a laboratory cross in which the genotypes of the parents are known in advance. As a consequence of this, quite large deviations due to selection may have to occur if a significant deviation is to be found. The expected value of p after a generation of selection is $(p - s_1p^2)/(1 - s_1p^2 - s_2q^2)$, with the denominator representing all the surviving organisms at the time the sample is taken. If we knew the number of zygotes in the population before selection occurred, and the number surviving after selection, then we could have considered the denominator of the expected value of p to be 1. The observed number of heterozygotes would then be the same as the expected number. But we do not have this information, so it is quite possible that the observed number of heterozygotes is by chance greater than the expected number, reducing the size of the deviation and the size of the χ^2 which the experimenter calculates. A simple numerical example is shown in Table 7.2a. Obviously the second χ^2 could not be calculated in practice, since it depends on a precise knowledge of the zygotic mortality. If such knowledge were available, one would be in a much better position to say something about the nature of selection.

Because the true extent of zygotic mortality is not usually known, a reduction in survival of only one genotype can also lead to apparent heterosis. A simple example of this rather discomfiting phenomenon, which is an alternative explanation for much of the data gathered from natural populations which indicates apparent heterosis, is given in Table 7.2b. Such an alternative explanation for an excess of heterozygotes can be ruled out if the expected genotype frequencies are known in advance, as in a laboratory cross between known genotypes, or if the genotype ratios are known at an earlier stage in the life cycle before selection has altered them.

Note the size of the selection coefficients in these examples. The difficulty of detecting heterosis can be considerable if they are smaller. If we use one of the

Table 7.2. An illustration of two difficulties in determining the amount of single-gene overdominance

(a) How early mortality can mask the true extent of overdominance

Genotype	*AA*	*Aa*	*aa*	Total
Initial numbers of zygotes	25	50	25	100
Numbers surviving after selection ($s_1 = s_2 = 0.28$)	18	50	18	86
Expected numbers based on the observed total number	21.5	43	21.5	
	$\chi^2 = 2.28$, not significant for 1 d.f.			
Expected numbers based on the *initial* total numbers	25	50	25	
	$\chi^2 = 3.92$, significant at the 0.05 level.			

(b) How selection against one homozygote can lead to apparent overdominance

Genotype	*AA*	*Aa*	*aa*	Total
Initial numbers of zygotes	25	50	25	100
Numbers surviving after selection ($s = 0.6$)	25	50	10	85
Expected numbers based on the observed total number	29.3	41.3	14.4	
	$\chi^2 = 3.85$, significant at the 0.05 level.			

favorite selection coefficients of theoretical population geneticists, $s_1 = s_2 = 0.01$, then for a population at equilibrium the observer would need to score a sample of over 150 000 to be even 95% confident that selection is operating.

As the above example shows, if selection is acting on a stage of the life cycle which is unavailable to the experimenter, it may remain undetected. For example, a population may appear to be in Hardy–Weinberg equilibrium, but the homozygotes may be far less fertile than the heterozygotes. This may result in very strong heterotic selection, but if the alleles are already at their selective equilibrium this will not show up either as a deviation from Hardy–Weinberg or as a secular change in allele frequencies. Differences in fertility in Drosophila made homozygous for wild-type chromosomes have been investigated in detail by Sved (1971), Tracey and Ayala (1974), and Anderson and Watanabe (1974), and have been shown to contribute an important component to fitness—much larger, in some cases, than that due to variations in egg-to-adult survival in the same experiment. But because of the difficulty of performing experiments to measure fertility compared with those that measure survival to adulthood, the nature of the genetic variability affecting fertility has not been investigated as fully.

A heterozygote excess may be caused by associative overdominance. It may also be generated by other factors. Neel and Ward (1972) found a slight, though not quite significant, heterozygote excess within Yanomama villages, even though there was inbreeding which should have led to an excess of homozygotes. They suggested that one way in which this could be generated was by different allele frequencies in males and females of the previous generation. The difference could have arisen through capture of women from another village. The excess may also have been due to lowered survival of one homozygote, as they suggested and as was pointed out earlier (p. 161). It is to be expected that if heterozygote excess were to be observed in human populations, it is precisely in populations living under primitive conditions that it should be most visible. The significant deviation from Hardy–Weinberg equilibrium observed for the sickle-cell allele in some West African tribes is a case in point. In the surveys of New Guinea tribes by G. A. Harrison and his co-workers (1976), dealt with on pp. 127–28, such deviations may easily be discovered upon more extensive analysis of the data.

As will be developed below, overdominance may not be a very large factor in those components of fitness which we usually measure in experimental populations. But that does not mean that it is not a large factor in other components of fitness, or that it may not be a large factor under more stringent environmental conditions. But as we have seen, if single-gene overdominance is common it cannot be unconditional, for otherwise large numbers of polymorphisms with intermediate frequencies would be observed.

Selection in spatially and temporally diverse environments

It is apparent from Chapter 5 that more than one type of selection is operating to maintain the polymorphisms, particularly those in the human population, about which something is known. In these cases heterosis, when it can be seen directly or inferred, is *conditional*—that is, it depends for its expression on a particular environment, and most notably in the case of resistance to malaria on a particular severe environment. Heterosis cannot be ruled out in other polymorphisms, such as the ABO system, but it must be, or have been in the past, sufficiently powerful to overcome the effects of selection against the heterozygote as a result of incompatibility. Further, it is striking that the selection that has been observed at the ABO locus with regard to disease susceptibility is not obviously heterotic.

Selection for and against different alleles in different environments, as a result of spatial or temporal environmental heterogeneity, can be invoked to explain several of these polymorphisms, for example the polymorphism for the Duffy-negative allele. The conditions for the maintenance of such polymorphisms are not especially stringent, provided that the polymorphisms have particular properties.

Selection in spatially and temporally fluctuating environments has received a great deal of attention from theorists. The current theoretical models have been ably summarized by Felsenstein (1976). Another excellent review of the theory and of some of the experimental work which has been done to investigate this phenomenon is that of Hedrick et al. (1976).

I will try first to summarize here some phenomena which selection in spatially-diverse and temporally fluctuating environments have in common, and to suggest some probable properties which variation subject to such selection should exhibit. The models currently in the literature are based on multiplicative selection. After dealing with them, I will turn my attention to a simple truncation-selection model employing diverse environments, and show that the advantage of truncation models is found here as well.

Mutiplicative models: selection in multiple niches. The first model to be proposed for selection leading to polymorphism in a spatially-diverse environment was that of Levene (1953). The model assumed that the heterozygote was exactly intermediate in fitness in each environment. As was shown later by Dempster (1955), Levene's model depends on selection being density-dependent in order for it to lead to the maintenance of polymorphism. If the heterozygote fitness is greater than the arithmetic average of the homozygotes, it changes into a kind of heterotic model.

A characteristic of Levene's and similar theoretical models is that they lead to the maintenance of polymorphism even under conditions in which, if there is a single niche available, there would be fixation of one or another allele. A great deal of effort and ingenuity has been directed towards investigating such extreme models, with the result that the properties of some less extreme but perhaps more likely models have not received the attention that they should.

Levene's model is as follows. Suppose that a population of organisms is living in *i* different environments, and that a large number of young organisms settle into the environments at random, the numbers in each environment being a function of the environment's size. Then a constant mortality occurs during development to sexual maturity, so that the number of adults surviving is also proportional to the relative size of the environment. It will be seen that if mortality acts differentially on different genotypes in the different environments, then the intensity of selection on a particular genotype will be a function not only of the selection for or against that genotype but of the size of the environment or environments in which the selection occurs. This is because, in Dempster's terminology, a *constant fertile adult number* is surviving in each niche.

Some examples will make this plain, and will illustrate some interesting properties of selection in two or more environments. Suppose W_1 and W_2 are the fitnesses of the *AA* homozygotes in environments 1 and 2, and V_1 and V_2 are the fitnesses of the *aa* homozygotes in the same environments. Table 7.3 shows what happens for various values of these fitnesses when environments are the same size.

We will assume that the final population size in each environment is 1000, giving a total population of 2000. Thus, in the Levene type of selection, the numbers of genotypes in each environment must add up to 1000. We can imagine, in contrast, another type of selection in which a large number of zygotes are acted upon by selection to give a final adult population size of 2000, but the numbers in each environment are not constrained to be equal. Then, survival in each environment is dictated by the fitness of each genotype, and is not dependent on the density within the environment. Thus, the proportions of gen-

Table 7.3. Various cases of the Levene model—equal environment size, $N_1 = N_2 = 1000$

$W_1 = 0.9$ $W_2 = 1.1$ $V_1 = 1.1$ $V_2 = 0.9$	Environment 1			Environment 2			Total frequency	
(a) No dominance	*AA*	*Aa*	*aa*	*AA*	*Aa*	*aa*	*A*	*a*
Before selection	250	500	250	250	500	250	0.5	0.5
Fitness	0.9	1.0	1.1	1.1	1.0	0.9		
After selection, normalized to:								
Total population	225	500	275	275	500	225	0.5	0.5
Individual environment	225	500	275	275	500	225	0.5	0.5
(b) No dominance								
Before selection	10	180	810	10	180	810	0.1	0.09
Fitness	0.9	1.0	1.1	1.1	1.0	0.9		
After selection, normalized to:								
Total population	9	180	891	11	180	729	0.1	0.9
Individual environment	8	167	825	12	196	792	0.10075	0.89925
(c) Complete recessiveness of the selected genotype								
$W_1 = 0.9$, $W_2 = 1.0$ $V_1 = 1.0$, $V_2 = 0.9$	Environment 1			Environment 2			Total frequency	
	AA	*Aa*	*aa*	*AA*	*Aa*	*aa*	*A*	*a*
Before selection	10	180	810	10	180	810	0.1	0.9
Fitness	0.9	1.0	1.0	1.0	1.0	0.9		
After selection, normalized to:								
Total population	9	188	845	10	188	760	0.1035	0.8965
Individual environment	9	180	810	11	196	793	0.104	0.896
(d) Overdominance								
$W_1 = V_1 = 0.9$, $W_2 = V_2 = 0.9$								
Before selection	10	180	810	10	180	810		
After selection, normalized to:								
Total population	10	196	794	10	196	794	0.108	0.892
Individual environment	10	196	794	10	196	794	0.108	0.892

otypes surviving can either be normalized to the total population size or to the population size in each environment.

In Table 7.3a, the fitnesses are symmetrically distributed around 1 and the frequencies of the *A* and *a* alleles are each 0.5 initially. Regardless of whether the normalization is to the size of the niche or to the total population size, the allele frequencies do not change after selection.

In Table 7.3b, the fitnesses remain the same but the frequency of *A* is now 0.1 and that of *a* is 0.9. If the surviving genotype frequencies are normalized to the entire population, once again the frequencies do not change. Despite strong selection the alleles will act as if they were neutral, and one or the other will eventually be lost by drift. But if the surviving genotype frequencies are normalized to the niche sizes, there is a slight (but very slight!) increase in the frequency of the rarer allele. This is because more mortality of the *aa* genotype has occurred in niche 2 than in niche 1.

Given that selection is dependent on the density in each niche, it is therefore quite possible for polymorphisms to be maintained. A set of fitnesses which would not lead to the maintenance of polymorphism in an undifferentiated environment *can* do so if the environment is differentiated into niches.

These examples have assumed that the heterozygote is precisely intermediate in fitness between the homozygotes in each niche. This is unlikely to be the case at most loci. In Table 7.3c, we consider complete recessiveness of the selected homozygote, so that the heterozygote has the same fitness as *aa* in niche 1, and the same fitness as *AA* in niche 2. Here, we can see that regardless of whether the surviving genotype frequencies are normalized to the total population or to the individual niche sizes, the rarer allele increases in frequency. For the particular set of fitness values chosen, the increase is about the same for both types of normalization. Thus, for sufficiently small partial dominance of the genotype that is recessive in the particular niche, the requirement that selection be density dependent is no longer necessary.

In Table 7.3d, unconditional overdominance is introduced. Here, the type of normalization makes no difference since fitnesses are the same in both niches. Note that the gain in frequency of the rarer allele is twice what it was for the complete recessiveness case. The complete recessiveness case with multiple environments can be thought of as a kind of diluted overdominance, in which selection against the homozygotes is carried out in only part of the environment each generation. While the heterozygotes are only as fit as the fitter homozygote in each environment, they are the fittest genotype when the fitnesses are averaged over all environments. This situation has been named 'marginal overdominance,' and was discussed by Wallace (1968).

Note also that for the example given here, the unconditional overdominance case is more than ten times as effective at increasing the frequency of the rarer allele than is the no-dominance case with density-dependent selection. The case of no dominance is delicately balanced, depending as it does solely on the density-dependent effects of selection in the different environments. Stability can thus easily be destroyed by a relatively small shift in the parameters.

Consider a no-dominance case in which environment 1 is three times the size of environment 2, given in Table 7.4. Once more, if the number of organisms surviving is normalized to the total population size, there is no selective change in allele frequencies and the polymorphism will eventually be lost through drift. But if they are normalized to the individual environments, then a situation which was stable for equal environment sizes has been transformed into an unstable one. The rarer allele *A* is selected against in this particular case. In fact, the *A* allele is always at a disadvantage regardless of its frequency. This is because more selection is taking place against the *AA* homozygote than against the *aa* homozygote, since the region in which the *AA* homozygote is disadvantageous is larger.

When dominance is introduced (Table 7.4b), then (depending on the degree of dominance and the relative sizes of the niches) the polymorphism can be

Table 7.4. The effect of changing the sizes of the individual environments

	Environment 1			Environment 2			Total	
(a) Unequal environment sizes, no dominance								
	AA	*Aa*	*aa*	*AA*	*Aa*	*aa*	*A*	*a*
Before selection	15	270	1215	5	90	405		
Fitness	0.9	1.0	1.1	1.1	1.0	0.9		
After selection, normalized to:								
Total population	13.5	270	1336.5	5.5	90	364.5	0.1	0.9
Individual environment	12.5	250	1237.5	6	98	396	0.09625	0.90375
(b) Unequal environment sizes with dominance								
	AA	*Aa*	*aa*	*AA*	*Aa*	*aa*	*A*	*a*
Before selection	15	270	1215	5	90	405		
Fitness	0.9	1.0	1.0	1.0	1.0	0.9		
After selection, normalized to:								
Total population	14	276	1241	5	92	372	0.1015	0.8985
Individual environment	14	270	1215	5	98	397	0.1015	0.8985
(c) Equal environment sizes with underdominance								
	AA	*Aa*	*aa*	*AA*	*Aa*	*aa*	*A*	*a*
Before selection	10	180	810	10	180	810		
Fitness	0.9	1.0	1.2	1.2	1.0	0.9		
After selection normalized to:								
Total population	9	173	934	11	173	700	0.0965	0.9035
Individual environment	8	155	837	13	195	792	0.098	0.902

stabilized. As in the previous examples, the necessity for density-dependent selection is eliminated if dominance is sufficiently great.

Instability can also result, even for equal environment sizes, when the selection coefficients exhibit underdominance, as is shown in Table 7.4c. In the example given, this occurs even when the numbers of organisms are normalized to the total in the population, because there is relative selection against the heterozygote. The selection against the rarer allele is less strong when the numbers of organisms are normalized to the individual environments, again because of the effect of density-dependent selection.

As we have seen, selection in multiple environments with dominance becomes, even in the absence of density-dependent selection, a kind of dilute overdominance. Levene's contribution was to show that in the presence of density dependent selection even fitnesses distributed symmetrically around unity could lead to stability—although this stability can be remarkably fragile. The requirement for stability in a two-allele system is that the harmonic mean fitnesses of the homozygotes, weighted by the size of the environments, are both less than the harmonic mean fitness of the heterozygotes similarly weighted. Because of the environment weighting, small shifts in the size of the environments can lead to instability if there is little difference between the harmonic means. Further, it can lead to very large shifts in the selective equilibrium even if stability is maintained.

Levene showed that the change in frequency of an allele is given by

$$\Delta p = pq \, \Sigma c_i \frac{(1 - V_i) + (W_i + V_i - 2)p}{V_i + 2(1 - V_i)\,p + (W_i + V_i - 2)p^2} \tag{7.1}$$

where c_i is the weighting factor associated with the i th environment and W_i and V_i are the fitnesses of the homozygous genotypes in that environment. Fig. 7.1 shows the effects of different degrees of partial dominance of the selected genotype on the strength of selection driving alleles towards an equilibrium of 0.5, in two environments of equal size. Fitnesses were chosen to keep the fitness differences between the homozygotes approximately constant. It will be seen that selection in the intermediate dominance case is far weaker than that in the complete recessiveness case. The case of complete dominance gives Δp values close to that for heterotic selection with selection coefficients of half the size.

The same pattern, of stabilization of gene frequencies by dominance, is seen when the fitnesses in the different environments are dissimilar. Fig. 7.2 shows the effect on the equilibrium frequency given two environments of equal size, in which the spread of fitnesses in the second environment is always twice that in the first. As dominance of the heterozygote decreases, the equilibrium frequency, which for complete recessiveness of the homozygote is at 0.66, shifts rapidly towards fixation. The equilibrium value has reached fixation even before the dominance of the heterozygote is lost.

Even with the stable, completely recessive case leading to marginal overdom-

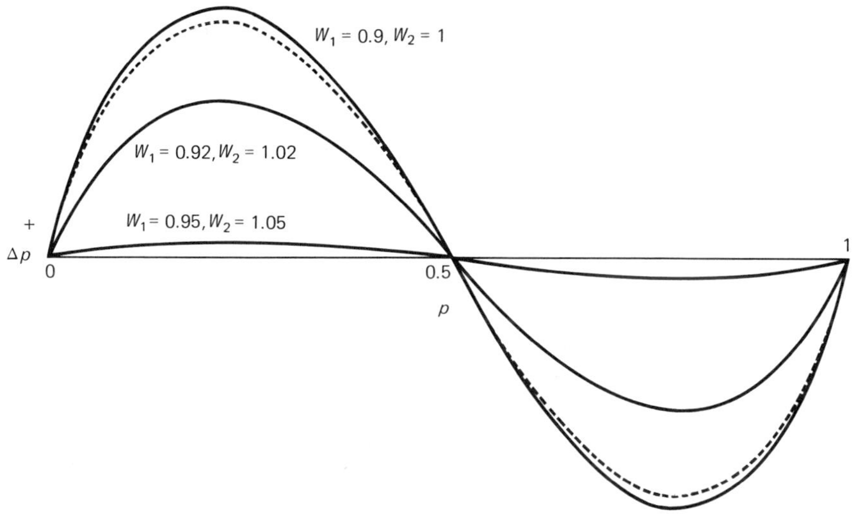

Fig. 7.1. The change in allele frequency per generation (Δp) as a function of frequency when the two subenvironments are of equal size. In all cases, $V_1 = W_2$, and $V_2 = W_1$. The dashed curve shows the change in frequency for unconditional heterotic selection when $s_1 = s_2 = 0.05$.

inance, the equilibrium frequency is very easily disturbed from 0.5 by a change in the environment size. Fig. 7.3 shows the effect on such a system when the size of the first environment (c_1) is altered from 0.5 to 0.1—that is from the case when both environments are of equal size to the case in which the second environment is nine times the size of the first. The equilibrium frequency is also shifted, from 0.5 to 0.11.

Similarly, equilibrium frequencies can shift dramatically with a change in the partial dominance. In Fig. 7.4, the dominance of the heterozygote has been progressively reduced in a marginal overdominance situation with symmetrical selection coefficients and the first environment one third the size of the second. The equilibrium frequency for complete recessiveness of the homozygotes is 0.75, but this is rapidly moved towards fixation as dominance is decreased. The

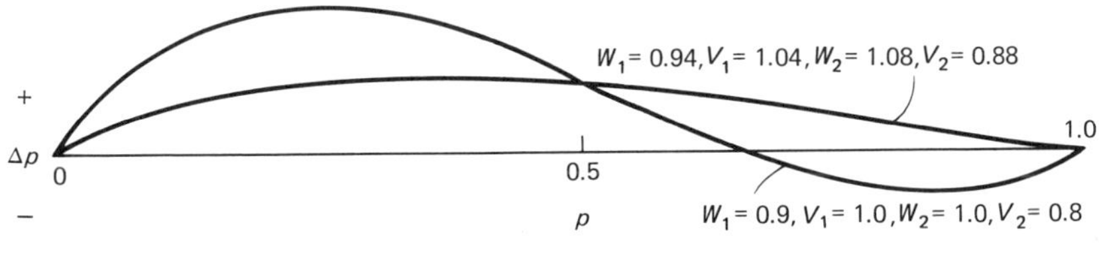

Fig. 7.2. The effect on Δp and $\hat{p}$ of changes in partial dominance in two environments of equal size. The spread of fitnesses in the second environment is always twice that in the first.

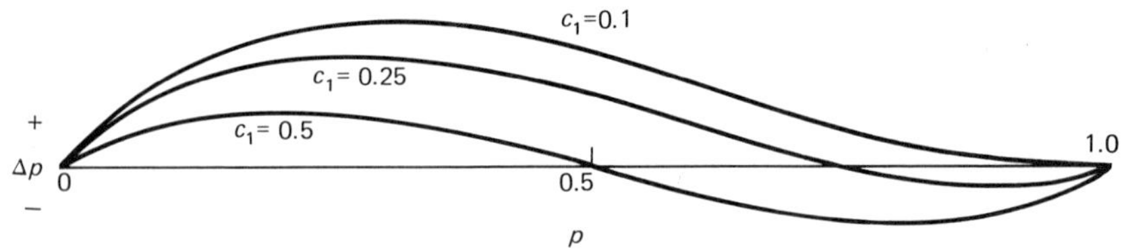

Fig. 7.3. The effect on Δp and $\hat{p}$ of changes in environment size. For each case $W_1 = 0.9$, $V_1 = 1.0$, $W_2 = 1.0$, and $V_2 = 0.9$.

system becomes unstable long before the condition of intermediate dominance is reached.

Dominance of the fitter genotype, complete or partial, converts a weakly heterotic system into a strongly heterotic one, though as indicated above, selection never becomes as strong as in a single-niche heterotic system. Further, if enough dominance is added to the system, as can be seen from Tables 7.2 and 7.3, the requirement that selection be density-dependent in each niche disappears.

A number of authors have worried about the apparent stringent requirements for the maintenance of polymorphism given intermediate fitness of the heterozygote. Maynard Smith (1966) pointed out that with density dependent selection in two different niches, the selection coefficients can vary over only rather narrow ranges, as we have seen from the examples above. He attempted to escape from this restriction by suggesting that there may be strong habitat selection, resulting in organisms carrying certain genotypes occupying particular niches and producing progeny which settle in the same niches. While this is certainly true for plants (see Allard 1975), the evidence for strong habitat selection is much less compelling for animals, particularly highly vagile ones. There is, however, another and perhaps more likely way for the strong restrictions on the range of selection coefficients to be relaxed, and that is if partial dominance is permitted in the system. Then, as we have seen, selection in differentiated spatial niches approximates a heterotic situation, although at least in multiplicative systems selection is not as strong. Such a system of fitnesses has been termed arithmetic mean overdominance by Felsenstein (1976), as contrasted with the much more restrictive situation considered by Levene

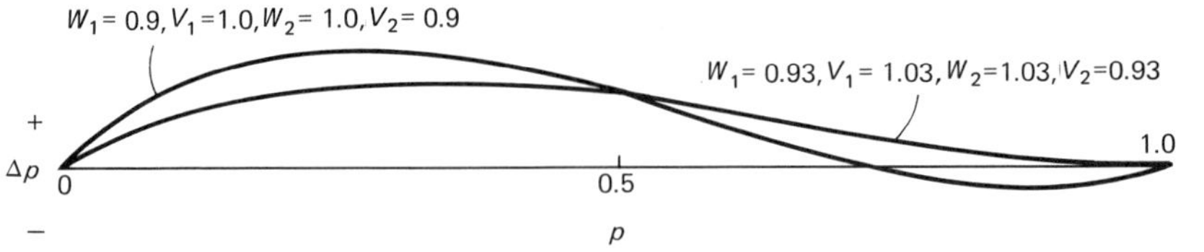

Fig. 7.4. The effect on Δp and $\hat{p}$ of changes in partial dominance when the first environment is one third the size of the second ($c_1 = 0.25$). In each case, $V_1 = W_2$, and $V_2 = W_1$.

which can be termed using Felsenstein's nomenclature 'harmonic mean overdominance.'

Multiplicative models: temporal variation. In a temporally fluctuating environment, it is again possible to maintain polymorphism for strictly intermediate fitnesses, but the conditions for this maintenance by multiplicative selection are slightly more restrictive than for the spatially fluctuating environment. Dempster (1955) was again the first to show that stability was possible, provided that the geometric mean of the heterozygotes averaged over time was greater than those of the homozygotes. Since the geometric mean is always equal to or greater than the harmonic mean, this is a more restrictive condition. The range of permitted selection coefficients was calculated by Hedrick et al. (1976) for the two-niche, two-allele case, and was found to be significantly narrower than that for spatial differentiation.

Again, however, the introduction of dominance allows these restrictions to be relaxed, as Dempster noted. Provided that the environmental 'switch' occurs at fairly regular intervals, so that one or another allele is not lost during long periods of environmental auto-correlation, introduction of dominance results in a marginal or arithmetic mean overdominance situation here as well.

Several other, perhaps less likely types of selection in fluctuating temporal environments have been examined. For example, Haldane and Jayakar (1963) envisaged a situation in which the heterozygote and one homozygote were equal in fitness, and the other homozygote fluctuated in fitness above and below this value. For a stable polymorphism to result, the geometric mean fitness of the fluctuating homozygote had to be less than the value for the other genotypes and the arithmetic mean fitness greater. These authors realized that this was a restrictive condition, and Hedrick (1974) showed that in a finite population this model was only effective in maintaining a polymorphism if the environment was negatively autocorrelated over time—that is, if the selection coefficients switched from positive to negative and back again in a regular fashion. If there was no autocorrelation or if there was positive autocorrelation, then regardless of the strength of selection at any one point in time, the rate of loss of polymorphisms was actually accelerated over the case in which only random drift was operating.

Wright (1975, pp. 54–5) has discussed the differences between stability in spatial and temporally fluctuating models in a very incisive fashion. In the temporal model one must consider the relative contributions of the different genotypes, so that the proportions of the various genotypes surviving over time are the products of their fitnesses at different times. Thus, even in the absence of 'soft' selection, the fitness of each genotype averaged over time is the weighted geometric average of the fitnesses at each point in time. This can produce 'cumulative overdominance' (Dempster 1955) even if the arithmetic means of all genotypes are all the same—for example, if the variance of the heterozygote fitnesses is less than that of the homozygote fitnesses. An increase in the dominance of the heterozygote, however, simply switches the system to an arithmetic mean overdominance, though as in the spatial fluctuation this is not as efficient in maintaining polymorphism as is strict overdominance.

Wright points out that in the spatial system the fitnesses are an average over space only, and so the weighted fitnesses of the genotypes are simply added up to give the overall fitnesses. If the arithmetic mean fitnesses of all the genotypes is the same, then the only way polymorphism can be maintained is with the introduction of density-dependent selection. As we have already seen, introduction of sufficient arithmetic mean overdominance eliminates the requirement for density-dependent selection.

There is no doubt that the requirement that selection coefficients switch signs in a regular way is a great restriction on temporal models. Spatial models require merely that the coefficients switch signs somewhere in the set of environments to which the species is subjected. Given sufficiently diverse environments, this may be a less restrictive condition than that imposed on strictly temporal models. Mixed models may be the closest to reality.

The expected distribution of polymorphisms. A number of complex variations have been proposed on these basic approaches to spatial and temporal variation. Some of the most interesting have been made by Gillespie and Langley (1974), and a following series of papers by Gillespie (1977 *a, b;* 1978 *a, b*). The mathematics involved in the model has become progressively more recondite, but if these authors will forgive my abstracting what I think are the essentials of the argument, it goes as follows:

1. There is no substantial evidence of 'overdominance' in enzyme activity as it has been measured *in vitro* (that is, for a greater activity of the heterozygote when enzyme activity is measured on a per unit protein basis) (Gillespie and Langley 1974). There is further good evidence from quantitative genetics (some of which we will review, see pp. 182–209) that heterosis is not a large factor in genetic variation in viability (Gillespie 1978 *a, b*). Thus it seems reasonable to assume that heterosis—at least heterosis acting under the conditions in which it is usually looked for—is not a large factor in maintaining genetic variation.

2. It is obvious that fitness and enzyme activity are not linearly related, but there is perhaps some fairly simple relationship that links them. Gillespie (1976) suggests such a relationship

$$\phi\,(X) = \frac{(1 + x)\,n}{\alpha + n} \tag{7.2}$$

where $n \geq 0$, using the following reasoning. If x is the enzyme activity and $\phi(x)$ is the fitness, then for a suitably chosen α one can generate fitness curves that correspond approximately to those obtained from experimental measurements. If α is fairly small, for example 0.05, then the function (plotted in Fig. 7.5) has properties that correspond to observation. Since h, the partial domi-

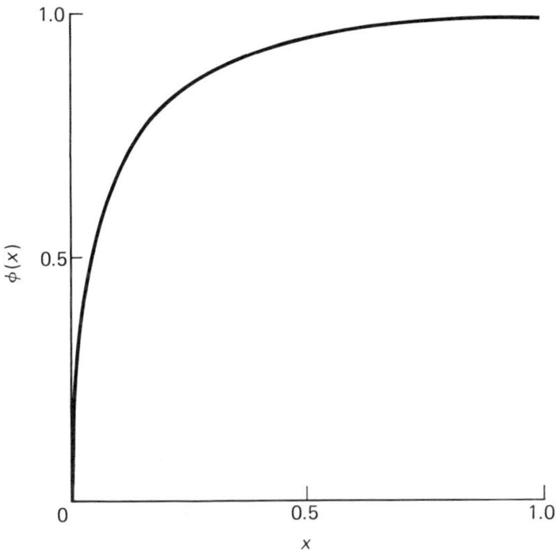

Fig. 7.5. Plot of Gillespie's (1976) fitness function for $\alpha = 0.05$.

nance, is equal to $\alpha/(2\alpha + s)$, the heterozygote effect of a deleterious allele will be inversely proportional to its homozygote effect. This is in fact what is seen; Gillespie points out that for *newly induced* lethals, partial dominance is low, about 5% of the homozygote effect, while the extensive experiments of Mukai suggest that partial dominance for viability polygenes of smaller homozygous effect is much higher, as much as 40%.

3. There is much evidence, to be given later in this chapter and cited in Chapter 4, that genes contributing to viability exhibit a strong genotype–environment interaction, suggesting that large changes in fitness could result if the environment fluctuated spatially or temporally.

Gillespie (1977*a*) shows that if fitnesses are related to activity in the way suggested above, and if selection coefficients fluctuate at random, with the activity of the heterozygote always being intermediate between those of the homozygotes, then polymorphisms can easily be maintained.

It will be seen from the shape of the curve in Fig. 7.5 that even though activities may be symmetrically distributed around one, fitnesses are not. Because the curve is concave, fitnesses will show arithmetic mean overdominance, so it is not surprising that polymorphisms can be maintained easily in this model even in the absence of density-dependent selection.

Gillespie has more recently turned his attention to the question which bedevilled us at the end of Chapter 6. If overdominance is common, why is the distribution of allele frequencies in natural populations U-shaped? One possibility is that dominance is very small, selection coefficients are varying in a random fashion, and selection at each locus is small. Then (Gillespie 1977 *a;* Gillespie, personal communication) the distribution of alleles takes on a U-shaped distribution indistinguishable from that expected from neutrality with very large population size, mutation rate or both. However (Gillespie, personal communication), if the partial dominance is increased the distribution becomes unimodal, just like the one we were confronted with at the end of Chapter 6. No consideration of unequal subenvironment sizes was taken into account in these calculations.

Multiple alleles and overdominance

The general disinclination on the part of many population geneticists to attribute much importance to overdominance was perhaps reinforced by a further factor. This has recently been examined by Gillespie (1977*b*), who quantified an earlier conjecture of Kojima. This is that it becomes progressively more difficult to establish a stable overdominant polymorphism in a population if more than two alleles are involved.

It is instructive to ask how often in the normal course of events such stable situations will arise. For a two-allele polymorphism, stability will ensue if the two homozygotes are less fit than the heterozygote. For a three allele case, at least two heterozygotes must be fitter than all the homozygotes, but one heterozygote can be much less fit. It cannot, however, be less fit than the least fit homozygote. For more alleles, the conditions become progressively more restrictive. Gillespie (1977*b*) used computer simulations to obtain the probability of a stable polymorphism when selection coefficients were drawn at random from a uniform distribution. The results, for increasing numbers of alleles, are shown in Fig. 7.6. A rapid decline in the likelihood of stability can be seen.

These calculations assume that the population is panmictic, and that the selection coefficients are constant over time and space. If these conditions are accepted, it is difficult to see how multi-allele polymorphisms could possibly be maintained in populations for any time.

Certainly, multiallelic polymorphisms that are difficult to explain on a selective basis do exist. Consider the XDH locus of *D. pseudoobscura*. Here there are a large number of alleles, perhaps larger than the first electrophoretic surveys indicated because of cryptic variation. But if a three-allele polymorphism has a restricted set of fitness relationships leading to stability, consider how restricted the set for a 37-allele polymorphism must be! Even assuming that alleles at a frequency less than 5% are deleterious and held in the population by mutation pressure, the requirements for selective maintenance of the three alleles at this locus that are more frequent than 5% are far more stringent than for a two allele system. It is possible that this is a case of multiple replacement of one allele by others, but this would have to be proceeding extremely slowly in order for it to be occurring at the same apparent rate in different parts of the population.

On the other hand, we have seen that the enormous multiplicity of G6PD deficiency alleles in man are maintained by strong localized selection in a

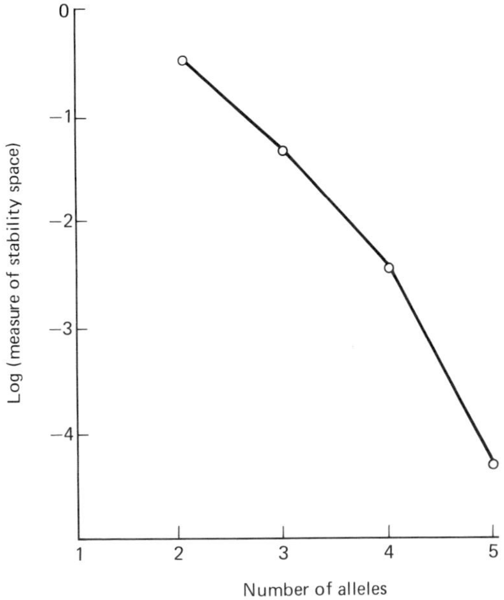

Fig. 7.6. Relationship between number of alleles and the proportion of randomly varying sets of fitnesses that lead to a stable internal equilibrium. From Gillespie (1977*b*).

patchy environment and that HLA alleles are most likely maintained by frequency-dependent selection. Who is to say that similar influences may not be operating on the *Drosophila* XDH locus?

Gillespie (1977*a*) has extended his model of intermediate enzyme activities and a convex fitness function to multiple alleles, though there are problems with his analysis. Given certain, quite modest assumptions about the nature of selection and the relative sizes of selection coefficients, he suggests that essentially unlimited numbers of alleles can be held in a population in a subdivided environment by marginal or arithmetic mean overdominance.

As we have seen, in the Gillespie and Langley model the enzyme activities of the heterozygotes are intermediate between those of the homozygotes, but because the fitness function is convex the homozygote fitnesses will average lower than those of the heterozygote fitnesses, leading to arithmetic mean overdominance. The set of fitnesses is therefore more restricted than that used to generate Fig. 7.6, which shows the likelihood of stability for the strict overdominance case. In their model, however, some of the homozygotes will be fitter than the heterozygote average.

Gillespie (1977*b*) assumed a small activity difference δ between homozygotes with the most closely matched activity. He could then show (given some approximations) that if $\delta < \frac{2\sigma^2}{n^2}$, where σ^2 is the variance in activity over all homozygotes and n is the number of alleles, then the polymorphism is stable.

This formula, comparing as it does an absolute difference in activity with a variance in activity, is unfortunately dependent on the scale on which activity is measured. If activities lie between 0.9 and 1, no polymorphism will be stable; if they lie between 90 and 100, all polymorphisms will be stable.

Nonetheless, multiallelic polymorphisms subject to heterotic selection do exist.

The difficulties with multiple-allele balanced polymorphisms can, at least intuitively, be surmounted by abandoning the requirement that the population be panmictic. We have already seen in the case of the β-thalassemias how an extensive series of alleles can be held in a population by localized heterosis, so that when the population as a whole is sampled, the number of rare thalassemia alleles appears to be very large. In a subsample of sufficiently reduced size the number would, of course, be much smaller, and if that subsample were isolated from the rest of the population and allowed to mate at random all but two or three alleles might be eliminated.

However, it is encouraging to note, as Hiorns and Harrison (1977) have indicated, that if selection acts very differently in different subpopulations even very high levels of migration will not swamp the differences. Environmental subdivision may be a powerful generator of diversity.

The distribution of unusual environments

We are still clearly in a dilemma. To recap, what we know or suspect about the mode or modes of balancing selection leading to the maintenance of polymorphism is as follows:

1. If the heterozygote is always being selected for, and the environment is not spatially or temporally subdivided, then strong and effective selection to maintain polymorphisms can result if selection is of the truncation type. The difficulty here is that one would expect, particularly in populations that have remained stable for long periods of time, that there would be far more polymorphisms at 0.5 than at frequencies closer to fixation. There is no sign of such a distribution, and in fact what we observe is just the opposite.

2. Selection to maintain polymorphisms can also occur in a spatially or temporally subdivided environment, even if the heterozygote is intermediate in fitness, if certain non-stringent assumptions are met. If dominance is low, then this selection has the interesting property that selective equilibria are extremely susceptible to small changes in selection coefficients or niche size in the case of spatial heterogeneity or to autocorrelation in the case of temporal heterogeneity. Indeed, if dominance is low then small changes in the conditions can easily lead to *instability* of the polymorphism. While quantitative arguments would be intensely dependent on the assumptions made, it would appear that if such selection were widespread it might not lead to a unimodal distribution of allele frequencies. Working against this is the expectation that, because stability is so easily lost, polymorphisms with low dominance should not be

retained in the population as easily as those with high dominance. As Gillespie has shown, this will again result in a unimodal distribution of allele frequencies.

3. Even if polymorphisms with high dominance tend to be retained, this need not lead to an excess of polymorphisms with intermediate frequency if the environment is divided into patches of unequal size. It seems likely that the most extreme environments to which a population is subject, in which conditional heterosis or marginal overdominance would be most likely to occur, would be limited in extent either spatially or temporally. This is because they represent the limits of the current adaptation of the species. If extreme environmental patches were distributed in this way, it might lead to a non-unimodal distribution of allele frequencies.

One would expect, on this view, that unusual alleles would tend to be found in unusual environments. There is some evidence that this is so; at subspecies boundaries in mice (Hunt and Selander 1973) and snails (Gould and Woodruff 1978), unusual alleles have been found at high frequency in a number of enzyme systems.

4. Because the conditions for the maintenance of multiple allele polymorphisms become progressively more demanding as the number of alleles is increased (Gillespie 1977*a*), then either such polymorphisms would appear to be maintained by balancing selection with perhaps the additional aid of new alleles produced by recombination, as has been hypothesized for the HLA locus, or by some localized selection for different alleles in different habitats, accompanied by a restriction in gene flow.

We have already seen in Chapter 5, that when heterosis is detectable it tends to be associated with extreme environmental conditions, and that under less extreme conditions one or more alleles are often selected against. Such a mixed set of selection conditions will still lead to polymorphism, but as in the case of marginal overdominance the equilibrium attained is extremely dependent on the size of the region in which heterotic selection is taking place. Figure 7.7 shows an example of spatial heterogeneity, in which the equilibrium frequency is plotted as a function of the relative sizes of the two environments in which heterotic selection and selection against one allele takes place. It can be seen that the equilibrium frequency drops rapidly as the size of the heterotic environment diminishes. However, unlike the case with intermediate dominance,

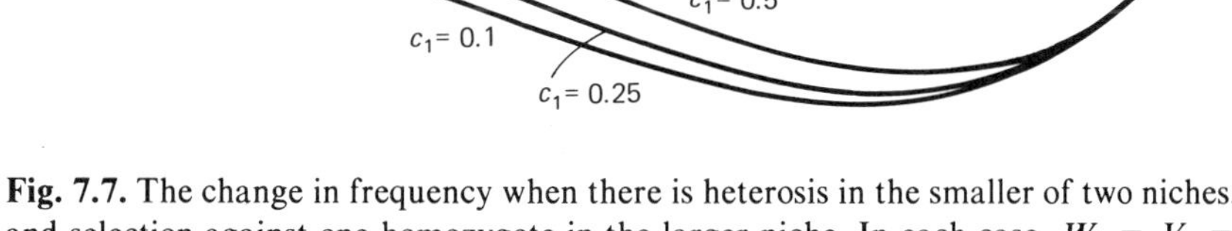

Fig. 7.7. The change in frequency when there is heterosis in the smaller of two niches, and selection against one homozygote in the larger niche. In each case, $W_1 = V_1 = 0.9$, $W_2 = 0.9$, and $V_2 = 1.0$.

the polymorphism does not become unstable before the size of the heterotic environment equals zero.

These calculations are for panmixia. In a population which is not panmictic, frequencies of heterotic alleles could be even higher locally in spite of strong selection against them elsewhere, as in the sickle cell case.

Such mixed types of selection, were they to be prevalent, might help to explain the large number of alleles near fixation. Stability due to conditional overdominance merely requires that an allele which is normally selected against is, under certain unusual and spatially or temporally restricted circumstances, selected for when it is heterozygous. Given that the unusual circumstances are sufficiently restricted, the allele will be balanced at a low frequency. Such selection is occurring for the sickle-cell allele and the thalassemias, and possibly for G6PD deficiency and hemoglobin E as well.

It is interesting to note that it is exactly this sort of polymorphism that is likely to be discovered in the human population. Since the majority of work on human polymorphisms has been in connection with diseases or clinical abnormalities, it is not surprising that those to which selection can be attributed are those in which the balancing selection is only expressed in unusual circumstances. No such detailed study of the clinical manifestations of a polymorphism has been carried out on any other organism, with the exception of some studies on maternal–fetal incompatibility in other primates and in domestic animals. The prevalence of such polymorphisms is an open question. What does seem to be ruled out, except perhaps in the case of human blood-group polymorphisms, is overdominance throughout all or most of a species range.

Frequency-dependent selection

In this type of selection, the frequencies of the genes, or more properly of the phenotypes, determine to at least some degree their fitness. Frequency-dependent selection can be very strong, and indeed can be more effective than simple overdominance at maintaining polymorphisms (Kojima 1971). But such selection has certain properties that are very different from those of the other modes of balancing selection we have considered. Frequency-dependence is undoubtedly responsible for some striking examples of organization of the genome, but it may not be the most common mode of selection. This is because frequency-dependence is a function not only of the frequencies of the phenotypes being selected, but also of the qualitative differences between them. The greater the qualitative difference between phenotypes, the more effective true frequency-dependence will be.

Nothing illustrates this more vividly than the first case of frequency-dependence to be investigated theoretically, that of Batesian mimicry (Fisher 1930). In this type of mimicry, most thoroughly investigated in butterflies (for review see Wickler 1968) the mimic imitates a distasteful model and is itself often (but not always) polymorphic. In a case of polymorphic Batesian mimicry, there are two possibilities. In the first, the mimic species is polymorphic for a mimetic and nonmimetic form. When the model is common, the mimic form has the advantage since predators have learned to avoid the model. When the model is uncommon, naive predators are more frequent and the non-mimetic form (which may exhibit protective coloration) is now at an advantage compared with the often conspicuous mimetic form. In the second type of polymorphic Batesian mimicry, the mimic is polymorphic for genes which result in phenotypes similar to two or more distasteful model species. This has been beautifully illustrated for *Pseudacrea euryatus*, which mimics five species of the genus *Bematistes* (Sheppard 1959, employing data collected by Carpenter 1920, 1949). The frequencies of the five mimic morphs track those of the models, since a more plentiful model confers an advantage on the butterflies mimicking it.

It is interesting that Batesian mimicry can sometimes lead to monomorphism, as in the case of the Monarch and Viceroy butterflies. The Viceroy *(Limenitis archippus)* looks remarkably similar to the Monarch *(Danaus plexippus)*, and there is no nonmimetic form of the Viceroy. Since the Monarch is an extremely abundant species and the Viceroy is less so and has a much smaller range, there is presumably no point at which the Monarch becomes sufficiently rare for the mimetic form to lose its advantage.

Müllerian mimicry, as Fisher pointed out, will not ordinarily result in polymorphism. In this type of mimicry, the mimetic species is similar to a model which is not distasteful but is dangerous, and which itself exhibits warning coloration. Thus, the more like the model the mimic can look, the greater its protection.

The most striking thing about the polymorphic mimics which have been investigated genetically, notably those of the butterfly *Papilio dardanus* (Clarke and Sheppard 1960 *a,b,c,*), is the extent to which the genes involved have been organized into a compact series, a supergene with limited recombination and consequently a high degree of linkage disequilibrium. Since several genes controlling different aspects of pigmentation and wing shape are involved, it is not surprising that they should be so organized; if the genes were free to assort with respect to each other a whole range of phenotypes would result, most of them not resembling the model, thus greatly weakening the mimetic advantage.

We might therefore expect frequency-dependent selection, if it were widespread, to contribute far more to the organization of the genome into supergenes than does selection which is not frequency-dependent. It will be seen in the next chapter that the genome of outbreeding organisms does not appear to be organized into supergenes, or if it is, the supergenes extend for only short distances along the genome. We will see that there are theoretical reasons for the expectation that, in outbreeding populations with long genetic maps segregating for large numbers of loci, supergenes cannot be maintained unless they are very short or unless there is some mechanical restriction on recombination. The most likely pressure maintaining such supergenes will be frequency-dependent selection, since the greater the qualitative difference between phenotypes the more effective frequency-dependent selection will become.

A further expectation is that if frequency-dependent selection has operated in a population for a very long time in a relatively constant environment, the number of supergenes that is selected in this fashion will be confined to the number of clearly different aspects of the phenotype that are subject to frequency-dependent selection. The supergenes for banding and color morphology in *Cepaea,* those for pattern, color and morphology in many butterflies, and those for histocompatibility polymorphisms in man and mouse all have profound effects on the phenotype. It might be argued that there are not a large number of other *totally different* aspects of the phenotype available for frequency-dependent selection in these organisms. Supergenes selected in a true frequency-dependent fashion may be very important in fitness, but may not be very numerous compared with genes that contribute to fitness in a frequency-independent fashion.

Because frequency-dependent selection becomes progressively stronger as allele frequencies deviate from equilibrium, it has been suggested as an alternative mechanism for circumventing the genetic load problem (Kojima 1971). In a population in which all allele frequencies are near their equilibrium, the frequency-dependent genetic load would be very small. But if large numbers of loci are being maintained by frequency-dependent selection, drift will tend to increase the load. As with other modes of multilocus balancing selection, multiplicative selection coefficients will lead to too large a load. Balancing frequency-dependent truncation selection is the only type of selection that can maintain very large numbers of frequency-dependent polymorphisms without a large load.

Rank-order truncation selection and conditional heterosis

When we compare the properties of truncation and multiplicative selection in subdivided environments, it is apparent once again that the two differ markedly. As we have come to expect by this time, the truncation model is again more efficient; the strength of selection is slightly reduced by subdivision of the environment, but the reduction is far less than that seen with multiplicative selection.

The marginal overdominance model has proved difficult to investigate theoretically, so I will briefly examine the simpler model of conditional heterosis.

Suppose that each generation a synchronously-reproducing population settles at random in n different equal-sized environments. In each environment, conditions are sufficiently different that a different subset of the loci in this population is subject to symmetrical heterotic balancing selection. We will suppose, for simplicity, that there are no shared conditions between environments, so that each subset is discrete. Further, each subset of loci is equal in number.

The sum of the selected alleles in all the environments is L. In subenvironments in which the loci are not heterotic, the alleles are neutral.

It will be seen that even though the equilibrium frequencies for the heterotically-selected alleles will be 0.5 in the subenvironment in which they are being selected, their equilibrium frequencies in the population as a whole will be $0.5/n$. It is this frequency which must be taken as $\hat{p}$.

Given that the selective criterion C is small, we can calculate σ_H^2 *within a subenvironment*. It is this variance on which truncation must act. Using the notation of Chapter 6,

$$\sigma_H^2 = \frac{\dfrac{2p_H q_H}{n}\left(1 - \dfrac{2p_H q_H}{n}\right)}{L} \tag{7.3}$$

This value for σ_H^2 can now be used in conjunction with the methods developed in Chapter 6 to determine the number of loci at which polymorphisms can be maintained to a given C.

Figure 7.8 shows the efficiency of selection in one, two, four and eight environments, for selection when $\hat{p} = 0.5$ within a subenvironment. Strength of selection is 20% truncation and replacement. It can be seen that subdivision into more than one environment has a relatively small effect on the number of

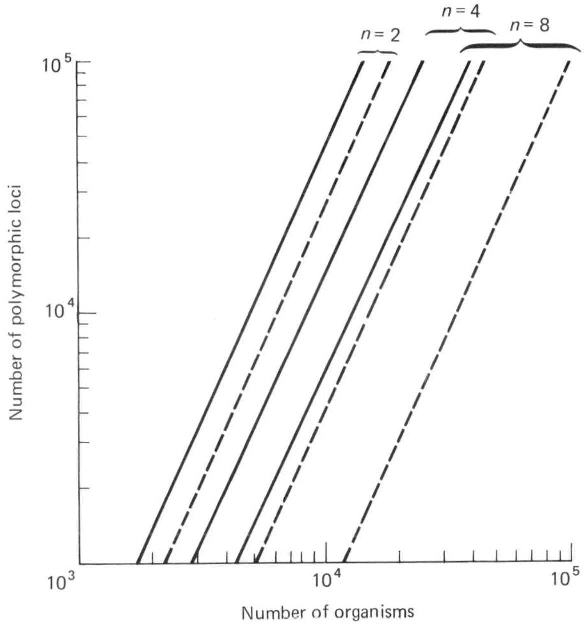

Fig. 7.8. Effect of conditional heterosis on the maintenance of polymorphisms. The number of subenvironments is given by n. In each case the solid line gives the number of polymorphisms that can be maintained in the absence of subdivision, and the dotted line gives the number that can be maintained when the environment is subdivided.

polymorphisms that can be maintained by rank-order truncation selection. For example, with eight subenvironments the number of organisms needed to maintain a given number of polymorphic loci is less than tripled, even though these loci are neutral 7/8 of the time.

The relatively small drop in the efficiency of truncation selection is due to the fact that, while subdivision of the environment results in a reduction in the number of loci being selected in each subenvironment, the total proportion of organisms being truncated remains the same. Thus, the selection coefficients associated with each locus become larger. By contrast, in a similar model with multiplicative selection, the rate of movement of alleles towards their equilibrium would be approximately halved as the number of subenvironments doubles, because the number of organisms removed by selection is halved. Of course, this difficulty could be circumvented by multiplying the selection coefficients associated with the homozygotes in each environment by the number of environments. This would keep the total reduction in fitness due to selection the same as in the single-environment case, and would approximately cancel out the effect of subdivision into multiple environments. Since the multiplicative model assumes that selection coefficients are known *a priori*, however, it is difficult to imagine a justification for such a procedure.

Conclusion

In this survey of some of the theoretical consequences of different types of balancing selection, I have emphasized the point that marginal overdominance and conditional heterosis will often be likely to lead to equilibrium frequencies for polymorphisms that appear to be close to fixation rather than intermediate. This is particularly the case if the dominance of selected genes is low, and if the more extreme subenvironments in which the strongest selection is operating are the most limited in extent. It may therefore be possible to explain on a selective basis allele frequency distributions that are very different from those expected with unconditional overdominance. As we have seen from Chapter 5, it is essential that the information about environmental diversity be added into the picture in order to understand the distributions of human polymorphisms. It is this information which is normally missing from theoretical treatments of the problem of allele-frequency distributions.

For the remainder of this chapter, we will turn to the question of whether there is evidence for widespread marginal overdominance, and whether there is evidence for conditional heterosis in organisms other than man.

Information from quantitative genetics on the nature of genetic variability

Since its founding by Fisher in 1918, the theoretical basis of quantitative genetics has been the partitioning of phenotypic variance into various components which can be attributed to additive genetics effects, dominance effects, epistatic interactions, environmental effects and genotype–environment interactions. These variances are additive, and phenotypic variation is the sum of these components. As Fisher showed, this way of approaching the problem of the relative contributions of genotype and environment is completely compatible with Mendelian inheritance. Fisher was able, therefore, to reconcile quantitative variation with the inheritance of Mendelian characters. Some very valuable information about the relative contributions of genotype and environment to the phenotypes of domesticated and experimental animals and plants has emerged from these studies.

In very simple terms, this partitioning of the phenotypic variance enables the worker to predict the breeding value of the genotype from the phenotype. If the genotypic component of the character is additive—that is, the heterozygotes are intermediate in their expression between the homozygotes at each locus contributing to the character—and if the environmental component is not too large, then when the breeder picks out animals or plants with desirable characteristics he can feel quite confident that the progeny of those animals or plants will also express the desired character and to nearly the same extent as the parents. He feels considerably less confidence if there is dominance involved, for homozygotes and heterozygotes for the dominant alleles will be indistinguishable and the effectiveness of selection will be correspondingly reduced. The same stricture applies to epistasis, which, regardless of its biochemical origin, can be thought of as dominance of one allele over an allele at another locus. The worst complication arises if there is an appreciable genotype–environment interaction component. If it is unpredictable—that is, non-linear—its effects can only be determined over the range of environments to which the experimenter or breeder has previously subjected the organisms. Because of its nonlinearity, it will be expected to produce unpredictable results if further environments are tested or if the environment changes in a way that is not under the control of the experimenter.

We have already seen in the first part of this chapter that non-linearity of genotype–environment interaction provides a mechanism for the maintenance of genetic polymorphism through balancing selection. The question is: does it? Is the genetic structure of populations simple or highly complex? If it is the latter, is the complexity such as to lead to the retention of many polymorphisms by balancing selection?

Predictions of the simple and complex models of population structure

The simple view of population structure, with most variants neutral and most of the remainder selected against, allows some prediction to be made.

1. A relatively small number of strongly-selected loci are involved in each continuously-distributed character, such as egg-to-adult survival of *Drosophila*. This is because a large number would produce an unacceptable genetic load.

2. Because of partial dominance of deviant alleles at these loci, there should

be a large component of the genetic variation for a character which is additive. This additive component and the genetic variance will both become progressively smaller as the harmful genes are removed from the population, which they will be most rapidly if the character is intimately concerned with fitness. If the character is not so concerned with fitness, deviant genes are freer to accumulate in the population through mutation pressure. There should therefore be a small genetic component of phenotypic variance in characters closely associated with fitness, and a larger component in characters less associated with fitness. Confusing this picture, however, is the added factor that new variation is being introduced by mutation. The larger the number of genes contributing to a character, the more rapidly variation will accumulate, regardless of the character's effect on fitness.

3. It should be possible, with any quantitative character determined by a small number of strongly selected genes, to select for a *single* genotype. This should occur regardless of whether selection is *directional,* resulting in a shift in the man of the selected character, or *stabilizing* selection against phenotypic extremes. Loss of vigor and adaptability will be due primarily to fixation of harmful genes as a result of unavoidable inbreeding during the selection process.

The more complex view also allows some predictions, some of which are in practice indistinguishable from these made above. Further, and unfortunately, the type of complexity which is postulated has a great influence on our expectation.

1. It is possible, because of the benefits of truncation selection, for a large number of strongly-selected loci to contribute to a continuously distributed character. Their effects may be strongly influenced by the environment, often in an unpredictable way.

2. If unconditional heterosis were acting on these loci, the breeding value of the genotype should not be directly predictable from the phenotype. The heterozygote should show the greatest expression of the character. In the next generation, segregation will once more produce an array of genotypes. There will be a very large dominance component to genotypic variation. However, as we have seen, other types of balancing selection are more likely to predominate.

(a) If there is much selection for conditional heterosis, one would again expect a large component of dominance variance—but this would only be the case if the overdominance is expressed. As we have seen, conditional heterosis tends to be associated with extreme environmental conditions, and of course most experimental measurements are made under far less stringent environmental conditions. Under these less extreme environmental conditions, loci should tend to revert to partial dominance and the genetic variance would appear to have a large additive component just as it does with the simple model.

(b) If there is much selection of the marginal overdominance type, then genetic variance in a single environment will again tend to be additive, since codominance or partial dominance will be the rule in a single environment. The only way to distinguish between this model and the simple model would be to change the environment and look for large and nonlinear genotype–environment interactions. Unfortunately, even if such interactions are found, it is usually not certain whether they are due to the *same* genes being expressed in different ways in the different environments or simply to the expression of different sets of genes. Evidence bearing on *this* thorny problem, as we will see, is exceedingly sparse.

3. Selection, particularly stabilizing natural selection acting under a variety of environmental circumstances, should not reduce the genotypic variance of a character as markedly if it is balancing as if it is purifying. Obviously, however, even with purifying selection the amount of genotypic variance remaining in the population will be a function of the number of loci involved, the mutation rate, the degree of dominance, and the epistatic interactions between loci that, like dominance, tend to slow down the rate at which harmful alleles are removed. Marginal overdominance, and possibly conditional heterosis if the environment is non-stringent, should both lead to the maintenance of quantities of apparently additive variance.

It seems likely, of course, that the true state of affairs lies somewhere in between these extremes. It would not take very many segregating overdominant genes to confuse the picture completely, so that the selection acting on a population might be predominantly purifying but appear to be balanced. This is most particularly the case if one bases one's arguments on *Drosophila* viability studies, which have for the most part dealt with the effects of entire chromosomes. Suppose there were a dozen loci exhibiting overdominance and affecting egg-to-adult survival scattered along the chromosome, and all the rest were monomorphic in fact or in effect (that is, carried neutral alleles) except for rare harmful variants. Then the complex model would appear to be the correct one, even though very few loci would be contributing to fitness differences. The prediction of this model that large numbers of loci are likely to be involved is not testable with such a system.

In fact, the case against overdominance seems strong enough to rule out even a small number of overdominant alleles. The partial dominance that has been observed for genes affecting viability suggests that overdominant alleles may be so rare as to make little, if any, real contribution to the genetic architecture of the population. They also seem to play very little part in the genetic architecture of domesticated plants that have been intensively investigated, notably corn. However, additivity by itself does not rule out the more complex model. Let us examine some ways in which the relative roles of purifying and balancing selection might be ascertained.

Do long periods of stabilizing selection tend to exhaust additive genetic variance?

The simple view of population structure predicts, with the caveats already noted, that genetic variability should be reduced as the 'ideal' homozygous phenotype is approached. Thus, if the environment remains stable for long periods, the genetic variance in a population should decrease relative to the total phenotypic variance. This will happen even if most genetic variance is due to dominance or epistasis, for while the approach to a uniform genotype will be much slower it will still proceed.

We must first define clearly what is meant by stabilizing selection. Mather (1953) divided selection into three types, stabilizing, directional and disruptive. The types are illustrated in Fig. 7.9, taken from Mather's paper. Mather was, of course, thinking about selection on the phenotype, so that stabilizing selection is selection against *phenotypic* extremes and says nothing about what is happening at the genotypic level. Unfortunately, the term stabilizing selection has been used (e.g. Goodman et al. 1975) in connection with selection on the genotype which can better be described as a subcategory of directional selection, purifying selection. That is, it has been used to describe selection towards a particular optimum amino-acid sequence at a particular period of time, with deviations from this sequence being deleterious and therefore selected against. The extreme stability of particular portions of proteins during long evolutionary periods is undoubtedly a result of purifying selection, but it remains to be seen how large a subset of stabilizing selection is made up of purifying selection.

It is generally considered that the largest amount of selection acting on natural populations is stabilizing. In view of this, it is surprising that more observations and laboratory investigations have not been made on such selection. By far the greatest amount of work has been done on directional selection, certainly in part because of its enormous economic importance, but also one suspects because the experimenter can actually *see* something happening.

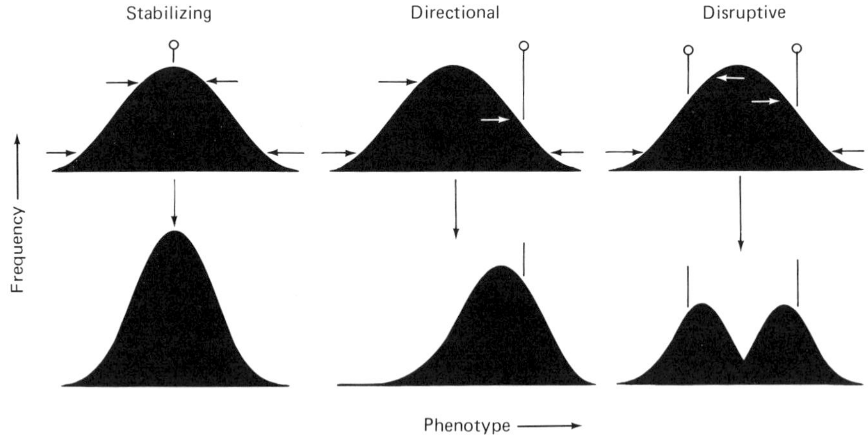

Fig. 7.9. The three types of selection. The phenotypic optima are located by the lines marked with an O. Mather assumed that selection increases in intensity with increasing distance from the optimum. From Mather (1953).

Observational evidence. The first and perhaps most famous investigation of stabilizing selection was carried out by Bumpus (1899), before the establishment of Mendelian genetics as a science. One hundred and thirty-six European sparrows, *Passer domesticus,* stunned by an ice storm, were taken to Bumpus' laboratory in Providence, Rhode Island, and 72 subsequently died while the remaining 64 recovered. Bumpus measured nine morphological characters, and found that birds with the more extreme measures were more likely to have died, while those closer to the mean were more likely to survive. In a careful reanalysis of Bumpus' data, Grant (1972) concluded that this was in fact the case for four of the characters—weight, length of humerus, length of tibiotarsus and length of sternum. Bumpus realized that his data, scanty though they were, provided strong evidence for Darwinian selection.

There is of course no indication of how much of the variability measured by Bumpus in these birds was in fact contributed by the genotype, though by analogy to man and domestic animals, particularly chickens (Lerner 1958), one would expect that bone length would have a large heritable component and weight a much smaller one. But certainly these birds possess a great deal of genetic variability which has expressed itself in one of the most rapid periods of evolution recorded for natural populations. Since its introduction into North America in 1852, it has differentiated into a number of different subspecies marked by differences in size, body proportions and color (Johnson and Selander 1964). Along the northern Pacific coast the birds have become distinctly darker, while lighter races are found in the Great Basin deserts. All this, however, is simply circumstantial and proves nothing about the effectiveness of stabilizing selection in removing or in maintaining genetic variability.

Another circumstantial case, one that has been discussed and reprinted widely and that has formed the basis for a good deal of clinical investigation, is given by Karn and Penrose (1951). These authors investigated the relative survival to age one year in a large sample of newborns in London from the years 1935–46. The graph of log survival against birth weights is shown in Fig. 7.10. Two things are apparent from the figure. The first is that, as might be expected, babies of very low birth-weight have a very low survival probability, since prematurity is a contributing factor. But so do babies with a very high birth-weight, a phenomenon which may be related to a variety of factors such as difficulties at birth and physiological abnormalities such as glandular imbalance in a very heavy babies or their mothers. There is, therefore, selection for an optimum birth weight—selection for an optimum phenotype. The second fact is more intriguing. It is that the weight for optimum survival is slightly but significantly displaced upward from the mean birth weight. There would appear to be a conflict here, presumably of long standing, between two selective optima. The first involves selection for heavier babies at birth, since their survival chances are better. The second involves selection for lighter babies, since the birth process is presumably easier.

This difference between mean and optimum birth weight is seen in other

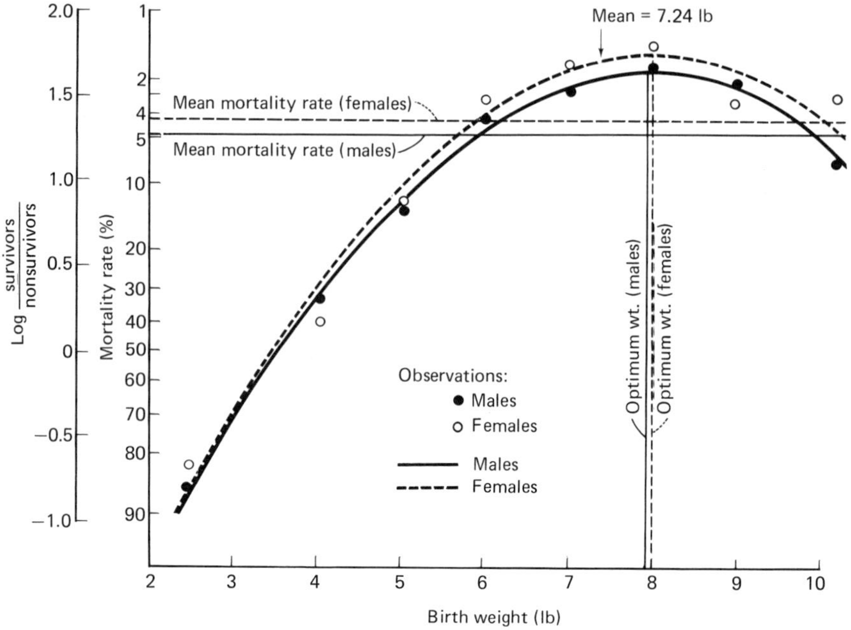

Fig. 7.10. Log of infant survival to 1 year plotted against birth weight. It will be seen that the mean birth weight is less than the weight conferring optimum survival. From Karn and Penrose (1951).

populations as well. For example, Saugstad (1977), in a study of normal neonates in a Norwegian population, found that among 135 695 infants the mean birth weight for males was about 3500 g and for females was about 3400 g. The birth weights for optimum survival, however, were about 4500 g for males (almost 10 pounds!) and 4000 g for females. The differences as well as the absolute weights were much greater than in the English sample (about 3200 and 3500 g for mean and optimal birth weights, respectively).

There seems, in these cases, to be a real conflict between two selective requirements. There is no single ideal phenotype possible in this case, since the two selective forces—slightly heavier babies have better survival and slightly lighter babies are born more easily—are in direct opposition to each other. There is also presumably no single ideal genotype as well, though the degree to which the genotype contributes to this character is problematical.

How much of birth weight is heritable? Twin studies indicate a broad sense heritability of about 0.61 (Osborne and deGeorge 1959). Estimates based on correlations between birth weights of parents and offspring are confounded by the rapidly changing state of nutrition in advanced human societies. Other correlations may be similarly affected. For example, Cawley et al. (1954) found, in a large study involving English babies from Smethwick and Birmingham, that there was a very small correlation (+0.1) of child's birth weight with height of the father, and only twice that (+0.2) with that of the mother. The difference between the effects of maternal and paternal heights was in the same direction as that observed for domestic animals by a variety of workers, though the correlation with maternal birth weight was higher in general for the animal samples. A mere lack of strong correlation does not, of course, mean that there could not be a large genetic component involved; it could be present chiefly as dominance and epistatic variance and therefore not be seen in an analysis such as that of Cawley et al. (1954). But as with the sparrows, the evidence for what is happening at the level of the phenotype is fragmentary at best. There seems from twin studies to be a substantial heritable component, but because of the fact that two optima seem to be selected for, it seems possible that disruptive rather than stabilizing selection might be occurring in these cases.

Another fascinating example of apparent stabilizing selection is found in the data of Conterio and Cavalli-Sforza (1960, discussed in Cavalli-Sforza and Bodmer 1971). These authors studied the distribution of heights in 20-year-old Italian conscripts (because of universal conscription, these conscripts were a very complete sample of Italian males of that age group). They found a significantly smaller variance, by a factor of about 1/6, among married than among unmarried men. This could be calculated to involve a selective elimination of 9.1% of the sample through presumed nonreproduction. Assuming that marriage is correlated with Darwinian fitness, we seem to have a clearcut case of stabilizing selection. Further, twin studies have shown that height in man has a large genetic component, giving heritability values of about 0.8 (Cavalli-Sforza and Bodmer 1971). The estimation of heritability using twins is an estimate of the *total* genotypic variation giving heritability in the broad sense, because identical twins have the same genotype and therefore the genotypic variation is expressed in the same way in each. Heritability estimates for height may vary from one human population to another. Roberts et al. (1978) found a significantly lower heritability (about 0.6) in a West African population, a difference they attributed to the rigorous and highly variable environment of the population they studied.

It would appear that only in this case of the three we have listed do we have evidence that stabilizing selection to a single optimum has not led to a disappearance of genetic variability, but note how tenuous the evidence is! We have only a rough idea of how strong or how effective the stabilizing selection may be.

Experimental evidence Data from experimental organisms are also difficult to interpret. A great deal of attention has been directed, particularly by English investigators, to the set of small bristles, usually about 25 in number, found on the sternopleural plates on the side of *D. melanogaster*. These were picked by Mather and Wigan (1942) as a likely object of selection, since their presence or absence seemed to have little bearing on the fitness of the fly. Indeed, it was shown early that the genetic variance in bristle number had a large additive component which made them ideal for directional selection.

They also showed signs, however, of having been subject to stabilizing selection. In his thesis, reported in Thoday (1958), Beardmore showed that there were different optimum numbers of bristles when *D. melanogaster* was adapted to four different temperatures. The numbers of sternopleural bristles decreased with increasing temperature. Unfortunately, this experiment was confounded with an environmental interaction; *any* flies raised at a higher temperature will produce fewer bristles on the average. The point of the experiment was that there was enough variation in the genes determining sternopleural bristle number to produce these different optimum numbers at different temperatures—though the figures for the amount of heritable variation exhibited by these lines were not published.

In a small-scale but more detailed study, Barnes (1968) showed that sternopleural bristle number also tended to stabilize at different numbers when the flies were raised at different temperatures—23.2 at 25 °C and 21.4 at 18 °C, which was the reverse of the trend seen by Beardmore. (It should be noted parenthetically that experiments of this sort are characterized by a remarkable lack of repeatability, usually attributed to ongoing genetic changes in the stocks being used. This lack of repeatability must certainly, in part, be due to the complexity of the phenomena being studied, and may tell us something about genetic variability in general.) Barnes went further than Beardmore and carefully studied several components of fitness, showing that there were indeed different fitness maxima at these sternopleural bristle numbers for the two different temperatures. The pattern appears to be similar to that for stabilizing selection for birth weight in man, with two different optima being selected for at the two temperatures tested (and presumably many other optima which could be selected for). In contrast to the birth weight case, however, we know that there is a large component of additive variance affecting sternopleural bristle number. The ability of temperature changes to affect the optimum suggests that it would be very difficult for natural selection to tend towards a single optimum genotype. Significant variance which is additive *in a particular environment* is being preserved, and somehow this variance is structured so as to result in different phenotypic optima in different environments.

Gibson and Bradley (1974) used this system to ask a very interesting question, and unfortunately obtained an ambiguous answer. The question was: is stabilizing selection effective in fluctuating as well as constant environments? A positive answer to this question would suggest that a single optimum genotype can be selected for despite the simultaneous selection of two phenotypic optima, in this case by two temperatures alternating during a diurnal cycle.

These authors found a reduction in the variance of the character, paralleled by a reduction in the size of the additive genotypic variance, in selected lines at both stable and fluctuating temperature by generation 19, while the control lines showed normal levels of additive variance. Unfortunately, from the standpoint of an easy interpretation of the data, subsequent to that point in the experiment the control lines showed a pronounced shift downward in their means (which the selected lines did not) and also showed a reduction in total and additive variance to levels the same as or even less than the selected lines. Because of the way the controls behaved, no obvious clearcut answer can be extracted from these very interesting experiments.

The question remains a central one, however. Will a population tend towards a single optimum genotype if the environment is kept uniform enough? Or will there be residual dominance variance representing heterotic genes that continue to segregate in the population? Dominance variance was not measured in the experiments of Gibson and Bradley. To make things even more complicated, will the final result of such an experiment depend on the nature and severity of the supposed uniform environment? And finally, how can one be sure that a 'uniform' environment is truly uniform? Certainly, as we will see, the fly bottles of the *Drosophila* experimenter do not constitute a particularly uniform environment. All these questions pose perhaps insuperable problems for the investigator.

What is the extent of genotype–environment interactions?

In the absence of widespread single-gene overdominance, if balancing selection is operating there should be a large amount of genotype–environment interaction. If the order of fitnesses of genotypes at a locus can be shown to be negatively correlated or even uncorrelated in different environments, such a situation is consistent with the maintenance of polymorphism. Indeed, the conclusion to which one is led from theoretical studies is that polymorphism will be expected from a wide variety of such fitness sets, as we have seen. Of course, genotype–environment interaction is not inconsistent with a model in which most selected variation is held in the population by mutation–selection balance. Like dominance and epistasis, this interaction will tend to slow the rate at which harmful genes are removed from the population. We can, however, ask whether the type and extent of interaction is consistent with balancing selection. Further, some limited data are available from Drosophila experiments at the single locus level which are consistent with a marginal overdominance model.

Information from plants. Estimates of genotype–environment interaction are far less common in the literature than are estimates of heritability in a single environment, because of the technical difficulty of obtaining the data. Some of the most elegant elaborations of experimental designs to obtain this information have been possible with plants. It is possible to grow plants under quite uniform conditions, and each plant can be considered to be growing in its own small microenvironment (Comstock and Moll 1963). Thus, accurate estimates of the genotype–environment component of variance should be obtainable. There are, however, difficulties in the way of obtaining reliable estimates of all components of variance, not simply the genotype–environment interaction component.

Robinson and Comstock (1955) reported on the results of extensive diallele crosses in maize, in which lines with known characteristics were crossed in many different combinations. The progeny plants were randomized with respect to environment. In one set of experiments the variation within lines of one male crossed with four females was measured; in another set, members of a segregating F_1 population were backcrossed to the two homozygous parental lines.

They pointed out that if there are genotype–environment interactions that cannot be detected because of the design of the experiment, for example if an apparently uniform environment is actually heterogeneous, this will tend to inflate the estimate of additive genetic variance. This is because the additive variance is considered to be what is left over after the other components have been subtracted from the total phenotypic variance. These experimental designs depend on a partitioning of the variance due to several causes. If other causes of variation are present, they will tend to confound the results.

In their paper, Robinson and Comstock concentrated on the relative sizes of the additive and dominance components of genotypic variance. Since in most cases the dominance variance was only about half the total, they concluded that there was enough additive variance for selection for yield to be effective. Indeed, the yields they obtained after one or two generations of selection matched prediction fairly well—in three out of the four strains used. The fourth showed a marked decline in yield, which they attributed to the observation that the total genetic variance was low in that strain.

In order to reduce the size of the dominance variance, however, they had to subtract the 'probable effect' of linkage. Linkage disequilibrium will tend to increase the dominance variance and make predictions of the genotype from the phenotype much more difficult, because the F_1 of plants or animals in linkage disequilibrium will exhibit 'pseudo-overdominance.' The F_1 will tend to express the characteristics being selected to a greater extent if whole blocks of recessive alleles are masked by dominant alleles at the same loci. The correction factor must be guessed at, since there are no good estimates at the moment of the amount of disequilibrium to be expected in corn populations. Their estimate of the increase in the amount of dominance variance due to disequilibrium was arrived at by carrying selected lines out for several generations and finding that the dominance variance appeared to decrease relative to the additive variance as selection proceeded and more and more loci became homozygous as a result of selection and inbreeding.

Perhaps the most intriguing aspect of this work from the standpoint of the present discussion is that they were unable to select for increased yield in one of their lines even though diallele crosses gave the information that additive variance was available for selection. There appears to be a real contradiction, in this case, between the information obtained from a diallele cross and that obtained from selection experiments.

Moll, Lindsey, and Robinson (1963) concluded from similar experiments that most or all of the overdominance observed for yield in corn could be attributed to the pseudo-overdominance due to linkage. This was because the overdominance tended to disappear after random mating of F_2 populations. This conclusion was reinforced by extensive experiments on the extent of genotype–environment interaction reported by Comstock and Moll (1963). In this paper, in which detailed models for genotype–environment interaction were proposed, the authors made the following points:

(a) As the universe of environments which is investigated expands, the genotypic variance will inevitably decrease compared with the genotype–environment interaction variance. If the range of environments employed in breeding experiments is sufficiently large, however, then for practical purposes all possible genotype–environment interactions will have been covered.

(b) If the *ratio* of components can be determined in several environments and can be shown not to alter significantly, then the genotype × environment bias should be the same in each environment and can for practical purposes be ignored. In other words, the additive variance will have the most practical importance.

One feels a little uncomfortable with this approach. If unusually wet or dry years happen not to occur in the course of the experiments, if a particular fertilizer should become unavailable in the future, or if a new disease or a variant of an old disease should attack the crop, it seems unlikely that the extent of genotype–environment interaction could be predicted from the data already in hand. The tendency to assume that a particular strain of wheat or rice will do well under a variety of conditions has led to problems when they are introduced to new conditions—see Chrispeels and Sadava (1977) for a survey of the fate of several 'green revolution' crops.

Still, the model constructed by Comstock and Moll (1963) is sophisticated—most particularly in that it avoids the assumption made in most models that the environmental variance under different conditions is comparable in size. In other words, one must avoid the error of simply cranking the *number* of replicates, years, or plots into an analysis of variance model as if they were independent. There is no guarantee, for example, that all the observations made in a single year might not be very similar to each other, but that the variation from one year to the next might not be very large. They investigated this possibility by looking at the variation in genotype x environment estimates from different subsets of the environmental variables.

The experimental data were collected from 60 half-sib families compared in two replicates in five locations over 5 years. The variance of variances estimated by a variety of methods was not high enough to be significant, and the genotype × environment interaction was also not high. Thus, these authors were led to the conclusion that, at least for the set of environments investigated, the genotype × environment interaction was not large enough to invalidate the predictive results of the partitioning of genotypic variability.

The information was gathered on plants grown under controlled conditions, with adequate nutrients and water, without appreciable competition from other species. Whether information gathered in this way is directly applicable to the situation found in nature is a moot point. Heritability may be high under natural conditions, however. A recent study of morphometric characters in Darwin's finches (Boag and Grant 1978), with data collected directly from nature, showed surprisingly high heritabilities for those characteristics which contribute greatly to species differentiation.

Even under laboratory conditions, the genotype–environment interactions may not be linear when environmental differences are great enough. We will see some examples of this later (pp. 200–207), but it may be worth mentioning now some results obtained by Butcher et al. (1972) in an investigation of clones of the mold *Aspergillus nidulans*. The characteristics measured were growth rate and protoperithecium formation, and 25 different clones were measured in twelve laboratory environments—two different media, two temperatures and three initial pHs in all possible combinations. Very strong genotype–environment interactions, not linearly related to the environments, were seen in all the strains investigated: similarities in interactions were used to group the strains tentatively into clonal types.

Growth rate was measured under essentially unrestricted conditions, with no competition. Nutrients were sufficient to ensure no slowing-down of growth during the time that the measurements were being made—and indeed such a slowing-down would have biased the results and made the complex model applied by these workers unusable. Thus even here natural conditions had not been approximated. The growth conditions were far more extreme than those usually employed in the laboratory, however.

Information from animals. When the estimates of different factors in inheritance are examined closely, some remarkable nonuniformities emerge. We have already seen that the controls in the experiment of Gibson and Bradley (p. 190) showed as great variation over time in the size of their additive genetic variance as did the experimental lines. Beardmore et al. (1975) discovered something even more discomfiting about additive variance as expressed by heritability. They found that the heritability of sternopleural chaeta number actually increased with the age of the parents. It increased 60–100% when the progeny of parents with mean ages of 3 days and 3–4 weeks were compared. This difference could not be due to differential mortality of the older flies, since these workers were careful to ensure that no appreciable mortality took place. The effect was less when despermed females that had been mated when young were used than when aged female virgins were employed. The effect was a highly consistent one, even though for at least part of the experiment the estimates of heritability themselves disagreed. Some values derived from half-sib comparisons were twice those obtained from regression of offspring on mid-parent, even though theoretically these two values should be the same—and indeed were the same for the last part of the experiment. The mechanism of the age effect is unknown. Essentially, the gametes produced by older flies gave rise to progeny more like the parents on the average than did those produced by younger flies. Is stabilizing selection acting on the pool of gametes over the lifetime of the fly?

As was pointed out earlier, sternopleural bristles were picked as a likely candidate for the investigation of polygenic inheritance by Mather and Wigan, because they appear to have little effect on the fitness of the flies carrying them. Further, phenotypic variation in these bristles has a large component of additive variance. But if experimental results of the sort found by Beardmore et al. (1975) are any guide, then even additive variation in a character not directly connected with fitness can be affected markedly by small changes in the environment, or in the experimental design, or both—changes so slight that they are not always under the control of the experimenter.

Spector (1956, pp. 111–13) has collected a large number of estimates of heritability for domestic animals. While some phenotypic characters tend to have higher average heritability than others, one striking aspect of the collected data is the amount of variation within characters. As with the correlations of polymorphisms with diseases discussed in Chapter 5, heritabilities tend to vary greatly from study to study, even when they are obtained using the same methods. In Table 7.5, I have listed characters in what seem to me to be descending order of importance to fitness (the reader may feel free to quarrel with me about the order chosen), along with the mean, standard deviation and number of heritability estimates. Of the three groups for which extensive data are available, only cattle show a trend of increasing heritability with decreasing importance to fitness.

It may also be that the amount of variation of a character itself increases the more closely it is related to fitness, though of course there are exceptions and again the data are fragmentary. Forbes Robertson (1957) determined for *D. melanogaster* the correlation between numbers of ovarioles—which determine, at least in part, how many eggs will be laid during the fly's lifetime—and body size as determined by thorax length. The former has a narrow sense heritability, estimated by selection, of about 0.14 when the effects of a major gene have been adjusted for, while the latter has a heritability of about 0.4. The coefficient of variation for ovariole number ranges between 11 and 15 percent, while that for thorax length is only 2–3%. Robertson and Sang (1944) showed that the rate at which eggs are laid is very much a function of slight differences in the diet of both larvae and adults, so there appears in this case to be a large genotype–environment interaction which is reflected in large variation in character.

Heritability, genotype–environment interactions, and viability measurements

Drosophila experiments

One of the important sources of data tending to support the simple view of population structure has been the exhaustive series of observations made by Mukai and his co-workers on the nature and partial dominance of viability

Table 7.5. Heritability estimates for various characters of domestic fowl (chickens and turkeys), cattle and swine. Characters are arranged in presumed order of descending importance to fitness, and heritability estimates using a variety of methods have been pooled. Data from Spector (1956)

Character	No. of studies	$\bar{x}$ ± S.D.
Domestic fowl		
No. of eggs produced	7	0.24 ± 0.093
Time to sexual maturity	3	0.53 ± 0.058
Total mortality	4	0.10 ± 0.021
Fertility	5	0.29 ± 0.15
Hatchability	6	0.19 ± 0.11
Egg weight	17	0.5 ± 0.18
Albumin weight	4	0.57 ± 0.47
Yolk weight	3	0.07 ± 0.11
Feathering	2	0.37 ± 0.064
Body weight	97	0.32 ± 0.22
Cattle		
Services per conception	1	0.05
Gain, birth to weaning	3	0.15 ± 0.072
Mastitis resistance	1	0.29
Gestation length	1	0.32
Milk production	1	0.32
Birth weight	4	0.46 ± 0.106
Mature weight	1	0.6
Swine		
Live pigs farrowed	6	0.18 ± 0.054
Litter size at birth	20	0.17 ± 0.115
21 days	3	0.20 ± 0.135
56 days	7	0.15 ± 0.134
% live pigs 56 days	1	0.4
Birth weight	7	0.017 ± 0.118
Body-weight gain	23	0.19 ± 0.137
Feed economy	12	0.25 ± 0.19
Body weight	45	0.18 ± 0.152

genes in Drosophila. The partial dominance was found to be as high as 0.4 (Mukai and Yamaguchi 1974), though it varied from experiment to experiment.

These and other experiments measuring viability in Drosophila were performed by using marker flies with crossover suppressors involving one or more chromosomes to 'extract' chromosomes from natural populations and pass then on to subsequent generations without recombination. The extracted chromosomes could be made homozygous or heterozygous with other chromosomes of known properties, and the egg-to-adult survival of flies carrying those chromosome combinations could be compared with that of marker flies segregating in the same cross.

The viabilities can be measured immediately upon isolation from a natural population, or (in an ingenious variation invented by Mukai) after they have been allowed to accumulate spontaneous variation for a number of generations while being prevented from recombining.

The homozygote and heterozygote viabilities in such an experiment can be looked upon as being equivalent to an experiment in which inbred lines are set up and crossed among each other to give F_1 hybrids. In such an experiment the regression of offspring on midparent yields the heritability as the coefficient of the regression line. The regression of heterozygote value on the average homozygote value gives the partial dominance of the viability genes being measured, and also should yield the heritability. In this case, the higher the partial dominance the higher the heritability should be. If partial dominance is high, then selection to increase or decrease viability should be highly effective since the expression of a gene in the homozygote is accompanied by its partial expression in the heterozygote. Dominance, and especially overdominance, should lead to a reduction in the size of the coefficient of regression.

Such a high heritability would appear to be in contradiction to the generalization accepted (with some caveats) earlier that characters closely involved with fitness tend to have low heritabilities as determined by response to selection. Here is a case of what appears to be high heritability in a major component of fitness in Drosophila, egg-to-adult viability.

Attempts to improve egg-to-adult viability in Drosophila by selection have not been made. It is indeed relatively straightforward to select *down* for a fitness character in an outbreeding organism like Drosophila, but at the end of the process one usually finds that one is investigating the effects of a small number of alleles with major deleterious effects on fertility or viability. Selecting up would appear *a priori* to be much more difficult.

Such an experiment could be done using marker chromosomes in Drosophila. The distribution of egg-to-adult viabilities of every sample of chromosomes from a natural population, when made homozygous, is approximately bell-shaped with a tail towards the lower end. In every survey, an appreciable number of chromosomes is found with zero or near zero viability, since they carry one or more lethals. When this range of viabilities is compared with that of heterozygotes for the same chromosomes, the homozygote range is found to exceed it at both ends (Fig. 7.11, data taken from Wills 1966). There is, in the terminology of Dobzhansky and Spassky (1963), an 'adaptive elite' of homozygous chromosomes conferring higher than average viability on their carriers. What would be the consequences of constructing flies using those chromosomes that confer high fitness? If, as the simple model suggests and the data appear to bear out, there is high partial dominance and therefore a high additive component to this character, then it should easily be selected for. In fact, however, experience in selecting components of fitness suggests that this will not be the case. Although the experiment has not been done specifically with Drosophila egg-to-adult viability, it seems unlikely that the apparent high additivity of the character would be translated into a high heritability.

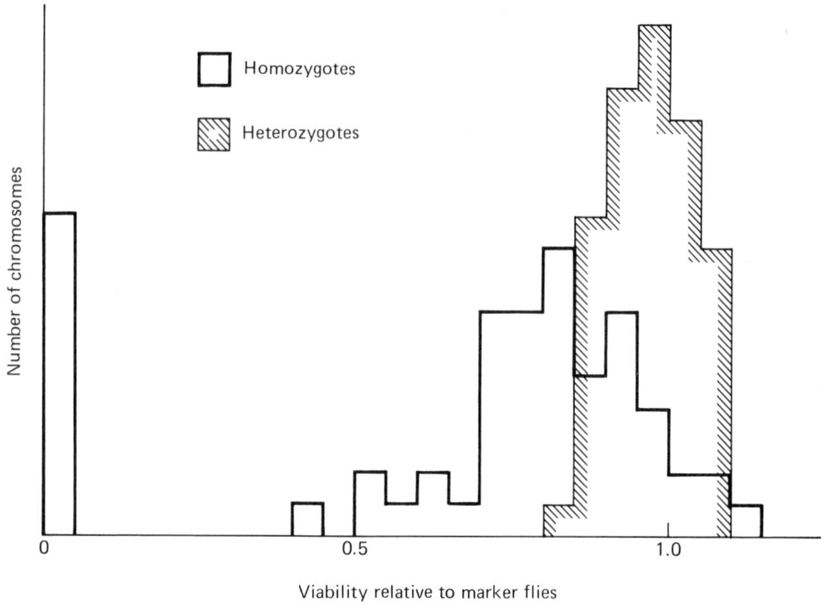

Fig. 7.11. Distribution of viabilities conferred by homozygous and heterozygous *Drosophila* chromosomes. Data from Wills (1966).

Why might such a discrepancy exist? Data collected from Drosophila viability experiments differ from the usual heritability estimates by being much more extensive. The viabilities conferred by a particular chromosomal homozygote or heterozygote are normally known with a high degree of precision. For example, the estimates of partial dominance made by Mukai are accompanied by standard errors, which gives the casual reader the impression that they are known much more precisely than they actually are. In fact, the partial dominances calculated for individual pairs of chromosomes vary greatly. At least in the case of chromosomes extracted from wild populations, the variance of partial dominances may be very large. It will be seen that if the variance is larger than the binomial variance to be expected entirely from sampling effects, this would have the effect of decreasing the effectiveness of selection.

An obvious source of increased variance is that due to genotype–environment interaction, which as we have seen decreases the size of the additive variance. Such an interaction may be quite significant even in data gathered in what the experimenter imagines to be a uniform environment.

That genotype–environment interaction may play a large role in data collections of this type is illustrated by some data of my own (Wills 1966), which I reanalyzed with this specific problem in mind (Wills 1975*a*). The experiment was of the usual type, with data gathered from third chromosomes of *Dro-*

sophila pseudoobscura carrying the Arrowhead gene arrangement and extracted by means of marker stocks from two wild populations. The viabilities of homozygotes were compared with those of heterozygotes for the same chromosomes. Viabilities were measured as the proportion of wild type flies emerging in crosses in which the Mendelian expectation was a 2:1 ratio of marker to wild flies. On the prescient advice of R. C. Lewontin, the viabilities of each chromosomal homozygote or heterozygote were determined to a high degree of precision by replicating each culture eight times.

Fig. 7.12 shows the viabilities of heterozygous flies compared with the average viabilities of the homozygotes for the same chromosomes. A clear partial dominance for genes with slight effects on viability is seen. There is, however, no difference in the mean viabilities of lethal-carrying and lethal-free heterozygotes, showing as have other studies that the partial dominance of lethal genes is very low. The scatter of viabilities is low because of the large number of replicates.

There is thus no indication of overdominance in the data, any more than in the data of Mukai. Because of the large number of replicates, however, it is possible to ask detailed questions about the role and extent of variation due to microenvironmental differences from one bottle to another within a set of replicates. While Wallace and Madden (1953) and Dobzhansky and Levene (1955) had considered this source of variation and attempted to factor it out of their experimental results, the small number of replicates in their experiments did not permit the following question to be asked. In an individual set

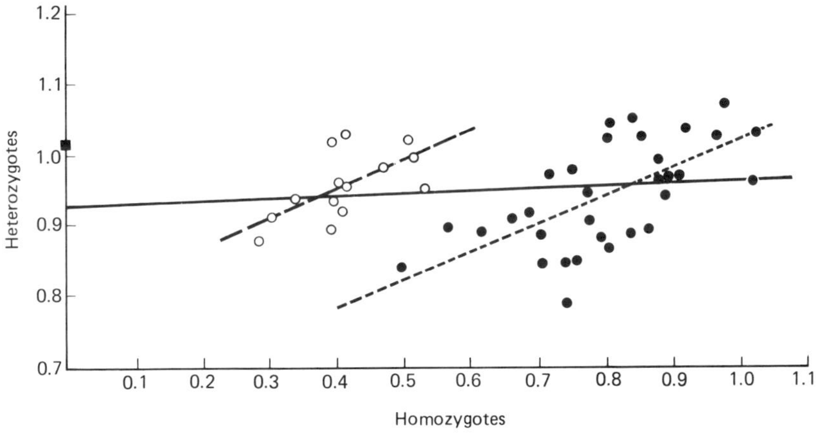

Fig. 7.12. Viability data for Arrowhead chromosomes extracted from a *D. pseudoobscura* population in Mesa Verde, Colorado. The heterozygote viability is plotted against the mean of the homozygote viabilities. The data are grouped into combinations of chromosomes without lethals (●), combinations in which one chromosome carries a lethal (○), and a single combination in which each chromosome carries a different lethal (■). From Wills (1966).

of replicates, how much of the variance can be attributed to genotype–environment interaction?

I first compared the variance between each set of replicate cultures with the binomial variance to be expected from a set of cultures containing the same number of flies but in which the viabilities of all flies are assumed to be the same. This was done for sexes within cultures and for total cultures. These variances were compared with an *F*-test.

Table 7.6 shows the distribution of significant *F*-values among classes of homozygous and heterozygous flies. It will be seen immediately that far more are significantly greater than would be expected by chance (23% of the comparisons, when about 2.5% would be expected to be significantly greater at the .05 level). Most of these significant comparisons are concentrated in the homozygotes, particularly those with low viability, but over a third are found in the heterozygotes, spread evenly across the entire range of heterozygote viabilities. Only about one fifth of the highly significant *F* values ($P < 0.01$) are found among the heterozygotes, however.

Is this truly genotype–environment interaction? It might be argued that certain genotypes are less well buffered against environmental alteration than others. Slight differences in the environment from one bottle to another will result in differences in apparent viability, but these differences are random and not correlated with the actual microenvironment. Thus, the greater variance of viability among some sets of replicates is simply a function of the tendency of some genotypes to give more variable results than others when tested in a given

Table 7.6. Comparison of observed between-replicate variances with those expected on the basis of sampling alone. Homozygotes and heterozygotes were arranged in order of increasing viability, then broken up into groups of approximately 20. From Wills (1975*a*, p. 185).

			No. of signif. *F* values		No. of *F* values	
	n	Mean viability	0.05 level	0.01 level	< 1	> 1
Homozygotes						
group 1	20	0.592 ± 0.087	1	8	0	20
2	20	0.755 ± 0.026	5	3	5	15
3	20	0.825 ± 0.023	2	3	3	17
4	20	0.895 ± 0.019	4	3	3	17
5	22	1.02 ± 0.074	2	2	6	16
Heterozygotes						
group 1	20	0.839 ± 0.037	3	1	10	10
2	20	0.905 ± 0.014	1	0	10	10
3	20	0.94 ± 0.0099	2	2	9	11
4	20	0.972 ± 0.012	3	1	7	13
5	20	1.011 ± 0.011	2	2	10	10
6	24	1.077 ± 0.040	3	0	11	13

set of microenvironments. Suppose the microenvironments could be precisely replicated. Then, if this were true, quite a different set of fitnesses would be obtained for a given chromosome even though the variance of fitnesses would be the same. Chromosomal homozygotes would be more likely to show such random variation in viability than heterozygotes, since they are presumably less developmentally buffered.

Fortunately, this alternative can be ruled out. Males and females within bottles are subject to the same set of microenvironments. If it is simply developmental 'noise' which is giving rise to the apparent variation in viability, the viabilities of males and females within the same set of bottles should be uncorrelated.

In fact, the viabilities of males and females are correlated. The correlations are very strong when there is significant between-replicate variance, but are still present when there is not. Table 7.7 shows the correlations of male and female viabilities, calculated within sets of replicates, for various subcategories of the data. It will be seen that all the correlations are positive, and all but one are highly significant. The most significant correlation (+0.807 for 54 degrees of freedom, significant at well below the 0.001 level) is that found between males and females in the same replicate when the tested chromosomes are homozygous and both sexes show significant between-replicate variance. To reiterate, both sexes are reacting in the same way to a given set of microenvironments—to such an extent, indeed, that developmental 'noise' seems to be playing a very small role.

Even when the between replicate variances are not significant, the effect of genotype–microenvironment interaction can be seen. When there is no detectable between-replicate variance above binomial expectation in either sex, the correlations are still positive—highly significant in the case of the homozygotes, and almost significant at the 0.05 level in the case of the heterozygotes (+0.106 with 292 degrees of freedom, with 0.113 significant).

Table 7.7. *Drosophila pseudoobscura* viability measurements: 'Within sets of replicates' correlations between viabilities of males and females in the same bottle. The stronger the correlation, the greater the presumed role of microenvironmental variation. From Wills (1975*a*).

Homozygotes			
	Significant between-replicate variance in:		
	(a) Neither sex	173	+0.296*
	(b) One sex	124	+0.286*
	(c) Both sexes	54	+0.807*
Heterozygotes			
	Significant between-replicate variance in:		
	(a) Neither sex	292	+0.106
	(b) one or (one case only) both sexes	138	+0.252*

*Significant at the 0.01 level.

In this experiment, each fly bottle contained the same amount of food, the same number of parents was allowed to lay eggs for the same period of time, and all cultures were raised at the same temperature. Even so, slight microenvironmental differences were enough to produce pronounced and highly significant genotype–environment interactions in the great majority of the genotypes tested, both homozygous and heterozygous.

Dobzhansky and his colleagues spent many years exploring the extent of genotype–environment interaction in *Drosophila*. Some rather vivid examples, summarized in Dobzhansky (1951, 1970), emerged. Some cases are directly relevant to the present discussion. Lewontin (1974*b*) replotted some data of Dobzhansky and Spassky (1944) on viabilities of particular fourth chromosomes of *D. pseudoobscura*, extracted from a natural population. The viabilities were measured at three different temperatures (Fig. 7.13). It will be seen that the chromosomes conferring the highest fitness in this sample show the

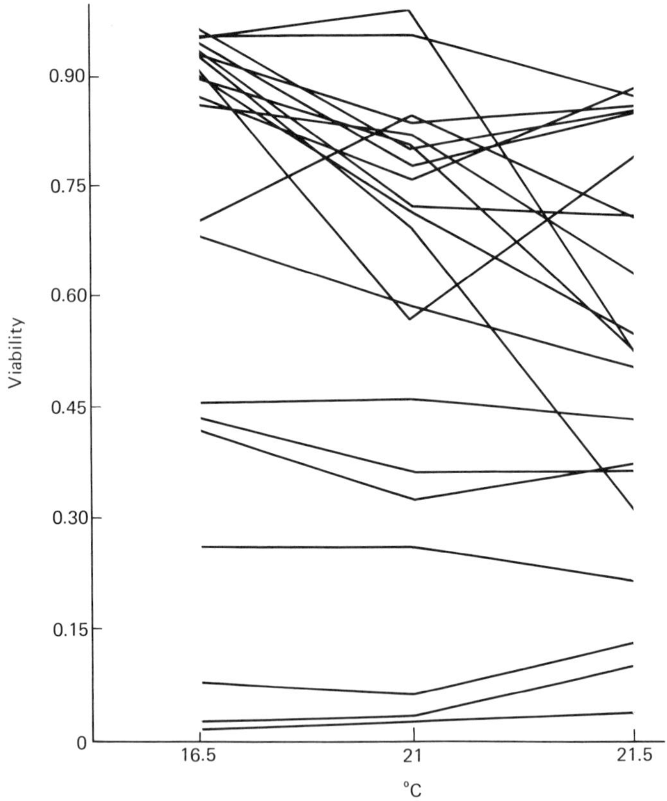

Fig. 7.13. Egg-to-adult viabilities of a set of fourth-chromosome homozygotes of *D. pseudoobscura* at three different temperatures. From Lewontin (1974*b*), using data from Dobzhansky and Spassky (1944).

greatest amount of genotype–environment interaction, quite the opposite of the third chromosomes in the study we considered in detail above. Examples with similarly indeterminate results can be multiplied many times from the literature.

It is possible to construct a model of such interactions, and I have done so in a very rough way in Fig. 7.14. In part a of the figure, I have assumed a set of eight homozygote and eight heterozygote microenvironments, each drawn at random from the universe of possible microenvironments. The microenvironments are different enough that they affect the viabilities of flies raised in them in quite different ways. In a large set, some of the homozygote and heterozygote microenvironments will be sufficiently similar to each other as to confer similar fitnesses, so that for any sufficiently large quantity of data there will be a correlation of homozygote with heterozygote fitnesses.

The range of fitnesses over a range of microenvironments for the homozy-

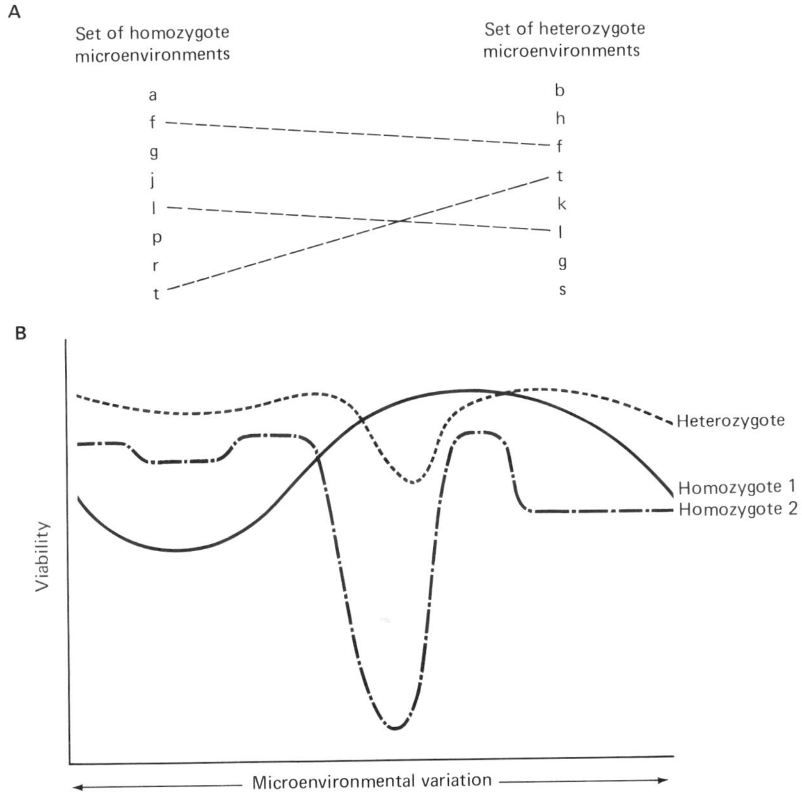

Fig. 7.14. A. Individual homozygote and heterozygote viabilities become progressively better correlated as the number of replicates increases. **B.** Hypothetical sets of fitnesses for two chromosomal homozygotes and their heterozygote over a range of microenvironments.

gotes and the heterozygote for two hypothetical chromosomes is shown in part b of the figure. Homozygote 1 shows little variation over the range of microenvironments, although there is enough to generate the sort of correlations between males and females that we saw in the set of homozygotes with little between-replicate variability. Homozygote 2, on the other hand, shows large variations, including a fitness 'valley' over part of the range. If the experimenter picked that part of the environmental range, this homozygote would show a low viability with a very high variability, since slight shifts in the environment would cause the viability to climb the sides of the fitness valley. Such a distribution of viabilities would explain the striking genotype–environment interaction exhibited by some (though not all) of the homozygotes with low viability.

The heterozygote shows a rough correlation with the viabilities of the two homozygotes, enough that partial dominance would be apparent in at least some of the environments. Note that over part of the environmental range homozygote 1 is actually fitter than the heterozygote. It would thus, over this part of the range, form part of the Dobzhansky and Spassky (1963) 'genetic elite', though it would be demoted from this exalted position were the environment to shift.

A chromosome carrying genes conferring low fitness over a range of environments would tend to be eliminated quickly, and few such 'unconditional' subvital genes would be retained in a natural population. But other classes of subvital genes, the harmful effects of which were highly conditional depending on the environment, would be retained longer—particularly if they were subject to balancing selection of some type.

I have suggested elsewhere (Wills 1975*a*) that the extent of genotype–environment interaction exhibited by subvital genes may prove to be a valuable diagnostic for the state of the population. If the interaction is low, the population may contain recently-induced variation which has yet to be eliminated by natural selection. If the interaction is high, as it seems to be in the case of natural populations of *D. pseudoobscura,* then the unconditionally deleterious genes may have been lowered in frequency. Many of the 'subvital' genes detected in natural populations may in fact be only conditionally subvital, maintained because of heterosis or some other form of balancing selection in environments not accessible to or not explored by the experimenter. Alternatively, the simple model would suggest that different loci are being expressed in the different environments.

Experiments employing balancer chromosomes should always be designed to examine the degree of genotype–environment interaction present. Data such as those of Mukai, collected both from natural population and from experimental populations in which newly arisen viability polygenes have been allowed to accumulate over balancer chromosomes, will be most valuable to reexamine in this regard.

Newly arisen viability genes should exhibit far less genotype–environment interaction than viability variants taken directly from natural populations.

Simmons, Preston, and Engels (1981) found a positive correlation between viability and competitive ability conferred by chromosomes that had been allowed to accumulate new viability mutants while heterozygous for balancer chromosomes. No correlation was found between these two characters for chromosomes taken directly from large cage populations, even though the genetic variance of viability conferred by these chromosomes was greater than that conferred by the newly-arisen variants. Newly-arisen and equilibrium viability variants are shown by this experiment to be *qualitatively different*.

Yeast experiments

Blatherwick and Wills (1971) examined genetic variation which had been produced by mutagenesis and by spontaneous events in yeast *(Saccharomyces cerevisiae)*. The purpose of the experiment was twofold: to determine how rapidly variability accumulates in this organism, and to measure the genotype–environment interaction associated with these newly-induced mutations and with mutations which had been subjected to selection.

Mutagenized and nonmutagenized cultures were grown to plateau in a variety of media (none of which made any difference to the results) in small culture flasks. After 10 days at plateau about 99% of the cells had died, and the survivors were sampled. Clones were established from random colonies, and the growth rates of cells from the clones were measured on a variety of media designed to be as stressful as possible compatible with regular and measurable growth.

The advantage of using yeast in an experiment of this type is that growth can be measured precisely under carefully specified conditions, rather as if each replicate in a Drosophila experiment could be assured to represent a specific microenvironment. One obvious disadvantage is that yeast is not an analogue of Drosophila or indeed of any organism that undergoes regular sexual reproduction. It was hoped, however, that some generalizations could be made from one system to the other.

In fact, the properties of newly-induced and selected variation were strikingly different. Most, but not all, of the newly induced variation resulted in a slowing of the growth rate, and strong positive correlations were observed between the growth rates of the same set of mutant clones on different media (Table 7.8a). The variation which had been subject to selection—variation examined at the end of one or more plateau periods—showed a strong interaction with the environment, to such an extent that the growth rates measured for a set of clones on one medium were uncorrelated with the rates measured for the same set on a different medium (Table 7.8b). One mutagen, ethyl methane sulfonate, could be observed to induce a mixture of the two types of genetic variability, since the variability with a high genotype–environment interaction appeared more rapidly in cultures treated with this mutagen than it did in untreated control cultures.

While it is, as was pointed out above, dangerous to generalize from one to

Table 7.8A. Correlations between growth rates on four stress media of yeast clones treated at zero time with nitrosoguanidine or ethyl methanesulfonate ($\bar{r}$ is the average correlation between growth rates on all different media). From Blatherwick and Wills (1971, p. 551).

(a) Nitrosoguanidine

	$YEPDE_2$	MVDEX	MVEtOH	MVN
$YEPDE_2$	0.953†			
MVDEX	0.330	0.971†		
MVEtOH	0.752†	0.352	0.936†	
MVH	0.556*	0.843†	0.460	0.949†

$\bar{r}$ = 0.549†, d.f. = 90

(b) Ethyl methanesulfonate

	$YEPDE_2$	MVDEX	MVEtOH	MVN
$YEPDE_2$	0.757†			
MVDEX	0.757†	0.971†		
MVEtOH	0.884†	0.794†	0.938†	
MVN	0.726†	0.985†	0.782†	0.946†

$\bar{r}$ = 0.821†, d.f. = 90

* = significance at 0.05 level.
† = significance at 0.01 level.

the other of these two very different organisms, it is worth noting that the yeast experiment and the Drosophila experiment of Simmons et al. (1981) bear out our expectation of what selection will tend to do in a population. Unconditionally deleterious genes with low genotype–environment interaction, genes which will tend to fall into the category of mutational load, were removed rapidly from a haploid population of yeast, while genes with a higher interaction were retained. Clones showing high genotype–environment interaction could show extremely different growth rates on different media, ranging from rapid to none at all. Further, while there were many instances of clones which grew more rapidly than the wild type control, none grew more rapidly on all the media used. Were such a clone to appear, it would carry an unconditionally advantageous gene, something which, as we have seen, does not appear to happen in a diploid population either.

The effect of the environment on partial dominance

Another observation which emerged from the *D. pseudoobscura* viability experiment (Wills 1966) was that the partial dominance of viability genes can be altered by the environment. This in itself is not remarkable—we have

Table 7.8B. Correlations between growth rates on four stress media of clones after 10 days of growth. (The significance levels are the same as those in Table 7.8A.)

(a) Untreated clones at 10 days:

	$YEPDE_2$	MVDEX	MVEtOH	MVN
$YEPDE_2$	0.729†			
MVDEX	−0.319	0.845†		
MVEtOH	0.014	0.071	0.874†	
MVN	0.252	−0.072	−0.138	0.223

$\bar{r} = 0.032$, d.f. = 90

(b) NG-treated clones at 10 days:

	$YEPDE_2$	MVDEX	MVEtOH	MVN
$YEPDE_2$	0.881†			
MVDEX	−0.363	.697†		
MVEtOH	−0.047	−0.222	0.820†	
MVN	0.316	−0.394	0.094	0.943†

$\bar{r} = 0.065$, d.f. = 90

(c) EMS-treated clones at 10 days:

	$YEPDE_2$	MVDEX	MVEtOH	MVN
$YEPDE_2$	0.881†			
MVDEX	−0.363	0.697†		
MVEtOH	−0.047	−0.222	0.820†	
MVN	0.316	−0.394	0.094	0.943†

$\bar{r} = -0.103$, d.f. = 90

already seen many examples of environmental effects on the expression of the genotype—but in light of the theoretical observations made earlier in the chapter this fact may take on added significance.

The first chromosomes to be examined were derived from flies trapped in Strawberry Canyon, California. Each culture was established with ten males and ten females of the appropriate genotype, and these flies (which had been kept together on fresh medium for several days beforehand) were allowed to lay eggs for 3 days before being transferred to a new bottle.

Productivity in these cultures was not good, and there appeared to be excessive larval mortality. Consequently, after seventeen Strawberry Canyon chromosomes had been examined, the protocol was changed and the flies were permitted to lay eggs for only 2 days. This protocol was followed for the remainder of the Strawberry Canyon sample and for the chromosomes derived from a population trapped in Mesa Verde, Colorado.

Productivity improved as a result; the mean number of marker flies eclosing per bottle rose slightly to 348 from 338. But the partial dominance, measured as the slope of the regression line of heterozygote on mean homozygote viability, also changed markedly, dropping from 0.3 to 0.18. Thus, at least for the genes being measured in this experiment, not only were strong genotype–environment interactions revealed that affected expression in the homozygote, but the degree to which the genes were being expressed in the heterozygote could also be altered, in this case by a relatively small change involving the stringency of the environment.

Now it will be recalled, from the arguments at the beginning of this chapter, that if alleles are being maintained in a state of balanced polymorphism by strong arithmetic mean overdominance, their equilibrium frequency may shift as a result of changes in niche size or differences in the sizes of the selection coefficients, but the internal equilibrium point will not be lost over a wide range of circumstances if the dominance of the selected allele is low. The system is least sensitive to such changes if the alleles being selected against are completely recessive. However, the system becomes highly unstable if the heterozygote shows intermediate fitness. Fluctuations in the severity of the environment could change a polymorphism from stable to unstable and back again. Thus, if many of the genes affecting viability in natural populations are polymorphic and maintained by arithmetic mean (marginal) overdominance, it might be very difficult, given the interplay and complexity of the forces involved, to predict the distribution of alleles at these loci.

Do other polymorphisms exhibit the same properties as viability genes?

The chromosomes dealt with in Drosophila viability experiments have a large effect, with a major impact on fitness even in a relatively unstressful laboratory environment. Are the viability alleles carried by these chromosomes and seen by the experimenter actually at one end of a continuum of polymorphisms which shades down through variants exhibiting smaller and smaller selection coefficients, and which would therefore be undetectable in experiments of this type? Do they represent a small or a large class of variants? Are they predominantly structural or regulatory genes? It is difficult to imagine ways to obtain answers to these questions.

The question of number, however, is important from the theoretical standpoint, for if there are many loci with small effect and only a few with large effect involved in fitness, then loci detected by such experiments may make up a relatively small part of the genetic variation in the population.

No experiments have yet been done to determine the actual number of segregating loci responsible for observed fitness differences in Drosophila chromosomes taken from nature. Such experiments would be very difficult to perform, particularly in view of the apparently strong genotype–environment interaction which many of these loci exhibit. Mukai (1969) has estimated that a fairly small number of polygenes which have been allowed to accumulate over balancer chromosomes (between about 7 and 40) are sufficient to produce as great a range of non-lethal viabilities as is observed in chromosomes taken from natural populations. The calculations are based on the assumption of equal effects of all the 'viability polygenes' and do not take into account gene–environment interaction (if this should occur to a significant degree in these particular experiments), but it would appear that not many genes are needed to produce at least the overt phenomena we are examining here. One experiment of particular interest would be to determine whether the chromosomes conferring low viability do so because of the effects of a single segregating gene. I suspect that this may often turn out to be the case, particularly when the chromosome lies in the bottom of a fitness valley. Otherwise one would have to postulate a number of genes which in combination on the chromosome result in a fitness valley with steep sides.

Should the chromosome-fitness measurements actually be dealing with a rather small number of genes of large effect, then any generalization from such experiments to genetic variability as a whole would be suspect. What would be measured in such experiments would be, not a 'chromosomal' fitness made up of the effects of hundreds or thousands of genes, but the effect of a small number of genes with perhaps unusual properties. Further, as emphasized earlier, genotype–environment interactions could be generated if different genes with large effect were expressed in different environments. To demonstrate that balancing selection is operating, we must somehow show that genotypes at a given locus are being selected differently in different environments.

Some possible examples of marginal overdominance at single loci in Drosophila

The human single-locus polymorphisms examined in Chapter 5 exhibit strong genotype–environment interactions. Do such interactions extend to other classes of polymorphisms, particularly to those electrophoretic polymorphisms about which little is known, and which appear from their allele frequency distribution to be neutral or subjected to purifying selection?

Some preliminary evidence is available on this point. I will mention at this juncture some experiments which my co-workers and I have performed to try to answer the question of what selective forces are acting on electrophoretic polymorphisms (Wills and Nichols 1971, 1972; Wills 1975*b*; Wills et al. 1975). Two loci of *D. pseudoobscura, esterase-5 (est-5)* and octanol dehydrogenase *(odh)* were investigated.

The experiments were started on the assumption that in a normally polymorphic outbreeding population, the selection coefficients (if any) associated with genotypes at most electrophoretically polymorphic loci would be quite undetectable. We also attempted to circumvent the problem that one cannot be certain whether it is the locus in question that is being selected or some other locus or loci nearby on the chromosome.

We reasoned that two conditions had to be met in order for selection to be detected. The first was that a specific selective pressure must be applied to the organisms. If a particular enzyme were being followed, an excess of its substrate or an analogue of its substrate in the medium might provide a suitable selective pressure. A control would be provided in the form of other substances added to the medium that would result in the same average reduction in survival of the flies. The control substances would not, however, be connected in any obvious biochemical way with the function of the enzyme being examined.

The second condition was that the flies must be inbred to reduce the effect of background heterozygosity. If rank-order selection is operating on these polymorphisms, then in an outbred background the contribution to fitness potential of a particular polymorphism would be small. If the flies were inbred but kept segregating for the polymorphism in question, the contribution of that locus to fitness potential—and consequently the selection coefficients associated with it—would grow larger. The use of selective pressures specifically tailored to the locus in question would, it was hoped, circumvent the difficulty that nearby loci in linkage disequilibrium might also contribute to differences in fitness.

Strains of flies recently isolated from nature were inbred by brother–sister mating for 38 generations. Separate lines were kept segregating for two common alleles at the sex-linked *est-5* and autosomal *odh* loci. Periodically during the inbreeding process, crosses were set up in which eggs expected to segregate in a 1:1 or a 1:2:1 ratio were collected and placed on various stress media. At the same time, eggs from the same type of cross but involving outbred flies were collected and placed on the stress media.

A variety of media were used. The ones expected to interact with the loci being followed were media containing 3 mM octanol and 6 percent ethanol, both substrates for the enzyme specified by the *odh* gene, and media containing tributyrin, a short-chain fat which was hydrolyzed slowly by the *est-5* enzyme. As expected, disturbances of Mendelian expectation in inbred flies segregating for *odh* were at their greatest when the flies were grown on medium containing octanol or ethanol. Disturbances of expectation in inbred flies segregating for *est-5* were found when the flies were grown on medium containing tributyrin. It is interesting to note, and central to the argument presented above concerning the probable small size of selection coefficients in outbred flies, that at no point in the experiment did the outbred controls segregating for the same alleles show any deviation from Mendelian expectation. Selective differentials were indeed enhanced by inbreeding.

Unfortunately, while the general pattern exhibited by the results was in accord with expectation, the details were complicated and extremely difficult to interpret. By generation 12 of inbreeding, the males of all three lines segregating for *odh* showed a significant excess of heterozygotes on medium containing octanol, as did the females of one of the lines. (Tributyrin medium was not employed this early in the experiment.) This pattern was taken to indicate that single-gene heterosis was being revealed by the inbreeding. Because the females are the homogametic sex the approach to homozygosity elsewhere in the genome must have been slower in the female lines so that the effect appeared in males first. This expectation appeared to be borne out by the 17th generation, by which time the females of two of the lines showed a significant excess of heterozygotes on octanol medium.

No more data were collected until generation 38, by which time the situation had changed dramatically. Now there was no sign of heterosis. Instead, in those lines where an effect was seen, the heterozygote was roughly intermediate in survival between the homozygotes. The heterosis seen at early stages of inbreeding seems likely to have been due, as Yamazaki (1972) suggested, to associative overdominance, since it disappeared on further inbreeding. The data for outbred flies and the 38th generation of inbreeding are summarized in Table 7.9.

Many complicating features of the data were apparent upon detailed analysis (Wills et al. 1975). The effects of the stress media were expressed more markedly in one sex than the other in most cases, though there were no cases in which the effect was expressed one way in one sex and the opposite way in the other. Further, there was a strong and consistent effect of octanol medium on male flies segregating for the *est-5* gene. No such effect was seen in these lines when they were grown on ethanol medium under conditions leading to approximately the same survival. No effect on any genotype was seen when these same inbred lines were grown on triacetin medium.

Could these effects on the inbred lines be attributed to the loci in question? The pattern is consistent with this interpretation for the flies segregating for the *odh* alleles. A pronounced effect on the inbred flies was seen on the two media containing 3 mM octanol and 6% ethanol, both of which are substrates for the enzyme. No effect on the pooled data was seen on tributyrin or triacetin medium. Survival of these flies on all media was low, with the lowest survival on tributyrin medium—on which no effect was seen.

Flies segregating for the *est-5* alleles showed a more complicated pattern. Again, no effect was seen on ethanol or triacetin. Complete hydrolysis of the latter compound yields acetic acid and glycerol, two common metabolites, so it may be that the stress acting on these flies to depress their survival occurs at some point other than the hydrolysis step. Strong effects were seen on both tributyrin and octanol medium, the former expected and the latter not. The octanol effect, puzzlingly, was confined to males. Wills et al. (1975) suggested that one of the products in the repeated series of beta-oxidations that shorten octyryl-CoA in two-carbon increments will be butyryl CoA, which might interact with the esterase-5 enzyme. Thus, the two enzymes might be more closely related biochemically than had been imagined at the start of the experiment.

The sex-limited effects indicate that genetic background is still playing a role in these crosses despite 38 generations of brother–sister mating. While the case for single-gene effects at the *est-5* locus is somewhat weak, the case for the *odh* locus appears stronger. The fascinating result from the standpoint of the pres-

Table 7.9. Fitness differences observed between octanol dehydrogenase (*odh*) and esterase-5 (*est-5*) genotypes of *D. pseudoobscura* upon inbreeding and specific stress. From Wills et al. (1975).

(a) Summary of data from 38th generation of inbreeding. Data from all inbred lines are pooled

		odh				*est-5* (sex linked):						
		Genotype			$\chi^2_{2\ \mathrm{d.f.}}$	Genotype (females)†			$\chi^2_{2\ \mathrm{d.f.}}$	Genotype (males)		$\chi^2_{1\ \mathrm{d.f.}}$
Medium		S/S	S/F	F/F		S/S	S/F	F/F		S	F	
Octanol	Obs.	179	314	124	10.0**	92	225	125	3.47	270	208	8.04**
	Exp.	154.25	308.5	154.25		107.5	221	113.5		239	239	
Ethanol	Obs.	100	261	146	8.79*	82	180	90	0.82	106	88	1.67
	Exp.	126.75	253.5	126.75		78.5	176	97.5		97	97	
Tributyrin	Obs.	124	279	111	4.42	148	218	132	9.8**	121	80	8.36**
	Exp.	128.5	257	128.5		121.5	249	127.5		100.5	100.5	
Triacetin	Obs.	186	322	166	2.52	108	260	102	5.33	196	217	1.07
	Exp.	168.5	337	168.5		120	235	115		206.5	206.5	

Fitness on octanol medium: S/S > S/F > F/F

Fitness on ethanol medium: S/S < S/F < F/F

†The expected female genotypes are not in a 1:2:1 ratio because the data had to be collected from two different types of cross, with parents S × S/F and F × S/F.

*significant at 0.05 level

**significant at 0.01 level

ent discussion is the different ways in which the two media with a pronounced effect, those containing ethanol and octanol, affected the survival of the three *odh* genotypes. As can be seen from Table 7.9, when the totals of all three lines for both sexes are pooled, there is a deficiency of S/S homozygotes and an excess of F/F homozygotes on ethanol medium and precisely the reverse on octanol medium. In each case, intriguingly, the heterozygotes are not quite intermediate, but show a slight partial dominance of the fitter genotype. The similarity to the arrays of fitnesses in the Levene-type models which we were investigating earlier in the chapter is striking and perhaps not entirely coincidental.

A different pattern was seen for the flies segregating for the *est-5* alleles. The fast hemizygote was at a disadvantage on both octanol and tributyrin medium. Of course, it seems likely that some medium could be found in which the slow hemizygote at this locus would have the disadvantage.

Despite some difficulties in interpretation, the pattern of variability revealed in this experiment conforms to our expectation. Selection coefficients in outbred flies are too small to be detected experimentally, even on specific stress media. With the one possible exception noted above, the greatest effects on inbred flies are seen on the media biochemically related to the enzyme in question. Finally, apparent overdominance in the early part of the experiment is replaced by partial dominance. In the case of the *odh* locus, the different arrays of fitnesses on the two selective media would be enough, if these two media constituted the environment of the flies, to maintain the polymorphism.

It is encouraging that both the *odh* and *est-5* alleles investigated conform in their properties to those of functional polymorphisms selected in a marginally overdominant fashion. Although inbreeding was not carried out, stressing of *D. melanogaster* segregating at the *adh* locus (Bijlsma-Meeles and van Delden 1974) and the amylase locus (de Jong and Scharloo 1976) showed consistent differences between genotypes. In the case of the *adh* experiments, the flies had to be grown in 20% ethanol before consistent differences in survival between genotypes were seen—perhaps a function of the fact that the lines were not inbred.

Do diverse environments allow more genetic variability to be retained?

If balancing selection is not primarily heterotic, it should be most effective under diverse environmental conditions. Powell (1971) attempted to approach this problem by exposing outbred and highly heterozygous populations of *D. willistoni* to uniform environments and to environments in which one of three factors (two types of food and temperature) were allowed to vary. He examined 22 electrophoretic polymorphisms in 13 populations which were exposed to these various environmental conditions for about 15 generations. Significantly more variability was retained in the diverse environments. The greater the environmental diversity, the greater the amount of variability retained.

This work was immediately criticized by King (1972), first on the grounds that *D. willistoni* is polymorphic for a large number of short inversions on every chromosome arm, and that Powell might have measured selection not on the alleles he was observing but rather on a few other strongly selected genes on the inversions carrying them. Powell (1972) attempted to counter this by arguing that no correlations had been found in this organism in samples from natural populations between particular inversions and enzyme alleles or between chromosomal heterozygosity and enzyme heterozygosity. It seems likely, however, that weak associations would be missed in such data, and weak associations might be all that are needed to give the effect.

King's second criticism, which is much more difficult to counter, was that the marker alleles followed in the experiment could themselves be neutral and be retained by associative overdominance with a few strongly selected loci nearby. The individual gene may not be the unit of selection in these experiments. This brings us back to the problem examined in the last section, which is that if balancing selection cannot be shown to be acting at the level of the single locus, apparent balancing selection could be nothing more than mutation–selection balance tempered by genotype–environment interaction. While on the surface Powell's experiment is consistent with conditional heterosis or marginal overdominance operating on these polymorphisms, it is open to alternative interpretations based on a simpler population structure.

Essentially the same experiment was repeated by McDonald and Ayala (1974), using *D. pseudoobscura*. These authors examined replicate populations, established from wild-caught flies, in one constant environment and three different fluctuating environments of increasing complexity. They found that in all cases, after the population cages had run for a year, there was more variability retained in the more complex environments than in the simpler one, but in their experiment the most complex environment retained less variability than the second most complex one.

For most of the polymorphic genes examined (17 out of 20) the first criticism levelled by King against Powell's experiment did not apply, since these genes are found on cytologically monomorphic chromosomes. The trend was seen, essentially unchanged, when only these 17 loci were examined. The second criticism, however, remains valid.

A similar experiment was reported by Minawa and Birley (1975). They employed two replicate populations in each of three stress environments, one constant and two variable. The founding population of *D. melanogaster* ('Texas') had been kept in mass culture in the laboratory for 200 generations. Despite this long period of culture, this population was still significantly heterozygous at eight of the loci assayed, giving an average heterozygosity of 8.2%.

Because of the experimental design (two replicates and two samples, at 27 and 54 weeks, in each of the three environments), it was possible to ask whether the heterozygosities at each locus showed significant variation among environments, and whether there was a significant secular trend from one sampling point to the next. The pattern that emerged was a complex one. Though only one locus showed a barely significant heterogeneity among environments after nine generations, after 18 generations four of the seven loci that were followed showed significant heterogeneity among environments and all seven, including the two that did not show significant heterogeneity, showed significant secular trends from the first sample to the second in at least some of the environments.

Clearly selection was going on. There was no significant difference between the heterozygosities of the populations in the two variable stress environments compared with the constant stress environment (7.71% heterozygosity on the average for the two variable environments and 7.00% for the constant environment). Because of the long period of laboratory culture before the start of the experiment, the population may already have lost much of the variability that would have been retained in diverse environments. Thus, there was little change in average heterozygosity under any of the conditions. Each population was started with 2000 flies and the authors suggest that this, in conjunction with the very long period of mass culture, should have been enough to eliminate linkage disequilibrium as a factor in these allele frequency shifts. The strongest (and I think irrefutable) evidence that selection was bringing about the changes in allele frequencies comes from examination of the replicate cultures. Only 3 out of 39 comparisons showed significant differences within environments. The presumption is that selection is in most cases affecting the replicate cages in the same way.

But are the selective forces acting directly on the loci that are being measured, or is there sufficient residual disequilibrium in the population to drag these alleles along in the wake of closely linked loci that are the ones that are *really* being selected? No satisfactory answer is forthcoming from these experiments because the stress environments—changes in food composition, fluctuating and suboptimal temperatures, etc.—have no obvious connection with the functions of the polymorphic loci being investigated.

It would appear, however, that in two out of the three experiments cited above, fluctuating environments did tend to retain more variability than a constant environment. In the third experiment, because the population was already adapted to laboratory conditions, there is a possible selectionist interpretation for why they did not. Whether the variability retained included the polymorphic genes that were used as markers, or whether it was primarily invisible variation closely linked to these marker genes, remains unresolved.

The question of whether variable environments allow the retention of more variation than do uniform ones appears to be answered in the affirmative. Such variation, as we have seen, need not be unconditionally heterotic. There may be a subclass of heterotic variability which is retained by selection in uniform environments as well, though great experimental difficulties lie in the way of deciding whether an environment is indeed uniform. Even replicate bottles in a *Drosophila* experiment vary significantly in their interaction with a particular genotype.

It seems clear that a thorough understanding of the nature of the forces that are retaining balanced genetic variability can only come from a study of the interaction of genotype with environment. Aside from unconditional lethals, few if any genotypes can be assigned unvarying fitnesses. It will be most instructive to investigate the nature of genetic variability, particularly the amount of overdominance, under extreme environmental conditions. I will venture the prediction that overdominance will be found to be much more common under such conditions than under the less stressful conditions prevailing in the laboratory or under cultivation and domestication.

As we have seen, however, the essential question facing us as we try to disentangle the nature of genetic variability is not whether balancing selection can act in the absence of unconditional overdominance. We have seen that it can, provided that genotype–environmental interaction is significant. Rather, the question is whether such selection can be shown to be acting at the single locus level. If it cannot, then the alternative explanation of mutation–selection balance is not ruled out, since genotype–environment interaction may be nothing more than the expression of alleles at different loci in different environments.

Two promising approaches to this problem have been explored in this chapter. In the first, specific environmental manipulations have been examined for their effect on single loci in inbred populations. In such populations the selective pressures have been shown to be enhanced by a reduction in the overall level of heterozygosity. The second approach involves subjecting populations to constant and fluctuating environments, in order to see how this affects the rate at which they lose variability. With both approaches, unfortunately, the experimental design is weakened by the possibility that alleles at loci closely linked to those actually being measured are in reality responsible for the effect. We will see in the next chapter how this possibility can be ruled out, provided that sufficient care is taken in the design of the experiment. Conditions under which the gene actually being measured is the unit of selection may not be so difficult to attain as has been imagined.

8. The conditions under which the gene is the unit of selection

The problem of associative overdominance in the interpretation of population genetics experiments is shown to be minimized in a large outbred population. Under these conditions associations between loci resulting from linkage can be shown to be small and primarily due to sampling. The sampling effect is demonstrated for the human MNSs and Rh loci. Experiments are suggested which take advantage of this phenomenon

In this chapter and the next, I will explore some of the ways in which drift and different types of selection can act on finite multilocus populations. I will try to integrate the work of many theoreticians who have examined these problems, and to suggest that there may be some important and previously unsuspected principles that can be extracted from what, to an outsider, must appear to be a terribly complex and hermetic set of theory.

The models presented here are crude and oversimplified, and will undoubtedly be refined greatly in the future. But they can be used as the basis for modifying experimental design, and this will of course be their chief value. Too often in this field, the experimental population-geneticist produces data which he imagines demonstrates the effect of selection on a particular allele, inversion or other measurable character in the population. Then the theoretician destroys his hypothesis by postulating that the effect may be due to other, closely-linked genes which are the ones really subject to selection. This adversary situation, examples of which we have already seen, leads to discouragement and a feeling of frustration. Might it not be possible to examine multilocus theory to see how this difficulty can be circumvented, and then to design experiments with the theoretical restrictions already firmly in mind? I will suggest ways in which this can be done.

Linkage disequilibrium and the interpretation of population-genetics experiments

So far in this book, we have considered the consequences of different types of selection acting on both finite and infinite populations, many of which are assumed to be segregating for large numbers of polymorphisms or which carry many loci capable of mutating to harmful alleles. It has become apparent, as was seen in the last chapter, that a full understanding of the role of these different types of selection and the classes of genes on which they act will continue to elude us, unless we can examine the effects of selection on individual genes; a very difficult task.

Our assumption in the calculations we have done up to now has been that all the loci are free to assort at random with respect to each other. This is obviously not the case, since genes are linked on chromosomes. If they *were* free to assort at random, then experiments such as those of Powell (1971) discussed in the last chapter would be more straightforward to interpret.

Leaving linkage aside, two interpretations of such experiments are possible.

(a) The reduced rate of loss of polymorphisms observed in a diverse environment is due to balancing selection of the marginal overdominance or conditional heterosis type.

(b) The polymorphisms observed are held in the population by mutation–selection balance, but this is complicated by dominance, epistatic interactions and genotype–environment interactions, all of which will tend to reduce the effectiveness of selection against the harmful alleles and reduce their rate of loss. The last will be especially apparent in complex environments, so the rate of loss will be slowest under such conditions.

The effect of this reduced rate of loss, due essentially to a weakening of the selective pressures acting on each locus, will be a higher average equilibrium frequency than would otherwise be achieved.

Linkage adds a new dimension of complexity to the situation, because the physical proximity of genes on the chromosome means that they can no longer be treated as individual units. Even though two polymorphic loci may each be in Hardy–Weinberg equilibrium, the frequencies of the gametes in the population may not be simply the cross-products of the individual allele frequencies, but may exhibit significant excesses of coupling or repulsion types. This excess has been known by a number of terms, but we will follow Lewontin and Kojima (1960) and refer to it as coupling or repulsion linkage disequilibrium.

Disequilibrium may be present even if the genes are on different chromosomes, particularly if the founding population is small. For example, if alternative alleles A and a are on one chromosome, and B and b on another, and if the population consists entirely of offspring of the cross $AB/AB \times ab/ab$, then the offspring will all be in complete linkage disequilibrium for the coupling gametes AB and ab. One further generation of random mating will restore linkage equilibrium in this case.

There are various measures for disequilibrium. The commonest is the coefficient of linkage disequilibrium, D, given by

$$D = g_{00}g_{11} - g_{01}g_{10} \tag{8.1}$$

where the g values are the frequencies of the gametes. The two alternative alleles at each locus are given here in binary notation, and by convention the first part of the equation is the product of the coupling gametes and the second part the product of the repulsion gametes. When each allele is at frequency 0.5, D can range between ± 0.25, but as Lewontin (1964) showed, this will not normally be the case and the maximum possible D will therefore usually be smaller.

Geiringer (1944) showed that D will be reduced by the factor $(1 - r)$ for each generation of random mating, where r is the frequency of recombination between the loci, provided that each locus is initially in Hardy–Weinberg equilibrium. For closely linked genes, any disequilibrium introduced by sampling or by selection may take many generations of random mating to fall to undetectable levels. If the reintroduction of disequilibrium by sampling or selection is great enough, very high levels of disequilibrium may exist at steady state between its introduction and its breakdown by recombination. Thus, there is a real possibility that an allele we are able to detect may be changing in frequency in the population, not through its own effects, but through the effects of other loci in linkage disequilibrium with it.

There are many problems with the treatment of disequilibrium. Universally agreed-upon methods for handling more than two alleles at a locus have not been developed, and in the subsequent discussion we will confine ourselves to two alleles at a locus. This may not be too great a simplification, since there often are two major alleles at a polymorphic locus, but it is a worrying limitation on current theory. Further, the best way of handling situations in which the two alleles are not equally frequent is not yet apparent. A number of authors have followed Hill and Robertson (1968) in using

$$r^2 = E\left[\frac{D^2}{p_1p_2q_1q_2}\right] \tag{8.2}$$

where r^2 is the square of the correlation of the frequencies at the two loci and the p and q values are the allele frequencies. This measure has convenient mathematical properties, particularly when the statistical consequences of the buildup of disequilibrium through drift are being investigated. I wish, however, to investigate populations in which all the loci are maintained by balancing selection at or near a particular equilibrium frequency (not necessarily 0.5). These are admittedly artificial cases, but perhaps not more so than the mathematically more tractable case investigated by a number of authors in which all loci start at 0.5 and then begin to drift and to build up disequilibrium by sampling (e.g. Sved 1968; Ohta and Kimura 1969, 1970; Hill and Robertson

1968; Hill 1974*a, b*). A convenient measure for cases in which all loci are at or near the same frequency is Lewontin's (1964) statistic D':

$$D' = D/D_{max} \tag{8.3}$$

where D_{max} is the maximum possible disequilibrium that can be attained between the two loci. As will be explored below, D_{max} may alter dramatically depending on whether the disequilibrium between two loci is coupling or repulsion.

This leads us to another problem. In the case of wild type alleles, how are we to decide which are coupling and which repulsion gametes? Here I follow the convention of Langley and Crow (1974), who define coupling disequilibrium as an excess of gametes carrying the two commoner or the two rarer alleles, and repulsion disequilibrium as an excess of gametes carrying one common and one rare allele.

If the effects of linkage are added to the interpretation of Powell's experiment, the possibilitity that balancing and directional selection are acting on the population must be modified by the possibility that most of the polymorphisms in the experiment are themselves neutral or nearly so. They may happen to be linked to a relatively small number of polymorphic loci subject to balancing or directional selection which are scattered throughout the genome.

The influence of strongly selected genes on neighboring genes can take many forms. As we have seen, it is impossible to interpret the results of most Drosophila viability experiments with any certainty when the fitness effects of entire chromosomes are being examined. Any number of genes affecting viability, from one to thousands, may be locked within an inversion. If there are few, all the rest of the genes in the inversion may simply be dragged along under the influence of a small number of genes with major effect. The advantage exhibited by inversion heterozygotes in many experimental situations may be due not to single-gene heterosis but to the enforced heterozygosity of a few genes of harmful effect. In a population free of inversions, such as was the case with most of the genome in the experiments of McDonald and Ayala (1974), a simple interpretation might be that a small number of loci are being retained by some form of heterozygote advantage, and that many other neutral loci in coupling disequilibrium with them are being retained by what Frydenberg (1963) called 'associative overdominance'.

Ohta and Kimura (1970, 1971) examined the likelihood that associative overdominance will have an effect on the rate of fixation of neutral genes, and found that (as might be expected) the most pronounced retardation of the rate occurred when the overdominant gene and the linked neutral gene were both at a frequency near 0.5. These workers made the point that the overall rate of fixation of neutral genes would not be affected by associative overdominance, but that certain neutral loci might be retained for long periods if they were closely linked to overdominant loci and at similar frequencies.

If we could be confident that the gene were in fact the unit of selection, or nearly so, then the effect of associative overdominance could be largely discounted in the interpretation of experimental results. Of course, we would still be unclear about the relative roles of balancing and purifying selection. Some ways in which the effects of these can be distinguished will be suggested at the end of the chapter. Meanwhile, we will examine the conditions under which the gene can be treated as the selective unit, and we will discover that these conditions may be quite easy to meet.

How linkage can influence the genome

The crystallization of chromosomes

The question of whether the gene is the selective unit was addressed directly in these terms by Franklin and Lewontin (1970). These authors showed, using computer simulations, that strong interactions could be set up between linked genes which at the outset of the simulations were in linkage equilibrium. These interactions resulted in the genes being inherited as blocks rather than individuals.

In a typical simulation, a population containing a relatively small number of closely linked genes with two alleles, representing a small segment of a hypothetical chromosome, was subjected to symmetrical balancing selection. The selection acting on each locus in the simulation was very strong, $s_1 = s_2 = 0.1$, so that the loci were driven strongly towards their equilibrium of 0.5. Thus, D could be as large as 0.25 for loci in coupling or as small as -0.25 for loci in repulsion.

Since coupling and repulsion disequilibrium are equivalent when both loci are at 0.5, a measure of the degree to which the loci were associating in groups was taken as the average of the absolute values of the disequilibria between adjacent pairs of loci.

The buildup of $|\overline{D}|$ for a typical pair of simulations is shown in Fig. 8.1. The population has an effective size of 667, and there are 36 loci closely linked on a small segment of chromosome. Because the recombination between adjacent loci is only 0.003, the total length of the segment is 10.8 map units (cM). To keep this in perspective, it should be remembered that a typical chromosome may be 100 or more map units long, and that even this represents only a small fraction of most genomes.

The simulation was started with all loci at or close to $\hat{p} = 0.5$, and with the loci uncorrelated with each other so that only the very small amount of linkage disequilibrium expected by chance was present. For the first 270 generations in one run, the buildup of disequilibrium was slow and appeared to reach a plateau at about 0.08. Then, from generations 270 to 310, the buildup was extremely rapid and suddenly reached a new plateau of 0.245, very close to the maximum possible. In the other run the buildup was much more gradual, extending over about 200 generations.

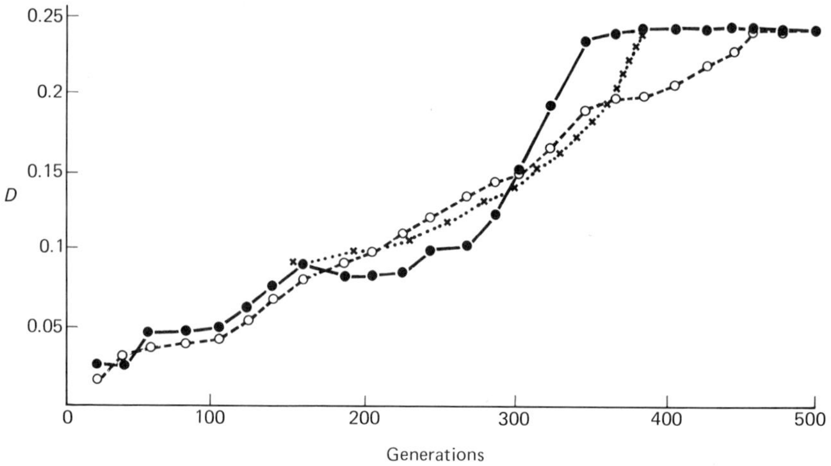

Fig. 8.1. The buildup of disequilibrium with time in typical computer runs of Franklin and Lewontin (1970, their Fig. 2). In these runs, $N_e = 667$, $s_1 = s_2 = 0.1$, $r = 0.003$, and $L = 36$. ● and ○ denote computer runs and × denotes the line derived from the theory of crystallization presented in this chapter.

What had happened was that two complementing types of gamete had come to predominate in the population, so that a heterozygote for the two types was heterozygous at most loci. As a consequence, the distribution of fitnesses in the population had become pronouncedly bimodal. About half the population had fitnesses close to one and half had fitnesses close to $0.9^{36} = 0.0225$. Because of this great disparity, virtually all the chromosomes contributing to the next generation were derived from organisms with a fitness close to one and were of the two complementing types.

Franklin and Lewontin termed this process, culminating in the appearance of the complementing chromosome types, *crystallization.* It can be seen from the figure that this is a two-stage process. The final rise to high disequilibrium values is driven by selection, but the achievement of the lower plateau, they realized, was due to quite another process.

Sved (1968*b*), in an important paper, examined in detail the effect of sampling on the generation of disequilibrium. He was able to show that disequilibrium would be generated in a finite population, even in the absence of selection, as a result of the simple fact that some gametic types would be represented more often in the next generation entirely by chance. Eventually, the buildup of disequilibrium by sampling would be counterbalanced by the breakdown of disequilibrium owing to recombination. The precise amount of disequilibrium to be expected at steady state for this process was difficult to determine, and as we will see is complicated by the fact that sampling may act differently when loci are at different frequencies. To obtain an approximation of this steady state disequilibrium, Sved was forced to make the assumption that all loci were at a frequency p of 0.5, something which would not be the case if the population were free to drift. Given this assumption, he was able to show that at steady state

$$D^2 = 1/[16(4N_e r + 1)] \tag{8.4}$$

where r is the recombination fraction between the loci. Other approaches to the problem with very similar results were made by Hill and Robertson (1968) and Ohta and Kimura (1969, 1970). The formulas are operationally very close to Sved's, even though they are not bound by the requirement that all loci be at 0.5.

It happens that under the conditions of the Franklin and Lewontin simulation, all the loci are driven very close to $\hat{p} = 0.5$, so that the restriction on Sved's formula is met. In the simulation shown in Fig. 8.1, the absolute value of the steady-state disequilibrium between adjacent loci is expected to be 0.0833, and it will be seen that the population reaches this level just before the sudden rise in disequilibrium to the new higher level. Franklin and Lewontin pointed out this close correspondence to Sved's expectation.

We can thus precisely define the crystallization of the genome in the Franklin and Lewontin sense. It is the appearance in the population of disequilibrium between loci of a strength greater than that predicted to occur when steady state is reached between the generation of disequilibrium by sampling and its destruction by recombination. Further, this excess disequilibrium must have appeared without any accompanying change in the rate of recombination between loci.

If such crystallized blocks of genes were widespread in the genome, Franklin and Lewontin suggested, then it would be most risky to assume that the behavior of a particular gene is due to selection on that gene itself. It may be due to selection on a much larger block of genes, invisible to the experimenter.

The extent of disequilibrium in natural and simulated populations

The extrapolation by Franklin and Lewontin from their data leads to the prediction that very strong disequilibrium, particularly between closely linked loci, should be common in natural populations. In fact, quite the opposite has been found. In various surveys of *Drosophila* populations (Kojima et al. 1970; Mukai et al. 1971; Langley et al. 1974) little disequilibrium has been found even between closely-linked loci. The only appreciable disequilibrium, and that not at very high levels, was between some electrophoretic polymorphisms and inversions segregating on the same chromosome.

In man, little evidence for disequilibrium has been found except when the loci involved are very closely linked—within about a map unit of each other. Appreciable disequilibrium has been observed for hemoglobins β and δ (Boyer et al. 1963) and HLA (Thorsby 1974). The strength of disequilibrium seen *may,* however, be due to sampling rather than selection. The most extensive sets of disequilibrium data currently available, those between the MN and Ss loci and the pseudoalleles of the Rh locus in man, are as we will see totally consistent with a large role for the sampling process. In the mouse, disequilibrium has been found between the *t* and H2 groups of loci, which are physically far apart on the chromosome, but this is accompanied by a localized reduction in recombination in this region (Snell 1968; Klein 1974).

Returning to computer simulations, disequilibrium has also not been found above the levels to be expected from sampling, in a number of simulations performed with populations somewhat different from those of Franklin and Lewontin. Some of these have been carried out by my coworkers and me (Wills et al. 1970; Wills and Miller 1976). These simulations were performed on rather small populations ($N = 100$), but in contrast to those of Franklin and Lewontin they involved rather long genetic maps with large numbers of loci. Most of the genetic maps were 100 cM long, carrying either 100 or 400 loci. Selection was of the rank-order truncation type, with as many as 30% of the population with the lowest mean heterozygosities being removed by truncation and a corresponding proportion of the organisms with the highest mean heterozygosities being assigned a fitness of two. As we have seen, this results in far greater per-locus selection than can be obtained by multiplicative selection involving the same average decrease in fitness, so that with symmetrical selection the loci were driven strongly towards $\hat{p} = 0.5$.

The populations were run for a sufficient number of generations to achieve approximate steady state, in which the rate of loss of polymorphisms due to drift was very low. At various points on their approach to steady state, the distribution of disequilibria between loci at different distances from each other on the chromosome was calculated.

There was no sign of crystallization. Rather, as can be seen from Fig. 8.2, the average disequilibrium monotonically decreased with increasing separation of the genes along the chromosome. The same shape of curve was seen regardless of whether or not selection was applied. In one set of runs (Fig. 8.2*c*), an attempt was made to generate complex interactions between loci. Different values of $\hat{p}$, uniformly distributed between 0.2 and 0.8, were picked at random for adjacent loci and the contribution to fitness potential of the genotypes at each

Fig. 8.2.A. Buildup of disequilibrium over fifty generations in a population with a uniform initial distribution of frequencies. $N = 100$, there are 400 loci per chromosome, and there is no selection. Dotted lines show the disequilibrium to be expected at steady state for sampling and recombination.

Fig. 8.2.B. Buildup of disequilibrium over fifty generations in a population similar to that of (A) to which 20% selection has been applied.

Fig. 8.2.C. Buildup of disequilibrium in a population similar to that of (A) to which 20% asymmetrical selection toward a uniform distribution of equilibrium frequencies has been applied.

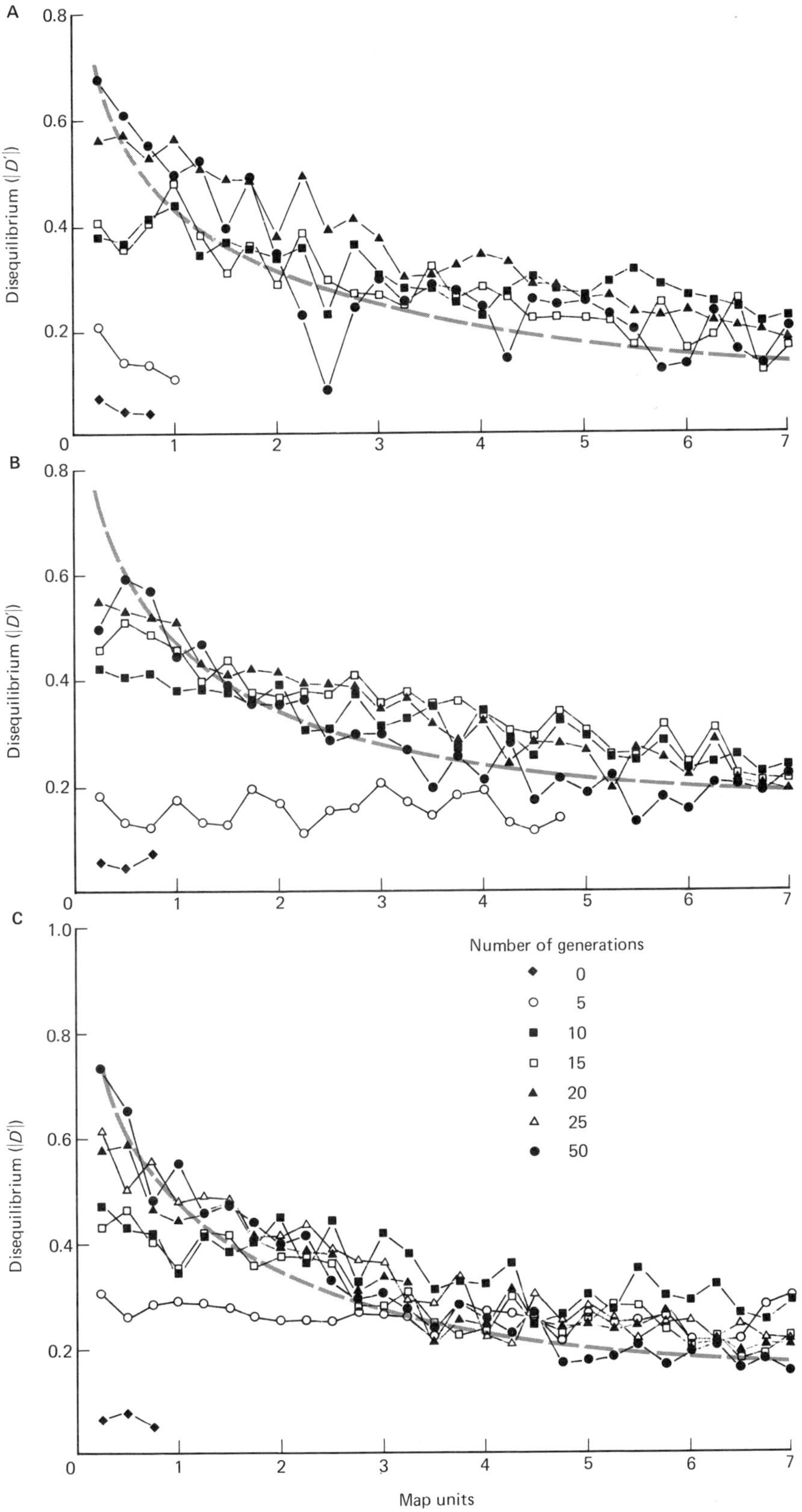
A
B
C
Disequilibrium (|D'|)
Map units
Number of generations
0
5
10
15
20
25
50
0
0.2
0.4
0.6
0.8
1.0
1
2
3
4
5
6
7

locus was adjusted accordingly. As will be seen from the figure, despite this complex set of conflicting selection pressures, the same disequilibrium curve was generated as in the other runs. Thus, while the *sign* of the disequilibrium—that is, which gametes tended to be in excess and which in deficiency—was undoubtedly strongly influenced by selection in these runs, its *strength* was determined by the process of sampling and recombination.

Because of sampling problems, which will be dealt with later (p. 238), loci were only picked for the disequilibrium calculation in which the allele frequency lay between 0.4 and 0.6. In the case of the runs made with an equilibrium frequency of 0.5, this was the majority of loci. The fit to the Sved expectation, given by the dotted line in the figures, was excellent. The dotted lines in Fig. 8.2 are calculated from Sved's formula on the assumption of $N_e = 100$ for no selection and $N_e = 80$ for 20% truncation and replacement selection. They can be seen to conform closely to the curves obtained after 50 generations.

To check this further, we fitted Sved's expectation to data from our simulations which closely approached steady state, and worked out the most likely value of N_e from a Gaussian least-squares approximation. As can be seen from Table 8.1, the correspondence between the predicted value and the number of organisms in the simulation is very close. The effective number of organisms in the simulations was determined approximately by subtraction of the number of organisms removed by truncation from N, the total number. This number, N_s, is approximately equivalent to N_e. Populations subjected to 20% truncation and replacement selection tended to yield an estimate for N slightly less than N_s, but this discrepancy may be due to the fact that the latter value is an approximation.

It will be seen shortly that multilocus populations subject to symmetrical truncation selection will not be expected to crystallize in any case, so that these simulations are a fair example of the effects of sampling.

Table 8.1. Estimates of effective population size in simulated populations subjected to different strengths and types of selection. All estimates made at generation 200. From Wills and Miller (1976, p. 394).

	Selection (%)	N	N_s
	0	106.88 ± 71.22*	100
Symmetrical	5	102.22 ± 20.7	95
	20	62.20 ± 8.71	80
	30	78.69 ± 8.75	70
Asymmetrical	5	105.90 ± 18.81	95
	20	63.86 ± 12.58	80

*Errors are 95% confidence intervals based on a linear hypothesis.

The theory of the crystallization process

Whether or not a population will crystallize can now be seen to be a central question. If it does, then selection contributes to the strength of disequilibrium in addition to determining which gametic types will be in excess. If it does not, then the amount of disequilibrium in the population can be predicted from sampling theory. This in turn would allow us to make predictions about the degree to which loci act independently even in the presence of linkage.

Simulated populations differ from natural ones in a variety of ways. They can be specified to remain at the same size for long periods of time, so that disequilibrium due to sampling will reach steady state. Natural populations can be expected to fluctuate in size over fairly short periods, so that they will not reach steady state. Further, unlike natural populations, simulated populations are isolated. Migration, particularly swamping migration, is a powerful generator of disequilibrium even in large populations. Thus, the calculations we perform here can only be applied to natural populations if they are known to be stable in size and not subject to large amounts of immigration.

We know that a finite isolated population of a constant size will reach steady state for sampling disequilibrium. To determine if it will crystallize, we must examine the population once it has reached steady state and determine whether the increase in disequilibrium due to selection is greater than the decrease due to recombination. If it is, then theoretically at least the population should crystallize. If the difference is very small, the crystallization may take a very long time and be unlikely to occur before the circumstances of the population change. If the increase in disequilibrium due to selection is less than the decrease due to drift, the population should not crystallize. I will show that, for a finite isolated population of constant size subjected to multiplicative balancing selection, the likelihood of crystallization will be determined by only four variables: the effective population size N_e, the recombination between adjacent loci r, the strength of selection at each locus, and the number of heterotically selected loci in the population L.

Let us start with a population at steady state for sampling disequilibrium, in which all loci are at $\hat{p} = 0.5$. The average gamete frequencies g_{00}, g_{01}, g_{10} and g_{11} for a pair of loci in coupling can be obtained by calculating D from equation 8.4 and noting that $g_{00} = g_{11} = 0.25 + D$, while $g_{01} = g_{10} = 0.25 - D$. These average values should not change over time if the population is not going to crystallize. To find the new value D^* of disequilibrium after one generation of selection, we must determine at least one of these gamete frequencies after selection, for example g^*_{00}. The value for g^*_{11} will be the same, and the values for g^*_{01} and g^*_{10} can be obtained by subtraction. The D^* can be obtained from equation 8.1.

Since disequilibrium is reduced by the factor r each generation, then ΔD, the change in disequilibrium due to selection each generation, will be

$$\Delta D = D^* - D - Dr \qquad (8.5)$$

I will determine g^*_{00} first for the two-locus multiplicative case, then generalize to L loci. The approach will be to determine the frequency of double homozygotes f_{aa}, of single homozygotes f_{ab} and of double heterozygotes f_{bb}, both before and after selection. It is assumed throughout that the frequencies of the loci being selected are 0.5. As a consequence, we need only consider the case of loci in coupling, since the case in which the loci are in repulsion is equivalent.

Suppose that there is coupling disequilibrium of strength D between the loci. Then the gamete frequencies g_{00}, g_{01}, g_{10} and g_{11} can be determined from Eq. 8.1. Before selection,

$$f_{aa} = g_{00}^2 + g_{01}^2 + g_{10}^2 + g_{11}^2 \tag{8.6}$$

$$f_{ab} = 2g_{00}g_{01} + 2g_{00}g_{10} + 2g_{01}g_{11} + 2g_{10}g_{11} \tag{8.7}$$

and

$$f_{bb} = 2g_{00}g_{11} + 2g_{01}g_{10} \tag{8.8}$$

In this symmetrical case, $f_{aa} = f_{bb}$.

Because 00 gametes appear in each of these three terms, g_{00} can be rewritten as

$$\begin{aligned} g_{00} &= g_{00}^2 + f_{ab}/4 + g_{00}g_{11} \\ &= g_{00}^2 + f_{ab}/4 + g_{00}^2 \end{aligned} \tag{8.9}$$

singe g_{00} and g_{11} have the same value.

Now suppose that the fitnesses of the three categories are W_{aa}, W_{ab} and W_{bb}, specified by multiplicative selection coefficients. The coefficients W_{aa} and W_{bb} will not be distributed symmetrically around W_{ab}—for example, if the selection against homozygotes is 0.1 at each locus, then W_{aa} will be 0.81, W_{ab} will be 0.9 and W_{bb} will be 1. The asymmetry of selection coefficients around W_{ab} is an important factor in the generation of disequilibrium, as will be developed shortly.

If the population is subjected to selection and then normalized, the new frequency of the double homozygote will be:

$$f^*_{aa} = \frac{f_{aa}W_{aa}}{f_{aa}W_{aa} + f_{ab}W_{ab} + f_{bb}W_{bb}} \tag{8.10}$$

Values for f^*_{ab} and f^*_{bb} are given by $f_{ab}W_{ab}$ and $f_{bb}W_{bb}$ divided by the same denominator. The normalized value of g_{00} after selection is now:

$$g^*_{00} = \frac{g_{00}^2 f^*_{aa}}{f_{aa}} + \frac{f^*_{ab}}{4} + \frac{g_{00}^2 f^*_{bb}}{f_{bb}} \tag{8.11}$$

The other gamete frequencies can be obtained by subtraction in this symmetrical case, giving the new value for coupling disequilibrium after one generation of selection:

$$D^* = g^*_{00}g^*_{11} - g^*_{01}g^*_{10} \tag{8.12}$$

Now, suppose that the disequilibrium value D initially observed in the population was the result of steady state for sampling disequilibrium. Then if ΔD from Eq. 8.5 is greater than zero, the disequilibrium in the population will continue to increase. If it is equal to or less than zero, the disequilibrium will remain at the steady state generated by sampling.

If the loci were in repulsion disequilibrium of strength $-D$, then Eq. 8.5 would become

$$\Delta D = D^* + D + Dr \tag{8.13}$$

and the condition for increased disequilibrium above steady state for sampling would be if ΔD were less than zero. But of course, coupling and repulsion disequilibrium are equivalent if the alleles at both loci are at 0.5.

Essentially the same approach can be used with the multilocus case, although there is a complicating factor. Here we assume, as is specified in the simulations, that all loci are equally spaced along the chromosome and all are subject to the same intensity of multiplicative balancing selection. If a pair of adjacent loci are picked at random, then if they happen to exhibit disequilibrium of strength D the frequencies f_{aa}, f_{ab} and f_{bb} can be calculated as before from equations 8.6, 8.7 and 8.8. But the average fitnesses of these three categories are not so easily calculated.

With all loci at 0.5, the fitness of category f_{ab} is simply

$$W_{ab} = (1 - s)^{L/2} \tag{8.14}$$

But W_{aa} and W_{bb}, while lower or higher than W_{ab}, are not lower or higher by the fitness differential at a single locus. A homozygote at the designated pair of loci will, on the average, also be homozygous at adjacent linked loci, so that the fitness difference will be greater than in the single locus case. To determine this fitness, one must determine the degree to which the loci exhibiting disequilibrium in a population are grouped together into blocks.

One approach to this problem is to determine the average number of runs of homozygous and heterozygous loci in an organism taken from a population in which the average strength of disequilibrium is specified. If there are J + 1 runs of homozygous and heterozygous loci, they will be separated by J junctions, and in the frequency class f_{ab} one of these junctions happens to coincide with our designated pair of loci. The average length of a run will determine the average increase in homozygosity needed to replace the heterozygous locus of the pair with a homozygous one. The stronger the average disequilibrium between loci, the longer will be the average length of a run, and the greater will be the difference between W_{aa} and W_{ab}.

It is possible to visualize the replacement procedure with a diagram (Fig. 8.3). If m_D is the average length of a block of loci, then in a long chromosome the removal of a heterozygous block and its replacement by a homozygous block has two possible outcomes, diagramed in the figure. In half the cases a single junction will be removed, and in the other half two junctions will be

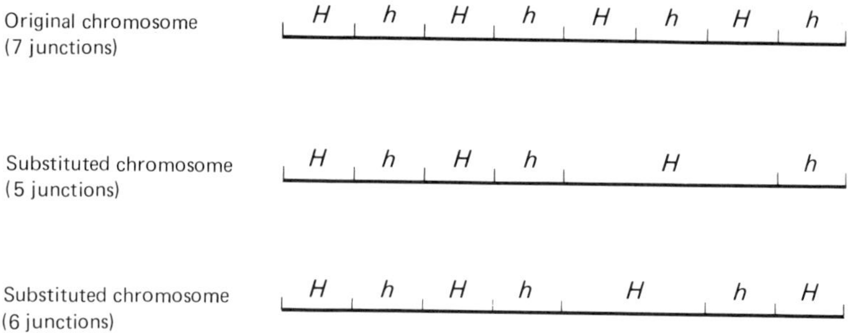

Fig. 8.3. The substitution of a heterozygous block of loci by a homozygous block in the case in which the length of the block is much less than the length of the chromosome. The substitution leads to two equally probable configurations in which the number of junctions between runs is reduced by one or two. *H* represents a heterozygous block and *h* a homozygous block. The increase in homozygosity needed to remove one junction is less than m_D, the average length of a homozygous block of loci.

removed. So, the increase in the number of homozygous loci needed to remove an average of one junction will be $m_D/1.5$.

If the chromosome is very short, or if m_D is a large fraction of L, the situation changes somewhat. If the average number of junctions is less than two, then the replacement of a heterozygous by a homozygous block will always remove the remaining junctions so that the increase is simply m_D or $L/2$, whichever is smaller. If the number of junctions lies between two and three, then the necessary increase will be $m_D/0.5J$.

It must be remembered that in these calculations we are working with *averages*. While the disequilibria between different pairs of adjacent loci will vary widely, and the lengths of individual runs of homozygous and heterozygous loci will also vary, taking their averages will allow estimates of the differences among W_{aa}, W_{ab} and W_{bb} to be made.

It is now necessary to calculate m_D. A run of homozygous loci will terminate at locus x if the zygote has the one of the following configurations at loci x and $x + 1$: 00/01, 01/00, 11/10 or 10/11. The probability of termination is the sum of the probabilities of these four configurations, as a fraction of the probability that locus x is homozygous:

$$T_{\text{hom}} = (2g_{00}g_{01} + 2g_{11}g_{10})/(1 - 2\hat{p}\hat{q}) \tag{8.15}$$

and by a similar argument

$$T_{\text{het}} = (2g_{00}g_{10} + 2g_{11}g_{01})/2\hat{p}\hat{q} \tag{8.16}$$

In the particular case we are considering, in which all loci are at 0.5, $T_{\text{hom}} = T_{\text{het}}$. It is possible, as will be seen later, to make some generalizations to cases in which all loci are at some internal equilibrium other than 0.5.

The average probability of obtaining a run of a given length b is distributed as a geometric series:

$$P_b = T_{hom}(1 - T_{hom})^b \quad \text{for } b = 1, 2, 3. \ldots \qquad (8.17)$$

The mean of this particular geometric distribution, with no zero class, is simply the inverse of T_{hom}, so

$$m_D = 1/T_{hom} \qquad (8.18)$$

The average length of a run of heterozygous loci can be calculated similarly.

In the absence of disequilibrium, the average length of a run of homozygous or heterozygous loci will be two. This length may increase markedly with strong disequilibrium. Consider an isolated, randomly mating population of size $N_e = 100$, with 50 equally-spaced loci subject to multiplicative balancing selection of $s_1 = s_2 = 0.1$ at each locus. The recombination coefficient r between adjacent loci is 0.01, giving a total map length of 50 centimorgans. Then the steady state disequilibrium as a result of sampling in the population will be (from eqn 8.4): $D = \pm 0.1118$, about half the maximum possible of ± 0.25. At the average pair of loci in coupling disequilibrium, $g_{00} = g_{11} = 0.3618$, and $g_{01} = g_{10} = 0.1382$. The probability of termination of a run of homozygous or heterozygous loci is 0.4, so $m_D = 2.5$, slightly longer than m_D for the case with no disequilibrium. Since the chromosome is much longer than m_D,

$$W_{aa} = (1 - s)^{L/2} + m_D/1.5 = 0.0602$$
$$W_{ab} = (1 - s)^{L/2} = 0.0718 \text{ and}$$
$$W_{bb} = (1 - s)^{L/2} - m_D/1.5 = 0.0856$$

These values can now be used to solve eqns 8.10, 8.11, 8.12 and finally 8.5, to yield ΔD for this population at steady state. This value is -4.3×10^{-4}, which, since it is negative, predicts that provided conditions do not change the population will remain at steady state for sampling disequilibrium indefinitely and will not crystallize.

Predicting the results of simulations

The value obtained from the numerical evaluation can be used to calculate D^*, and this in turn can be used in eqn (8.5) to calculate ΔD due to selection for any population for which D before selection has been specified. This calculation gives the change in the average value of coupling disequilibrium. Since the cases we are considering are symmetrical, changes in the average value of repulsion disequilibrium will be the same but the opposite in sign.

This tool gives us great flexibility. It is possible to specify a population with a given effective population size, number of loci, strength of selection per locus and recombination between adjacent loci, and determine its probable fate with regard to the crystallization process. We can ask the question: given that such a population has achieved steady state for sampling disequilibrium, is the disequilibrium generated by selection likely to result in an increase beyond the steady state? This will be a function of the steady-state disequilibrium itself, which will in turn be, as we have seen, a function of the population size and of the recombination between loci. But it will also be a function of the strength of selection acting on each locus and of the variance of heterozygosity. These latter two variables will determine the distribution of fitnesses in the population. If, for example, many loci are segregating simultaneously, then the variance of heterozygosity will be very small and even very strong per-locus selection will not change the disequilibrium greatly.

Fig. 8.4 shows the change in D as a function of D for the Franklin and Lewontin simulations discussed earlier, and also for some simulations of Yamazaki (1977) which will be examined below. In none of these simulations was ΔD ever negative, so that populations of any size with these sets of characteristics (small L and r and large per-locus selection) should eventually crystallize. The rate of increase of D becomes very great as D increases, since at high levels of D selection is acting on an entire chromosome and is very strong.

Let us turn to the cases investigated by Franklin and Lewontin and presented in Fig. 8.1. At first, sampling was responsible for the generation of disequilibrium, and both populations reached an average steady state of 0.0833. It can be calculated from the equations given above that, at this level, ΔD due

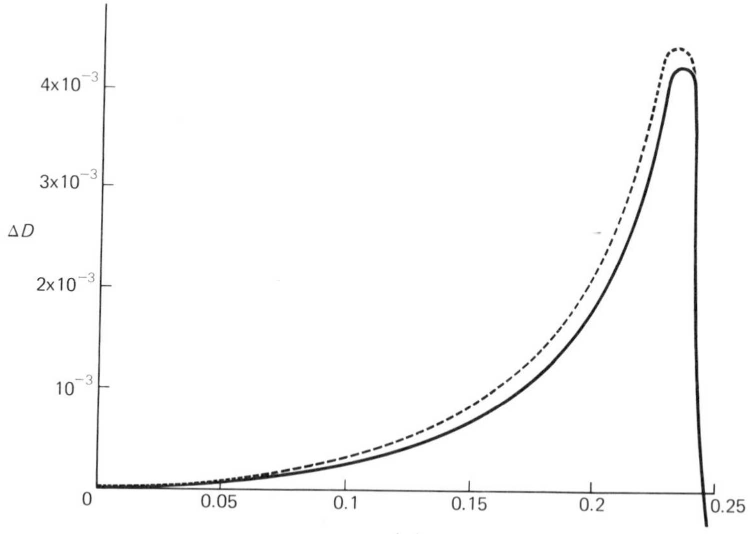

Fig. 8.4. The change in D as a function of D for the simulations of Franklin and Lewontin (1970, solid line) and Yamazaki (1977, dotted line).

to the difference between selection and recombination is 2×10^{-4}. The population should therefore crystallize, and the theoretical path of crystallization is given in Fig. 8.1. The theoretical crystallization curve lies between the curves for the two simulations.

The simulations of Franklin and Lewontin given in the figure lie at the boundary between populations that will crystallize and those that will not. They found that by keeping L, s_1, s_2 and N_e constant but increasing r from 0.003 to 0.005, a third simulated population did not begin to crystallize even though the simulation was continued for 500 generations. The present theory predicts this. The steady-state disequilibrium for such a population is $D = \pm 0.067$. In such populations the increase in D due to selection at this D level is 3.5×10^{-4}. When $r = 0.003$, the decrease due to recombination is 2.0×10^{-4}, so that the increase per generation is the difference between these values or 1.5×10^{-4}. When $r = 0.005$, however, the decrease due to recombination is $0.005 \times 0.067 = 3.4 \times 10^{-4}$. Now the difference is only 10^{-5} per generation, and to increase D from 0.067 to 0.077 will take approximately 1000 generations. A slight increase in r was enough to prevent crystallization from occurring for the duration of the simulation.

The simulations of Yamazaki (1977) are shown in Fig. 8.5. They show a

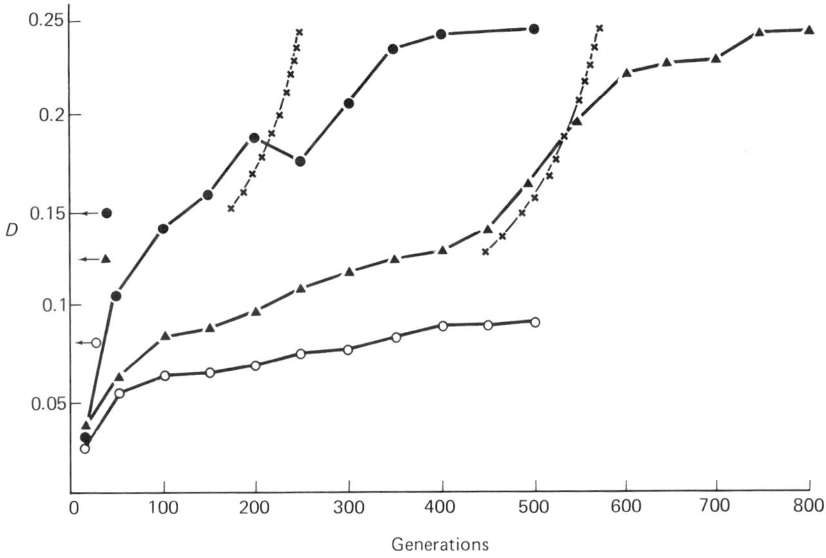

Fig. 8.5. Effect of population size on the time and rate of crystallization. Data from Yamazaki (1977, his Fig. 2). ● $N_e = 200$, ▲ $N_e = 400$, ○ $N_e = 800$. In all runs, $s_1 = s_2 = 0.1$, $r = 0.0022$ and $L = 88$. The arrows on the ordinate indicate the amount of steady-state disequilibrium to be expected from sampling alone in a given run. As in Fig. 8.1, the lines marked by × show the rate of increase in disequilibrium to be expected from crystallization theory.

strong effect of population size, with smaller population size leading to more rapid crystallization as would be predicted. From the curve shown in Fig. 8.4 the largest population ($N_e = 800$) should also eventually crystallize, but the run was terminated just after the population reached steady state for sampling disequilibrium and before the crystallization began.

It will be seen that the theoretical curves for these populations do not match the simulations well. The lack of precise correspondence could be due to approximations in the theory or to the fact that in a finite population the final crystallization must await specific recombination events. The simulations are marked by plateaus followed by sudden increases in D.

The conditions for crystallization

Crystallization and multiplicative selection

The close fit of the theory to published simulations suggests that we can extrapolate to situations that have not been explored by simulation, and that indeed could not be explored without an enormous expenditure of computer time. In Fig. 8.6, I have explored four different types of populations subject to multiplicative selection. For each population, the total map length has been allowed to vary. The question asked is: what is ΔD at steady state for sampling disequilibrium?

The two dashed lines show ΔD for populations in which $N_e = 100$ and $L = 100$. In the line showing the greatest effect of selection, s_1 and s_2 have been chosen to give a fitness $\overline{W}$ of only 0.01, which is very strong selection on this segment of the genome. In the line below it, the per-locus selection has been adjusted to give $\overline{W} = 0.1$. It will be seen that in either case ΔD is substantial when the total map length is short, but it drops rapidly with increasing total map length. This is because the steady state disequilibrium between adjacent loci becomes small as recombination between them increases. Beyond a total map length of 20 cM for the more weakly selected population and 54 cM for the more strongly selected population, ΔD actually becomes negative because the increase due to selection is now more than counterbalanced by the decrease due to recombination.

The two solid lines give ΔD when the other population measures are kept the same but the effective population size increases to 1000. ΔD is now much smaller, and the points at which ΔD become negative have been displaced to the left. This is because the steady-state strength of sampling disequilibrium has been reduced in the larger population size.

With multiplicative selection, it appears that if the population size is moderately large and the total map is long, crystallization will not occur even in strongly selected populations. There are, however, three unusual circumstances in which crystallization *could* occur in such populations.

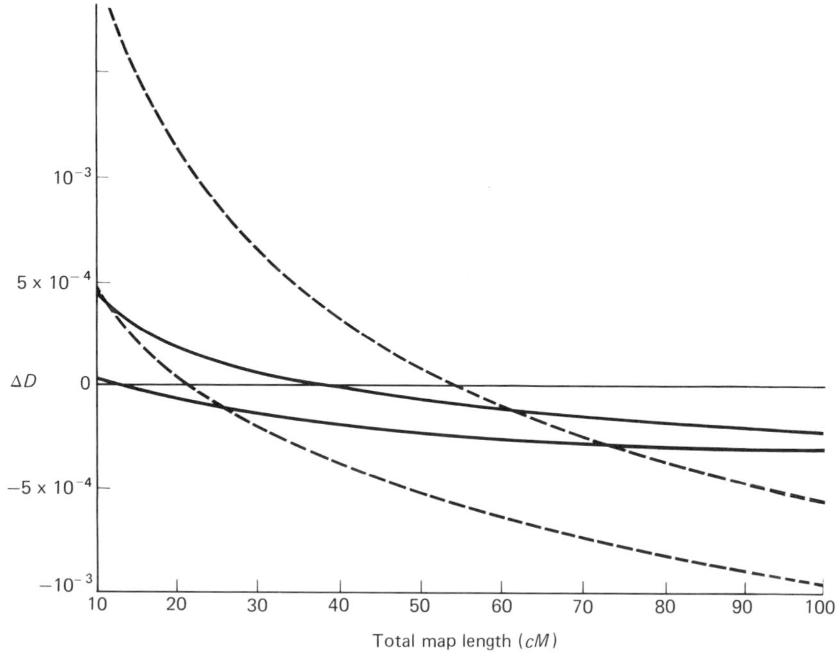

Fig. 8.6. The change in D due to the difference between the increase in D due to selection and the decrease due to recombination when the population is at steady state for sampling-generated disequilibrium. In all populations $L = 100$. The abscissa is total map length, governing the rate of recombination between the loci. For the two dashed lines, $N_e = 100$, and for the two solid lines $N_e = 1000$. The upper line in each set is generated from populations in which $s_1 = s_2 = 0.088$, and the lower line from populations in which $s_1 = s_2 = 0.045$, leading to average fitnesses of 0.01 and 0.1 respectively.

(a) If recombination were greatly reduced over a portion of the genome, the steady state disequilibrium in that region would markedly increase. Inversions, translocations and other mechanical restrictions on recombination could bring about such an effect, as could selection for genes modifying recombination in such regions.

(b) If the rate of recombination for the genome as a whole were to decrease, the entire genome could crystallize. As we will see in the next chapter, there are strong selective reasons for supposing that this will not occur.

(c) If the selective pressures were not uniformly distributed over the genome, but were concentrated in a small segment, then that segment might crystallize even without modification of the recombination rate. The segment would have to be circumscribed in extent, like the portions of the genome dealt with by Franklin and Lewontin and by Yamazaki, and it would have to contribute the majority of the reduction in fitness of the population. The rest of the genome would be much more weakly selected, and as a consequence the strength of disequilibrium in the rest of the genome would certainly be a function of sampling.

Crystallization and rank-order truncation selection

The simulations shown in Fig. 8.2 at the beginning of the chapter were subjected to balancing rank-order truncation and replacement selection. None of the simulated populations showed any signs of crystallization. Is this what theory would predict?

The truncation and replacement selection performed in the simulations is of a special type. All loci are at or near 0.5, and the two selection thresholds are placed symmetrically around the mean $\overline{H}$. Thus, so far as gamete composition is concerned, the two halves of the population above and below the mean are mirror images of each other. Removal of a fraction of the most homozygous organisms and replacement by the mirror-image fraction of the most heterozygous organisms will leave the gametic composition of the population, and therefore its disequilibrium, unchanged. It is therefore not surprising that the simulations of Fig. 8.2 exhibit levels of disequilibrium no greater than those predicted from sampling theory.

There are other circumstances in which truncation selection will leave the levels of disequilibrium unchanged. For this to be true, there appear to be two requirements. The first is that allele frequencies be at 0.5. The second is that W_{aa} and W_{bb} be symmetrically distributed around W_{ab}. If the difference in fitness between W_{aa} and W_{ab} is $-y$ and the difference between W_{ab} and W_{bb} is $+y$, then Eq. 8.11 can be rewritten as:

$$\begin{aligned} g^{*}_{00} = {} & \frac{g^{2}_{00} f_{\mathrm{aa}}(W_{\mathrm{ab}} - y)}{f_{aa}[f_{\mathrm{aa}}(W_{\mathrm{ab}} - y) + f_{\mathrm{ab}} W_{\mathrm{ab}} + f_{\mathrm{bb}}(W_{\mathrm{ab}} + y)]} \\ & + \frac{f_{\mathrm{ab}} W_{\mathrm{ab}}}{4\,[f_{\mathrm{aa}}(W_{\mathrm{ab}} - y) + f_{\mathrm{ab}} W_{\mathrm{ab}} + f_{\mathrm{bb}}(W_{\mathrm{ab}} + y)]} \\ & + \frac{g^{2}_{00} f_{\mathrm{bb}}(W_{\mathrm{ab}} - y)}{f_{\mathrm{aa}}\,[f_{\mathrm{aa}}(W_{\mathrm{ab}} - y) + f_{\mathrm{ab}} W_{\mathrm{ab}} + f_{\mathrm{bb}}(W_{\mathrm{ab}} + y)]} \end{aligned} \tag{8.19}$$

Given that $f_{\mathrm{aa}} + f_{\mathrm{ab}} + f_{\mathrm{bb}} = 1$, and in this symmetrical case $f_{\mathrm{aa}} = f_{\mathrm{bb}}$, this reduces to

$$g^{*}_{00} = g^{2}_{00} + f_{\mathrm{ab}}/4 + g^{2}_{00} \tag{8.20}$$

and symmetrical selection leaves the disequilibrium unchanged.

This situation does not hold for multiplicative selection. Even though the differences in heterozygosity between the aa and ab and between the ab and bb subcategories are the same, the fitnesses will be different because they are a logarithmic function of H. It is this asymmetrical distribution of fitnesses which results in the generation of disequilibrium by multiplicative selection, as we have seen.

As with multiplicative selection, the generation of disequilibrium by truncation selection when allele frequencies are not at 0.5 will be strongly dependent on the assumptions made about the distribution of frequencies in the population. Disequilibrium may be generated, since in these cases the frequencies of 00 gametes in f_{aa} and f_{bb} are not the same. This question will repay careful theoretical study, particularly with regard to the problem of what happens when the selective equilibria of the loci in the population are not 0.5. It may be remarked at this point, however, that the greatest differences in the frequencies of 00 gametes in f_{aa} and f_{bb} will be found when allele frequencies are close to fixation. Under these circumstances the variance of heterozygosity, and therefore the effectiveness of truncation selection, will be greatly reduced. Although the point is still an open one, it seems unlikely that truncation selection will prove to be more effective than multiplicative selection in producing crystallization when the number of segregating loci is large and the genetic map is long.

How likely is crystallization, and what will be its effects?

So far, we have dealt with balancing selection that is driving polymorphisms towards frequencies of 0.5 at all loci simultaneously. The likelihood of crystallization seems to be very small for large outbreeding populations with long genetic maps, regardless of whether the mode of selection is multiplicative or truncation. In fact, it can been seen that symmetrical balancing selection is the most favorable for crystallization. If polymorphisms are being driven to frequencies other than 0.5, then the variance of heterozygosity will be smaller and so will the variance of fitnesses in the population. This will reduce the likelihood of crystallization. It also seems probable, though this will bear further investigation, that the presence of multiple alleles at many loci will also reduce the probability of crystallization because of the great variety of chromosomal types that will be selected for. Yamazaki (1977) found that the rate and time of onset of crystallization were both delayed if there were more than two alleles at each locus at the start of simulations.

Still, crystallization has occurred in some habitually selfing plants. Complementing sets of gametes, apparently involving an appreciable part of the genome, have been found by Allard and his coworkers, as was pointed out earlier. Because the effective population size *from the standpoint of the generation of disequilibrium by sampling* is very small in these plants, it is not surprising that crystallization of the genome has occurred. The number of selected genes that were actually involved in the crystallization is of course an open question.

What is the effect of crystallization on the preservation of genetic variability? The effect may actually be that variability is lost. Consider the actual process of crystallization itself, the 'birth trauma' of the crystallized chromosome so to speak. Crystallization results in the takeover of the population by complementing blocks of genes. These complementing gametes are drawn from the population of gametes available at the start of the crystallization process. It is highly unlikely, especially when L is large, that perfectly complementing gametes will be present initially or will be generated by recombination during the process. Thus, the gametes that end up predominating in the population could carry the same allele at a number of loci, and in the process of their taking over the population these loci will be driven to fixation.

This effect has been observed by Yamazaki (1977), who found that even in cases in which the selection per locus was very strong, the process of crystallization increased the rate of fixation in the population to levels far higher than would be expected if the loci had been free to assort at random. We will explore this phenomenon at greater length in the next chapter, when we turn to the question of the maintenance of variability in the presence of disequilibrium.

Thus, crystallization has its costs. In inbreeding plants, the process of crystallization, at least initially, was probably accompanied by a marked loss of variability. New balanced polymorphisms would accumulate in such a population very slowly, since each new polymorphic allele is in effect a new chromosome. Each new chromosome would have to replace those already in the population, since there would be no recombination.

We can conclude from the above arguments that for large outbreeding populations the gene is essentially the unit of selection. If the chromosomes of a population are densely packed with balanced polymorphisms, disequilibrium between adjacent loci may be strong as a result of sampling, but the strength of disequilibrium between loci will rapidly diminish with increasing genetic distance between the loci. In such populations, the *sign* of the disequilibrium may be under genetic control, but its *strength* will be predictable as a function of sampling and recombination.

The pattern of disequilibria to be expected from sampling

The likelihood that the strength of disequilibrium can often be predicted from sampling theory opens up some exciting possibilities. We can, for example, get some idea of the extent to which associative overdominance will play a role in the outcome of experiments such as those performed by Powell (1971) and McDonald and Ayala (1974). To do this, however, we must ask a rather complex question. Since most polymorphisms are at frequencies other than 0.5, what is the distribution of disequilibria to be expected from sampling under these circumstances? Our intuition will tell us that it will vary as a function of allele frequencies. If, for example, both the loci at which the sampling is taking place are near fixation for the 1 allele, then g_{11} will be very large. Many organisms will be homozygous 11/11, and sampling will have no effect on this subgroup of the population. The effect of this will be to restrict sampling to a subgroup of the gametes in the population, and this will in turn have an effect on the amount of disequilibrium to be observed at steady state and the rate at which it builds up.

I will first quantify this phenomenon, then use it to examine a variety of situations in which pairs of loci have stayed near certain frequencies long enough to exhibit steady-state sampling disequilibrium. I am not concerned with the variance of disequilibria in an entire population (the only theoretically tractable solutions to this problem are restricted to cases in which the expectation of D is zero), but with a different question. Assuming a large outbreeding population, are there circumstances under which associative overdominance will have an appreciable short-term impact on gene frequencies?

Coupling and repulsion diseqilibrium are defined according to the convention of Langley and Crow (1974). These authors suggested that if there is no *a priori* reason for distinguishing alleles—for example, if all the alleles involved are wild type—then coupling disequilibrium can be defined as an excess of gametes carrying the two commoner and the two rarer alleles at the loci. Repulsion disequilibrium is exhibited by the loci when there is an excess of gametes carrying one rarer and one commoner allele.

It is assumed, further, that the two loci at which disequilibrium is building up each have two alleles, but we place no restriction on the allele frequencies. The loci are either being maintained at these frequencies by balancing selection, or are neutral but have stayed near these frequencies long enough to be at steady state for sampling disequilibrium.

The situation can now be markedly different from the $p = 0.5$ case. Selection or drift will determine whether the loci exhibit coupling or repulsion disequilibrium, and the 'half-distributions' of coupling or repulsion disequilibria possible at this pair of loci may be markedly different from each other.

The two half-distributions will be the same when both loci are at 0.5. In this case, all the gametes in the population are potentially substitutable as a result of sampling. If they are coupling gametes they can be substituted by repulsion gametes, and vice versa. If we define the proportion of gametes that is potentially substitutable as G_e, the *effective proportion of gametes,* then in the limiting case in which $p = q$,

$$G_e = 1 \tag{8.21}$$

When p_1 or p_2 or both are not 0.5, then the coupling and repulsion cases must be considered separately. Each forms a half distribution, and the proportion of gametes that is potentially substitutable will be $G_{eC}/2$ or $G_{eR}/2$, where G_{eC} and G_{eR} are the effective proportion of gametes for pairs of loci in coupling and repulsion disequilibrium respectively. Specifically, $G_{eC}/2$ is the proportion of repulsion gametes which can be substituted by coupling gametes as a result of sampling or selection in the course of the generation of coupling disequilibrium. This is $2p_1q_2$ or $2p_2q_1$, whichever is smaller. $G_{e_R}/2$ is the proportion of coupling gametes which can be substituted by repulsion gametes in the generation of repulsion disequilibrium. This is the smaller of $2p_1p_2$ or $2q_1q_2$.

Each of these values determines the possible size of a half-distribution. Thus,

$$G_{e_C} \text{ is the smaller of } 4p_1q_2 \text{ or } 4p_2q_1, \quad (8.22)$$

and

$$G_{e_R} \text{ is the smaller of } 4p_1p_2 \text{ or } 4q_1q_2 \quad (8.23)$$

This reduction in the effective proportion of gametes available for sampling is shown for a simple example in Table 8.2. In the case in which $p_1 = p_2 = 0.1$, if there is the maximum possible coupling disequilibrium, 18% of the gametes will have been converted from 01 or 10 to 00 or 11, as contrasted with 50% for the $\hat{p} = 0.5$ case. This is $2pq$. But since only half the locus pairs on average will exhibit coupling disequilibrium if there is no genome-wide bias favoring one type of disequilibrium over the other, then we are seeing only half the distribution of disequilibria. The other half is never seen because those pairs of loci are in repulsion disequilibrium and the proportion of convertible 1gametes in repulsion disequilibrium for the $\hat{p} = 0.1$ case is much smaller. Thus, the number of potentially convertible gametes yielding coupling disequilibrium must be multiplied by two, giving $4pq$.

Sampling will act on this reduced proportion of gametes in exactly the same way as it would if all the gametes were available for sampling. Provided both

Table 8.2. The effective proportion of gametes G_e available for sampling for two values of p.

	Equilibrium	Complete coupling disequilibrium	Complete repulsion disequilibrium
$p_1 = p_2 = 0.5$			
g_{00}	0.25	0.5	0
g_{01}	0.25	0	0.5
g_{10}	0.25	0	0.5
g_{11}	0.25	0.5	0
		$4pq = 1$	$4p^2 = 1$
$p_1 = p_2 = 0.1$			
g_{00}	0.01	0.1	0
g_{01}	0.09	0	0.1
g_{10}	0.09	0	0.1
g_{11}	0.81	0.9	0.8
		$4pq = 0.36$	$4p^2 = 0.04$

loci stay near the same frequency, the steady-state disequilibrium can be obtained from a modification of Sved's formula (8.4). Because the number of potentially convertible gametes is different for coupling and repulsion when $\hat{p} \neq 0.5$, two average values of disequilibrium will be obtained when N_e is multiplied in Sved's formula by the factor G_e. This D value, when multiplied by $4D_{max}$, will give Lewontin's D'.

$$D'_c = +\ 1/16\ (4N_eG_{e_C}\ r + 1)\cdot 4D_{max} \tag{8.24}$$

and

$$D'_R = -\ 1/16\ (4N_eG_{e_R}\ r + 1)\cdot 4D_{max} \tag{8.25}$$

Because fewer gametes are available for sampling, the average strength of steady-state sampling disequilibrium *as a fraction of the total possible* is greater for repulsion than for coupling disequilibrium if $\hat{p} \neq 0.5$ for either locus. Both will be increased over the case in which $\hat{p}_1 = \hat{p}_2 = 0.5$ for the same N_e.

Further, as we have already seen, the rate of buildup of sampling disequilibrium is an inverse function of effective population size. Thus, the smaller G_e becomes, the more rapidly a population will achieve steady state for sampling disequilibrium. This is an important consideration when we turn, as we will at the end of the chapter, to the question of the strength of disequilibrium to be expected between a selected and a neutral locus. If the buildup of disequilibrium is rapid, then the requirement that the loci must remain near a particular value in order for Sved's formula to apply is not likely to be seriously violated.

Figure 8.7 shows the results of a typical computer run in which 96 loci were selected towards an equilibrium frequency of either 0.2 or 0.8 in a population of size $N_e = 100$. The population was subjected to 20 percent truncation and replacement selection in order to ensure that all loci were kept near their equilibrium. Part (A) of the figure shows the distribution of the locus pairs with respect to their D_{max} values, and demonstrates that pairs in coupling and repulsion disequilibrium remain distributed around their expected D_{max} values of 0.16 and 0.04 respectively.

The expected values of D/D_{max} at steady state can be calculated from eqns. 8.24 and 8.25, yielding 0.53 and 0.78 respectively. In Part (B) of the figure, the approach of loci in coupling and in repulsion that are near the expected D_{max} values is followed for 50 generations. It can be seen that locus pairs in repulsion (dashed line) rapidly reach their expected steady state value (arrow), while locus pairs in coupling (solid line) approach their steady state more slowly. The theory is seen to be an adequate predictor of the buildup of disequilibrium in populations in which crystallization is not expected and loci are being driven towards internal selective equilibria other than 0.5.

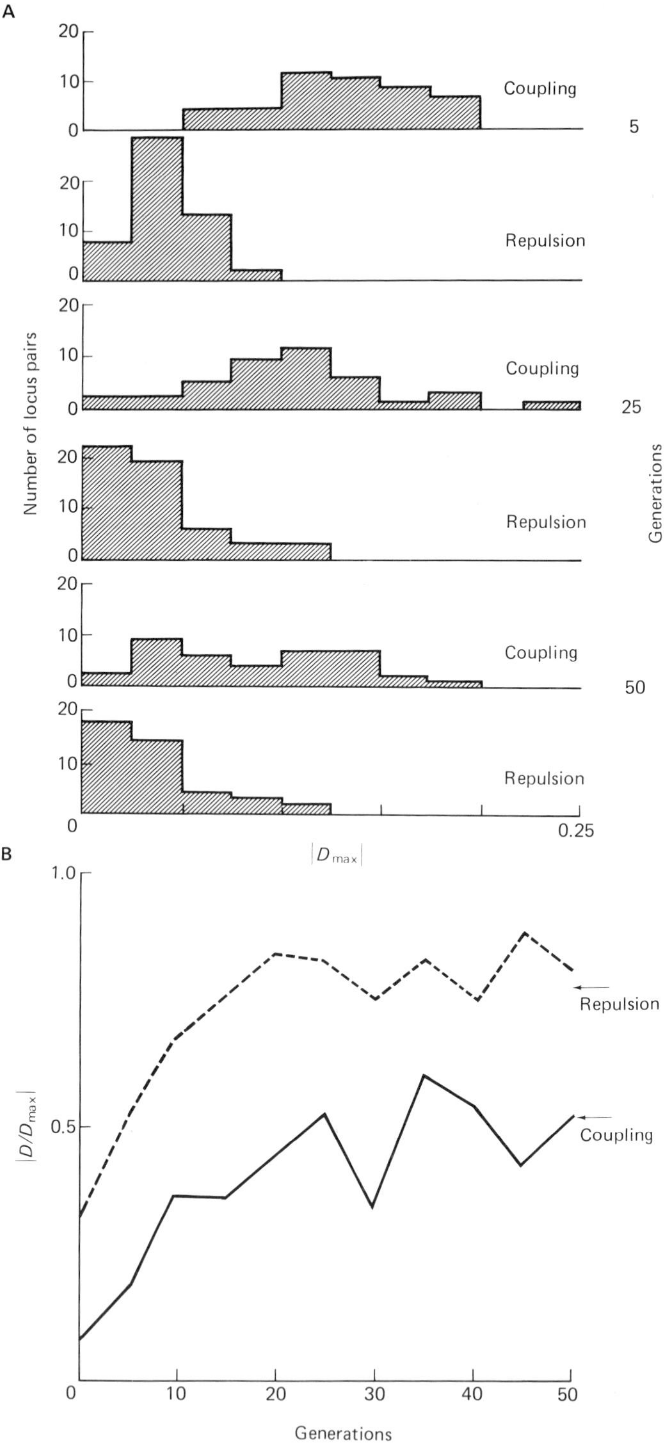

Fig. 8.7. The buildup of sampling disequilibrium in a population in which polymorphisms are being selected towards internal equilibria of 0.2 or 0.8. In this population, $N_e = 100$, $L = 96$, $r = 0.01$ and selection is 20% truncation and replacement. Part A of the figure shows how locus pairs continue to be grouped around their expected coupling and repulsion D_{max} values. Part B shows the increase of D/D_{max} with time to levels predicted by the present theory.

Data from humans illustrate the influence of allele frequency on sampling disequilibrium

The MNSs system

The MNSs blood-group system consists of two closely linked genes, each carrying two common alleles, which in most human populations are in pronounced linkage disequilibrium. The precise distance between these loci is unknown, but Gedde-Dahl et al. (1967) estimate it on the basis of rare recombinants to be at most one map unit.

Extensive information from virtually all human populations is available for these two closely linked blood groups. In Fig. 8.8, I have plotted the information from 200 studies compiled by Mourant et al. (1976, their Table 2.11.1). These data were gathered with the aid of anti-s antiserum, so that the frequencies of the four gametic types MS, Ms, NS, and Ns could be determined with a minimum of uncertainty.

On the horizontal axis is the absolute value of D_{max}, the maximum possible disequilibrium given the allele frequencies in the sample. The vertical scale is the absolute value of D/D_{max}, which can range from 0 to 1.

The distribution is the one we would expect if sampling were generating the disequilibrium between the loci. There is little disequilibrium, as a fraction of the total possible, when the frequencies at both loci are at or near 0.5. N_e for

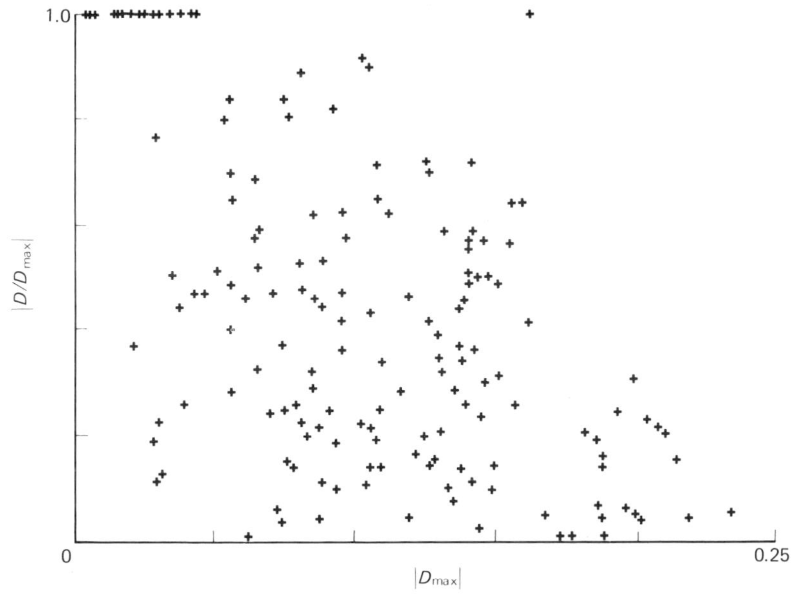

Fig. 8.8. Data from surveys of the human MNSs system, plotted to show the inverse relationship between the size of $|D_{max}|$ and the variance of D/D_{max}.

these samples, calculated from Sved's formula, averages around 1500 (assuming $r = 0.01$), which is perhaps not an unreasonable value for human deme sizes. But the variance of disequilibrium increases rapidly as alleles at one or both loci deviate from 0.5. This would be predicted from the calculations in this section, since as we move towards low values of D_{max}, G_e becomes small relative to N_e. The correlation of D/D_{max} and D_{max} is -0.564, which for 198 degrees of freedom is highly significant.

By the criterion of Langley and Crow (1974), there are 57 coupling and 143 repulsion cases among the data. It is impossible to say whether this represents a selective trend or a historical accident.

There is one alternative interpretation of the data. If the MNSs system is actually a multiple-allele system rather than two closely linked loci (the apparent recombinants being explained by mutation or uncertain parentage), then apparent disequilibrium could be generated simply by an excess of MS and Ns or Ms and NS alleles. But such disequilibrium should not be dependent on D_{max}. The lack of populations with high D/D_{max} when D_{max} is high would not be expected on this genetic model, since high apparent D/D_{max} could occur in a population even when M and S were both near 0.5. A multiple-allele model could only fit the data if there were heterotic selection acting on the locus such that the four alleles, MS, Ms, NS, and Ns, are all held at equal frequency when M and S are near 0.5. We have seen that the conditions for the selective maintenance of such a multi-allele polymorphism are restrictive. Further, such selection has not been detected (Morton and Chung 1959).

Because the sizes of the samples making up the data vary greatly, there is the further possibility that the relationship between D/D_{max} and D_{max} is an artifact of sample size. If it were, then large values of D/D_{max} would be found in small samples. The correlation of D/D_{max} with sample size is -0.011, however, which is not significant.

Thus, the distribution of disequilibria at this pair of loci is in accord with sampling theory, though this does not of course rule out a role for selection. There is certainly no sign of a 'supergene' involving this pair of loci, since the amount of disequilibrium seems to be no more than would be expected from the action of sampling.

The Rh locus

Precisely the same analysis can be carried out for the Rh locus. A good deal of evidence has now accumulated that the Rh locus is actually a complex of three pseudoalleles, as was first suggested by Fisher (1947). These alleles, designated C, D, and E, were inferred by Fisher on the basis of European data to be in the order DCE. His calculation was based on the ratio of presumptive recombinants to the heterozygotes for ancestral gametic types in the population; the higher this ratio the greater the recombination and therefore the greater the genetic distance between pseudoalleles.

There are various problems with the Fisher model, so that the alternative interpretation of a large series of multiple alleles at this locus is not ruled out. While antibodies can readily be produced to the major alternatives at the C and E 'loci', no antibody has yet been found that reacts with the major presumed alternative at the D locus, designated d. Further, there are some *'cis–trans'* effects or interaction antigens, particularly the *f* specificity which is found only when c and e are on the same chromosome. Even the 'deletion' phenotypes which have been discovered, which are again consistent with the order DCE, are susceptible to alternative interpretations (Boettcher 1964, 1965; Cavalli-Sforza and Bodmer 1971). Furthermore, populations elsewhere in the world do not always fit the Fisher model, since in many cases recombinant frequencies do not fit the order DCE (see Table 8.4).

When the DCE data are examined as were the MNSs data in the preceding section, however, Fisher's model is supported. Table 8.3 shows the correlations between D/D_{max} and D_{max} for the combinations D–E, C–E, and C–D, and the slopes of the regression lines. It will be seen that the strongest negative correlation with the steepest regression is found in the D–E data, supporting Fisher's hypothesis that D and E are the two outside markers. The next strongest correlation and next largest slope are found in the C–E data, and in the C–D data the correlation is not significant. On the assumption that the disequilibrium observed between these pseudoalleles is sampling-generated, the order would appear to be DCE, with D and C closer than C and E. Again, there is no significant correlation of sample size with D/D_{max}.

The alternative hypothesis, that the Rh polymorphism consists of a series of multiple alleles at a single locus, would require heterotic balancing selection to maintain eight alleles. While not impossible, this seems unlikely in view of the known modes of selection at this locus.

As was mentioned earlier, the data from many human populations are inconsistent with Fisher's model (in which he had initially suggested that C and E are closer together than D and C). Drift or differential selection in different populations may have rendered the data inconsistent with the pseudoallele

Table 8.3. Correlation coefficients (r) and slopes of the regression line (b) between D/D_{max} and D_{max} for the various pseudoallele combinations at the Rh locus. The correlation of D/D_{max} with sample size is also given. Data from Mourant et al. (1976).

	No. of sets of observations	D/D_{max} with D_{max}		D/D_{max} with sample size
		r	b	r
D and E	452	−0.458**	−4.15	+ 0.023
C and E	643	−0.141**	−0.36	+ 0.045
D and C	455	+0.068	+0.23	+ 0.041

**Significant at the 0.01 level.

model. In Table 8.4 I give the results of a series of calculations similar to those that Fisher (1947) carried out on the European data available at the time. These calculations have been extended to the data currently available in different parts of the world.

Fisher's method was to note that the three most frequent gametic types in the European population are CDe, cDE and cde. Assuming that these were the ancestral types, recombinants between them would yield the three types CDE, Cde, and cdE, and a fourth type cDe which, since it is produced by each recombinational event, should be equal in numbers to the other three. The type CdE can only be produced by two recombination events starting from the three ancestral types, and is indeed the rarest type in all human populations.

Fisher inferred the order DCE by noting that the greatest frequency of recombinants, relative to the frequencies of heterozygotes for the 'ancestral' types in the population, was found for recombinants between D and E, so that these appeared to be the outside markers. At that time, there appeared to be a higher relative frequency of recombinants between D and C than between C and E in the northern European population, though it will be seen from the table that more data has changed these relationships somewhat and it now appears from this group of data that the three pseudoalleles are equally spaced. It will be seen that Fisher's model is still perfectly consistent in that the sum of CDE, Cde, and cdE is still equal to cDe as it was when he first performed the calculations.

I have given in the table the frequencies of the heterozygotes for the three ancestral gametic types, the frequencies of the presumptive recombinants between them, and the ratio of these two values. The larger this ratio the greater the apparent recombination between the two ancestral types and the greater the distance between the pseudoalleles. In many cases the map so constructed is linear, but in others the two shorter distances sum to more or less than the distance between the two presumptive outside markers.

The internal consistency of the data with Fisher's model is also checked by comparing the sum of CDE, Cde, and cdE with the frequency of cDe. The order and spacing of the genes inferred from the data, and the degree of consistency with Fisher's model, are given.

It will be seen that the worldwide data are not particularly supportive of Fisher's model, except that when the sum of CDE, Cde, and cdE is equal to cDe the inferred gene order is DCE. If Fisher's model is correct, selection against the different recombinants of the ancestral gametic types must have varied in different populations. It is therefore interesting that the analysis of the pattern of disequilibrium given above, which is independent of Fisher's test, is supportive of his model. Mapping of the locus by cloning and sequencing, which will certainly be possible soon, will be the ultimate test of his hypothesis.

It will be very informative to develop sampling theory encompassing more than two alleles at a locus, in order to examine the patterns of disequilibrium at the HLA complex. This set of linked loci is known to be subject to selection,

Table 8.4. Fit of Fisher's (1947) model to different geographical groupings of the human population. It is assumed, following Fisher, that CDe, cDE, and cde are the ancestral gametic types, and the others have arisen by recombination. Data from Mourant et al. (1976)

		Frequency of heterozygotes for ancestral sequences			Ratio of recombinants to ancestral sequences = crossover between							
	n	A 2 × CDe × cDE	B 2 × CDe × cde	C 2 × cDE × cde	C & E CDE/A	C & D Cde/B	D & E cdE/C	CDE+ Cde+ cdE	cDe	Gene order	Consistency	Linearity
N. Europeans	65 792	0.117	0.334	0.108	0.035	0.038	0.083	0.026	0.024	D–C–E	Good	Good
S. Europeans	29 550	0.112	0.332	0.078	0.109	0.061	0.228	0.05	0.041	DC–E	Good	Fair
E. Europeans	43 942	0.118	0.335	0.105	0.013	0.052	0.035	0.024	0.021	D–EC	Good	Good
Middle East	7 294	0.128	0.276	0.072	0.044	0.07	0.055	0.029	0.07	D–E–C	Poor	Poor
N. India	8 316	0.13	0.204	0.031	0.099	0.077	0.081	0.032	0.06	Ambiguous	Poor	Poor
S. India	9 868	0.108	0.248	0.032	0.131	0.078	0.181	0.039	0.043	DC–E	Good	Fair
Far East	14 500	0.287	0.041	0.012	0.024	0.382	1.330	0.039	0.038	D–CE	Good	Poor
N. Africa	12 645	0.100	0.267	0.089	0.033	0.037	0.097	0.022	0.117	D–C–E	Poor	Fair
Cen. Africa	5 537	0.015	0.061	0.03	0.095	0.209	0.147	0.018	0.558	D–EC	Poor	Fair
S. Africa	5 151	0.019	0.052	0.041	0.188	0.311	0.17	0.027	0.541	D–E–C	Poor	Fair
N. American Indians	5 537	0.315	0.106	0.069	0.156	0.405	0.193	0.105	0.022	D–E–C	Poor	Good
Cen. Amer. Indians	2 020	0.329	0.065	0.043	0.168	0.143	0	0.064	0.042	DE–C	Fair	Fair
S. Amer. Indians	5 116	0.358	0.057	0.047	0.162	0.36	0.055	0.082	0.042	DE–C	Poor	Fair
	215,271											

and the question of whether their disequilibrium shows a pattern attributable to sampling or to selection is a highly interesting one. Certainly, the strength and sign of disequilibrium at the HLA loci varies from one population to another.

If patterns similar to those found in the MNSs and Rh systems should become apparent at other pairs of loci and in other loci of outbreeding populations, this would strongly reinforce the basic premises on which the calculations of this chapter are based—that the strength of disequilibrium is determined primarily by sampling and the gene can be considered the unit of selection.

Linkage disequilibrium and associative overdominance

We can now turn to the question posed at the beginning of the chapter. How do we know that when loci are observed to change in frequency, the effect may be due not to the loci being observed but rather to associative overdominance or a 'hitchhiking effect' caused by a small number of strongly selected loci invisible to the experimenter? Such effects would not be apparent if the loci in the population were all in linkage equilibrium, but such a situation seems extremely unlikely.

In outbred populations, the selection acting on individual selected functional polymorphisms will probably be very small. It is remarkable, for example, that in a sample of 1068 families from northeastern Brazil, Morton (1964) and coworkers (Morton et al. 1966) could find little evidence of selection acting on 14 polymorphic loci. In this population the mean family size is large and inbreeding is relatively low, with very high infant mortality. No significant differences were found between children and adults either in frequency or in proportion of heterozygotes at any of the loci. The only significant association with disease was the well-known one of nodular goiter with PTC nontasting. The only effect of incompatibility was a slight but significant deficiency of A_1 children from incompatible matings. Among these people diet is poor and public health measures are inadequate, so that one would certainly think that selection on these polymorphisms would be stronger than in most studies.

This situation can be contrasted with a number of reports in the literature of apparent heterozygote advantage with increasing age of the cohort. In shellfish, the general pattern in natural populations appears to be one of homozygote excess greater than expected under Hardy–Weinberg equilibrium. This is presumably due to inbreeding, which is rather surprising in view of the reproductive pattern of these organisms. But the oldest mussels (Koehn et al. 1973; Koehn et al. 1976; Tracey et al. 1975) or the largest oysters of a given cohort (Singh and Zouros 1978) show a smaller deviation from Hardy–Weinberg than do the youngest or the smallest. In the case of the oysters, this was apparent in five out of the six loci examined. The more heterozygous organisms survived better or grew faster.

In the case of partially inbred organisms this can almost certainly be attributed to associative overdominance. This is in turn traceable to the disequilibrium generated by inbreeding, as Singh and Zouros point out. The difference between the pattern observed in these organisms and that observed in human populations is striking, and almost certainly attributable to differences in population structure.

It is therefore difficult to rule out associative overdominance or hitchhiking in experiments of this type, and of the type discussed earlier at the end of Chapter 7. However, *if* the population measured, either in the field or in the laboratory, is large and outbreeding, so that the polymorphisms all conform to Hardy–Weinberg expectation, and *if* there are no mechanical restrictions on recombination such as inversions present in the population, then disequilibrium is unlikely to extend over long genetic distances. Perhaps the experiment which most closely corresponds to these conditions is that of McDonald and Ayala (1974, discussed on p. 214). If our calculations concerning the probable extent of disequilibrium in such populations are correct, then it seems likely that in order to explain the effect seen by these workers as associative overdominance, one would have to postulate a large number of strongly-selected genes scattered through the entire genetic map.

Such systems will repay careful theoretical study. Some rough calculations, however, are illuminating. On the assumption of steady-state disequilibrium produced entirely by sampling, I have worked out the degree of associative overdominance exhibited by a neutral locus when a nearby locus is subject to selection for the heterozygote. If p_1 and q_1 are the allele frequencies at the selected locus, and p_2 and q_2 the frequencies for the neutral locus, then H_2, the heterozygosity at the neutral locus after selection, is given by:

$$H_2 = \frac{\begin{array}{c}2(p_1p_2 + D)(p_1q_2 - D)(1 - s_1) + 2(q_1p_2 - D)(q_1q_2 + D)\\(1 - s_2) + 2(p_1q_2 - D)(q_1p_2 - D) + 2(p_1q_2 + D)(q_1q_2 + D)\end{array}}{1 - s_1p_1^2 - s_2q_1^2} \tag{8.26}$$

The disequilibrium at steady state can be calculated for the appropriate allele frequencies, effective population size and recombination frequencies from equation 8.24 and 8.25. This is only an approximation, since neutral alleles may not exhibit this steady-state disequilibrium value, but as we have seen the buildup of disequilibrium may be rapid enough that allele frequencies would not have had time to change greatly during the buildup.

Fig. 8.9 shows values plotted for steady-state disequilibrium given that $N_e = 100$ or 1000, $p_1 = p_2 = 0.05$ and s_1 and s_2 at the selected locus are both equal to 0.1. This produces an increase in heterozygosity at the first locus of 0.026 after selection. In this case, the curves for coupling and repulsion disequilibrium are the same. It will be seen that there is a pronounced effect of population size, since the average D values for the $N_e = 100$ case are much greater than those for the $N_e = 1000$ case. For the latter case, the associative

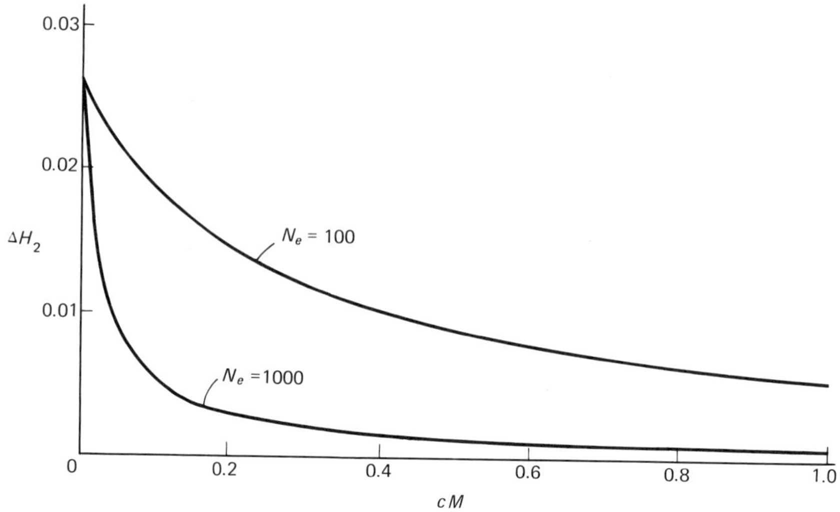

Fig. 8.9. Associative overdominance at a neutral locus as a function of distance from a selected overdominant locus. In each case $p_1 = p_2 = 0.5$ and s_1 and s_2 at the selected locus $= 0.1$.

overdominance is only one tenth of the overdominance at the selected locus when the loci are separated by 0.2 cM. This is equivalent to selection coefficients at the neutral locus of 0.01, too small to have much of an effect on allele frequencies during the few generations of an experiment. For the smaller population size, the associative overdominance is one-fifth of the selected overdominance, still a substantial amount, if the loci are separated by 1 cM.

Most loci, however, are not at 0.5. Figs. 8.10 and 8.11 show what happens when both loci are at the same frequency, 0.2 or 0.1. The associative overdominance is on the same scale as the preceding figure. The strength of selection against the rarer allele has been increased in these and subsequent calculations so that the magnitudes of associative overdominance in all the figures can be compared directly.

Fig. 8.10 shows the case for $N_e = 100$. It will be seen that coupling disequilibrium produces pronounced associative overdominance over long genetic distances, and that the rate at which it drops with increasing map length is a function of the frequencies at the two loci. The rate of decrease of associative overdominance is slowest for $p_1 = p_2 = 0.1$ in the figure, because for this set of figures G_{ec} is small and disequilibrium between the loci also decreases slowly with increasing map distance.

For repulsion disequilibrium, the case is quite different. At these frequencies, most of the heterozygotes at the first locus are matched with the commoner homozygote at the second, so that if the loci are in repulsion disequilibrium an increase in heterozygosity at the first locus results in only a small increase in heterozygosity at the second. Thus, the average associative overdominance between a selected and a neutral locus at these frequencies is likely to be small.

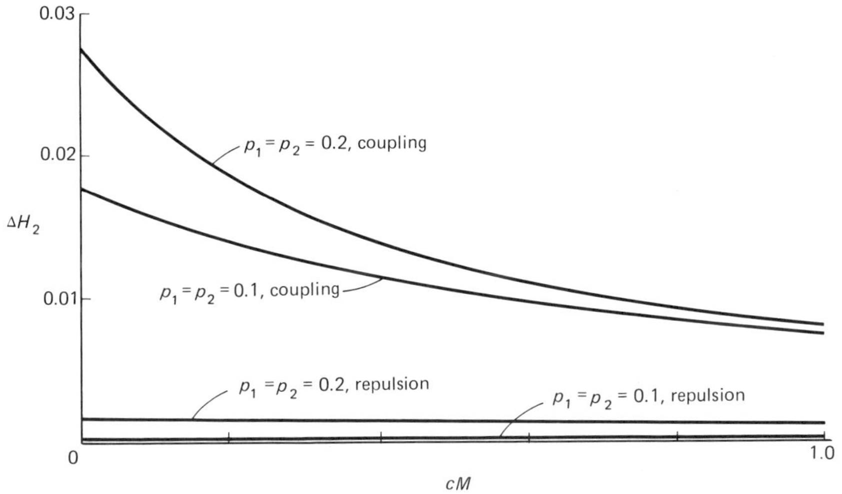

Fig. 8.10. Associative overdominance at a neutral locus when both loci are far from 0.5 and the population size is small. Note the great difference between coupling and repulsion disequilibrium. $N_e = 100$.

Fig. 8.11 shows the results for the same set of frequencies when $N_e = 1000$. It will be seen that if the loci are separated by 1 cM, there is very little associative overdominance exhibited by the neutral locus, whether it is in coupling or in repulsion.

Fig. 8.12 shows the effect on associative overdominance when the loci are at different frequencies. For the frequencies picked, 0.5 and 0.1, coupling and

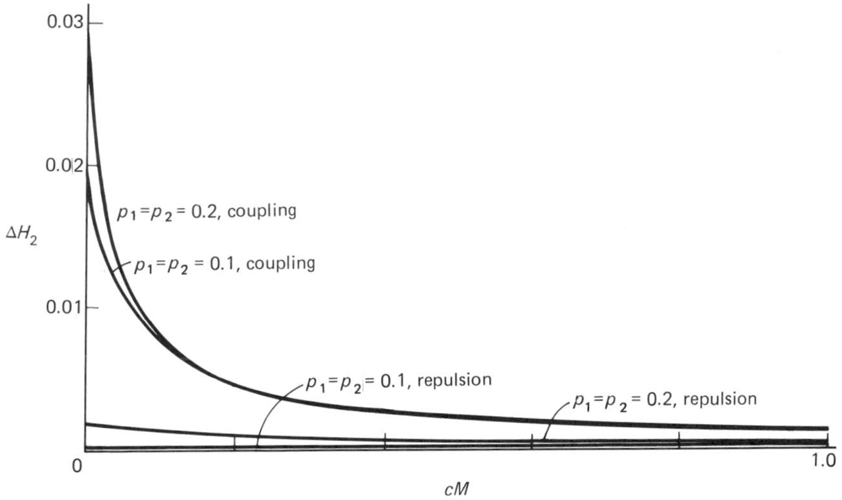

Fig. 8.11. Associative overdominance when both loci are far from 0.5, in a large population, $N_e = 1000$.

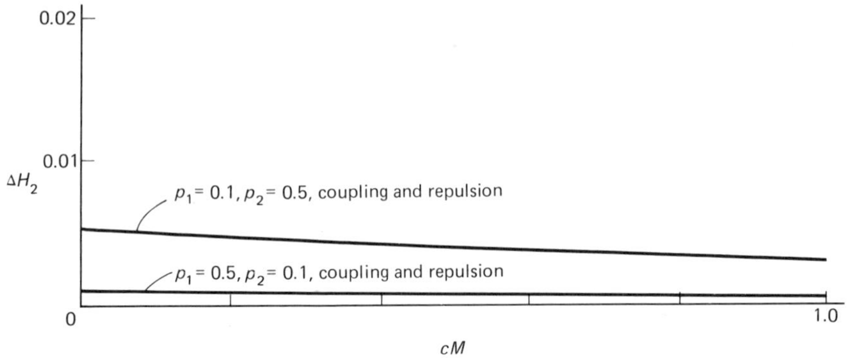

Fig. 8.12. Associative overdominance when the two loci are at different frequencies in a small population, $N_e = 100$.

repulsion disequilibrium are the same, and all show a very low value even for $N_e = 100$. Such situations, in which the linked loci are at different frequencies, must be very common.

If the strength of disequilibrium is due to sampling and recombination, then associative overdominance over distances greater than 1 cM will primarily be due to loci with rare alleles in coupling disequilibrium. The amount of associative overdominance is intensely dependent on the population size, so experiments designed to measure selection on the entire genome must be arranged to ensure that at no time does the effective population size become small.

Assuming that the results of McDonald and Ayala (1974) were obtained from populations that had not passed through pronounced size bottlenecks, then to attribute their results to associative overdominance it is necessary to postulate a large number of strongly heterotically selected loci in coupling with the marker loci. The density should be at least one per map unit. At least as many would be at the same frequency and in repulsion, thereby having little effect on observed polymorphisms far from 0.5. Several times as many would be at frequencies other than the frequencies of the observed polymorphisms, say conservatively four times as many. Then, if the total map length of Drosophila is about 250 map units, we must postulate 1500 strongly selected polymorphisms in order to give the observed effect. The chance that some of the observed polymorphisms are themselves subject to selection can therefore be seen to be quite high. Indeed, if the population size has been kept at 1000 or more throughout the experiment, so that a selected locus in coupling must be found at less than a map unit from each observed polymorphism, the total number of selected polymorphisms becomes so high that we must rely on a model such as that presented in Chapter 6 to explain them.

It is impossible to determine how much balancing selection was necessary to maintain the levels of heterozygosity McDonald and Ayala observed in their complex environments without a knowledge of the effective population sizes involved, however, so a more quantitative statement cannot be made at the present time. The proper design of future experiments of this type must include careful estimates of the effective population sizes throughout the experiments.

In short, if disequilibrium is distributed in populations as a result of the sampling process, there can be little associative overdominance in large populations unless the number of selected loci is also large.

The Powell (1971) and McDonald and Ayala (1974) tests for selection were much more sensitive than that applied by Morton and his coworkers, since the first two experiments followed the levels of heterozygosity over a number of generations, while Morton's group was looking for deviations from Hardy–Weinberg equilibrium and trends in allele frequencies over a single generation. It may be that extensive experimentation of the type first carried out by Powell will give us the most precise information about the nature of the mechanisms maintaining polymorphisms in general. Associative overdominance may not be as great a problem in the interpretation of such experiments as had been imagined initially.

We should thus be able, with careful experiments, to rule out the possibility (p. 218), that associative overdominance or hitchhiking are causing the effects seen in multilocus population experiments. Possibilities (a) and (b) (p. 218) will be much more difficult to distinguish. Suppose a series of Powell-type experiments is carried out with the appropriate safeguards, and it is found that certain polymorphic loci are preferentially retained in certain environments. Is this due to balancing selection of the marginal overdominance or conditional heterosis type, or to purifying selection complicated by epistasis and dominance? Either explanation or a mixture of the two would serve to explain the phenomenon, and it may be that further investigation will reveal that both play a role. Regardless of the actual forces acting on each locus, if many loci can be shown to be retained when the gene is the unit of selection, some form of truncation selection will have to be invoked.

It should be emphasized that it will be necessary to employ experiments of this type in order to detect the type and strength of selection acting on functional polymorphisms with unknown physiological effect. It is not possible to generalize to these polymorphisms from experiments designed to measure selection on lethal or severe-subvital genes.

9. The evolution of recombination

The contrasts between selection in which the fitness is determined directly by the genotype and selection in which it is determined only indirectly by the genotype are explored. It is suggested that for most of the cases investigated the former gives a selective advantage to reduced rates of recombination between loci, while the latter gives a selective advantage to increased rates of recombination

In this chapter, I will turn to a problem that has bedevilled population geneticists for fifty years. The problem, first stated in clear form in Fisher's (1930) landmark book *The genetical theory of natural selection,* is fundamental. Briefly, there seems to be no satisfactory reason why genetic recombination should be preserved in the face of selection for increased fitness. This difficulty has been set out in some detail in a recent book by Maynard Smith (1978), and it has particularly ironic overtones. It is perfectly obvious that high rates of recombination exist in most organisms, unless the process is hindered by mechanical restrictions such as inversions. Yet, there is no theoretical reason why such high rates should have an advantage. To be valid, such a reason must be applicable over a broad range of conditions and must result in high enough selection pressures to counteract the strong tendency for selection to reduce recombination rates.

I will deal first with what appears to me to be the essence of the problem, and then examine some of the solutions that have been proposed to it. Some of these solutions lend themselves to quantification using the concepts developed in earlier chapters, particularly Chapter 8. As a result, it is possible to make some estimate of the likelihood that these proposed solutions actually play an appreciable role. It will be seen that, as Maynard Smith also concluded, none of the proposed solutions are particularly convincing.

I will then turn to another solution that has hardly been dealt with by theoreticians attacking the problem, a solution rooted in the peculiar and relatively unexplored properties of balancing truncation-selection. This may be the most convincing solution from the standpoint both of the strength of selection for recombination that it can generate and the fact that this process can operate under a wide variety of conditions—including those for which alternative satisfactory answers have not yet been found.

The essence of Fisher's problem

Fisher initially pointed out that if certain gametic types confer higher fitness than others, they will of course be selected for. If recombination between the genes on the gametes is low, then these gametic types will increase in frequency in the population. Any reduction in the rate of recombination will be selected for, since it increases the rate at which the fitter types accumulate. Eventually, the genome should 'congeal', to use Turner's (1967) phrase.

While Fisher used a frequency-dependent two-locus model to illustrate his case, all other types of selection can lead to the same result.

Balancing selection

We have seen some examples of this in Chapter 8. Either truncation or multiplicative balancing-selection will lead to the crystallization of the genome if per-locus selection is strong enough and recombination is low enough. While we were able to show that this should not occur in large outbred populations given the rates of recombination which we know or can infer to be present in most organisms, the crystallization of the genome could certainly be brought about if a modifier gene appeared that reduced the rate of recombination and thus increased the length of runs of homozygous and heterozygous loci. This would in turn increase the variance of heterozygosity among the organisms carrying the modifier gene, increasing the range of fitnesses in this subgroup. Since, by definition the organisms with higher fitnesses contribute disproportionately to the next generation, the modifier should increase in frequency.

Purifying selection

Suppose we consider a population that has a high frequency of deleterious or conditionally deleterious alleles at many loci. Then purifying selection can also be more effective if these deleterious alleles can be organized by disequilibrium into groups. In this case it is the variance of homozygosity for deleterious genes that is increased by disequilibrium, thus increasing the effectiveness of both multiplicative and truncation selection.

Substitution selection

Selection in which a less favorable allele is being substituted by a more favorable one will also be enhanced in effectiveness if the more favorable alleles at different loci can be organized into blocks. The rate of substitution will be increased, again because the variance of fitnesses in the population will be increased.

All these situations have one obvious factor in common. It is that a reduction in the rate of recombination increases the variance of the selected character and therefore the variance of fitnesses in the populations. This will always be advantageous and therefore the genome should congeal. Are there any circumstances under which this will not occur? There are some, but as we will see it is not always obvious just how effective they are in counteracting Fisher's congealing process.

Mechanisms that reduce the variance of the selected character

The Muller ratchet and the Hill–Robertson effect

In his initial discussion, Fisher saw (though he did not quantity) a solution to the problem. This solution and several others have been elaborated by other workers, and like Fisher's original problem they also have a factor in common. This is that there are circumstances in which disequilibrium can reduce the variance of a selected character rather than increase it. This will in turn reduce the effectiveness of selection.

Fisher pointed out that if there were a constantly arising stream of new mutations, in a population in which recombination was absent or very rare, then the rate at which these favorable genes would be fixed would be slowed. This is because double mutants would be unlikely to arise and be fixed simultaneously. This phenomenon was explored at greater length by Muller (1964), who likened the process of fixation of favorable mutations in the absence of recombination to a ratchet. If two favorable mutations arose simultaneously in a population they would most likely arise in different organisms. If both survived the first few generations, only one of the two would be eventually fixed, and it would probably be the one conferring the higher fitness. Only if a second mutation arose in the same genome as the first could both become fixed, since favorable mutations arising in other gametes would be forced out of the population by the fittest gamete.

Fisher intuitively felt that this process was unlikely to have much selective importance unless the rate at which new mutations arose was high. We will see that this is indeed the case, and that the ratchet mechanism and modifications of it are likely to be swamped by much more powerful selective pressures.

It was pointed out in the last chapter that disequilibrium generated by sampling will tend to increase the variance of heterozygosity. The assumption underlying the calculations that were made was that as one proceeded along the chromosome and examined the disequilibrium exhibited by adjacent loci once steady state had been reached, about half the locus pairs would exhibit repulsion disequilibrium. It was reasonable to make this assumption for polymorphisms that have been held in the population for some time by balancing selection, since the disequilibria exhibited would be much more likely to be a result of the sampling history of the population than the early history of the polymorphisms.

The Muller ratchet, however, presupposes a very different type of distribution of disequilibria. For the ratchet to work, the favorable mutations must almost always arise in different gametes, which in turn means that they will exhibit repulsion disequilibrium. If the initial frequencies are low and the population size is very large, the amount of repulsion disequilibrium will be so small as to be undetectable. But if the population size is small enough to permit the generation of appreciable disequilibrium by sampling, then the mode of origin of the mutant alleles becomes very important. If most of the mutants originate in repulsion disequilibrium, then most of the disequilibrium generated by sampling should also be repulsion disequilibrium.

This phenomenon was addressed specifically by Hill and Robertson (1966), who were interested in the rate at which directional selection could proceed and how this would be influenced by linkage disequilibrium. In a series of simulations, they subjected two loci to strong directional selection in a finite population. The initial frequency of the first was always the same, and that of the second was permitted to vary. Fig. 9.1 shows the probability of a fixation of the first locus as a function of the initial frequency and strength of selection at the second. No recombination was permitted between the loci.

It can be seen that the effect of a second selected locus is always to retard the fixation of the first, and that the retardation is greatest when the second locus is initially rare and strongly selected. Hill and Robertson measured the disequilibrium in their populations and found it to be always repulsion disequilibrium—that is, the mutant alleles at the two loci tended to be on different chromosomes.

Felsenstein (1974) examined this 'Hill–Robertson effect' in some detail, and suggested that in situations such as this the Muller ratchet has been reinforced by disequilibrium generated by sampling. As was realized but not fully quantified by these authors, the type of disequilibrium generated was all-important.

What appears to be happening in such populations is this. The Hill and Robertson simulated populations were initially in approximate linkage equilibrium. But if the mutant allele at both loci was rare the majority of gametes carrying mutants in the populations would carry one mutant or the other and not both. This can be seen to be equivalent to the Muller ratchet, in which newly-arising mutations are almost certainly going to occur in different genomes. As a result of this preponderance of such gametes in the population, any disequilibrium generated by sampling will tend to be of the repulsion type, as defined in the Langley and Crow (1974) sense.

In a finite population, any disequilibrium generated by the Muller ratchet will be of the repulsion type. Disequilibrium generated by sampling will then reinforce this. This reinforcement of the Muller ratchet is then sufficient to

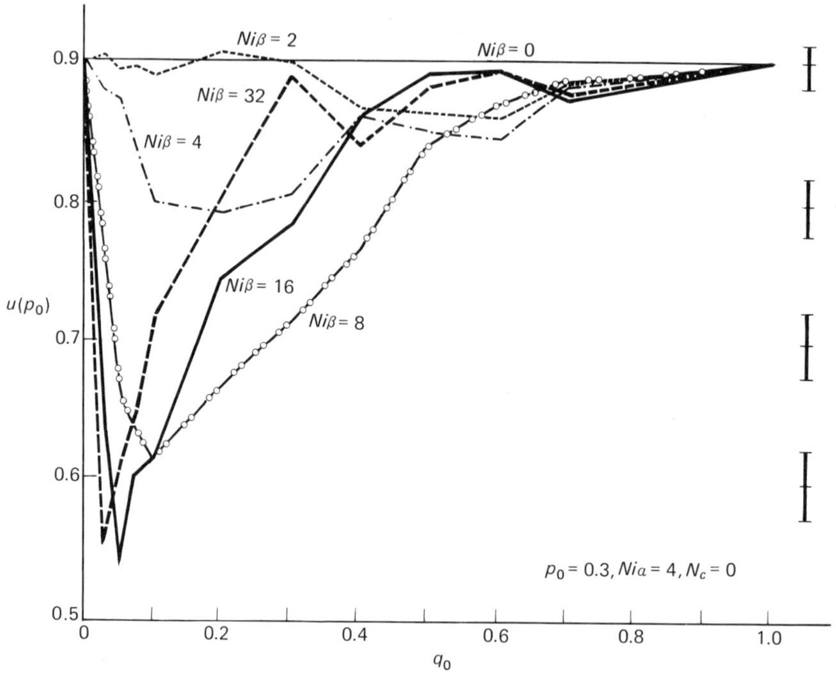

Fig. 9.1. The probability of fixation at one locus $|u(p_0)|$ as a function of the initial frequency and strength of selection at a second completely linked locus. The $i\beta$ value measures the initial frequency and relative intensity of selection at the second locus. The initial frequency of the first locus is always 0.3. From Fig. 1 of Hill and Robertson (1966).

retard the progress of linked loci towards fixation, since the variance of the selected character will be small. Repulsion disequilibrium will reduce the variance of the selected character to *less than* the binomial variance, since the selected genes will tend to be more evenly distributed among the gametes of the population than they would be by chance.

Quantification of the Hill–Robertson effect

Felsenstein and Yokoyama (1976) examined the effects of the reinforced Muller ratchet through the use of simulated populations which were allowed to accumulate either favorable or harmful mutations. Since new mutations were introduced primarily in repulsion, most of the new disequilibrium generated by sampling was presumably of the repulsion type, though this was not measured.

The most striking aspect of their simulations was the enormous mutation rate and strong selection that had to be introduced in order to give the advantage to genes increasing the rate of recombination. Table 9.1 gives their data

Table 9.1 Probability of fixation of a recombination locus, in populations with 20 segregating background loci at which favorable alleles occur by mutation. The selective advantage of each favorable mutation is given by s, and the mutation rate per locus per generation by μ. From Table 1 of Felsenstein and Yokoyama (1976).

s	0.25	0.5	1	3
(a) Recombination locus recessive				
μ		$N = 100$		
0.1	0.48	0.54	0.47	0.44
0.3	0.57	0.52	0.53	0.51
0.6	0.53	0.51	0.59	0.59
1	—	—	—	0.56
μ		$N = 200$		
0.1	0.56	0.53	0.54	0.43
0.3	0.59	0.55	0.49	0.58
0.6	—	—	0.61	0.58
1	—	—	0.63	0.61
(b) Recombination locus dominant				
μ		$N = 100$		
0.1	0.53	0.45	0.61	0.56
0.3	0.52	0.54	0.64	0.76
0.6	0.59	0.63	0.78	0.86
1	—	—	0.86	0.92
μ		$N = 200$		
0.1	0.47	0.50	0.55	0.67
0.3	0.52	0.58	0.79	0.89
0.6	—	0.77	0.96	0.94
1	—	—	—	0.98

for two population sizes, for either a recessive or a dominant recombination locus, for various degrees of advantage of mutant alleles and for various mutation rates in 20 'background' loci. The numbers in the body of the table give the fraction of the replicate populations that was fixed for the recombination allele.

In order for recombination to be favored, extreme conditions had to be chosen. Genes were completely linked unless the recombination gene was present, in which case they assorted at random. Mutation rates ranged from 0.1 to 1 per gamete per generation, which certainly provided the 'abundant stream' postulated by Fisher. The selective advantage of the mutant alleles ranged from 25% to 300%.

In spite of all this, if the recombination gene was recessive, only about 60% of the replicates were fixed for it under the most extreme conditions—one mutation per gamete per generation, and a 300% advantage for all mutant alleles. This was in spite of the fact that the recombination allele started at a frequency of 50% in all the replicates. If the recombination allele were dominant (part (b) of the table), selection was somewhat more effective, but extreme conditions were still necessary to produce a selective advantage for recombination.

When similar simulations were run with a very high mutation rate to deleterious alleles (selection coefficients as high as 0.32 and mutation rates as high as 1.28 per gamete per generation), alleles for recombination again tended to win out, but the effect was much less than that for favorable alleles and much less consistent.

Fisher appears to have been right, which is not surprising. The Muller ratchet, even when enhanced by sampling-generated repulsion disequilibrium, appears to have only a weak effect on genes influencing recombination. Further, purifying selection acting on harmful genes is less effective than directional selection acting on advantageous genes. The variance of the selected character is much smaller in the former case than in the latter, so that any changes in the variance introduced by disequilibrium will have a smaller effect.

We can quantify the Hill–Robertson effect further, by using the methodology introduced in the last chapter. It will be recalled that the average steady-state disequilibrium between pairs of loci in coupling and in repulsion will tend to be different if the allele frequencies at the loci are not at 0.5. These values, D_{C} and D_{R}, were given by eqns (8.24) and (8.25). As a consequence of the effect, the average lengths of runs of homozygous and heterozygous loci would be different in two populations in which all loci were in coupling disequilibrium and all loci were in repulsion disequilibrium.

Of course, the likelihood of such populations occurring is remote. Even if the Muller ratchet, supplemented by the Hill–Robertson effect, is operating, the majority of the locus pairs may be in repulsion disequilibrium but there will by chance be some in coupling disequilibrium. But let us imagine for a moment two such multilocus populations, one in which all the adjacent locus pairs are in coupling disequilibrium and one in which all the adjacent locus pairs are in repulsion disequilibrium. We can now obtain the average coupling and repulsion gamete frequencies, designated by the subscripts C and R, from eqns (8.24) and (8.25), and determine the probabilities of termination of runs of heterozygous and homozygous loci from modifications of equations 8.6 and 8.7:

$$T_{\mathrm{het_C}} = (2g_{00_C}g_{10_C} + 2g_{11_C}g_{01_C})/2\hat{p}\hat{q} \tag{9.1}$$

$$T_{\mathrm{hom_C}} = (2g_{10_C}g_{11_C} + 2g_{01_C}g_{00_C})/(1 - 2\,\hat{p}\hat{q}) \tag{9.2}$$

$$T_{\mathrm{het_R}} = (2g_{00_R}g_{10_R} + 2g_{11_R}g_{01_R})/2\hat{p}\hat{q} \tag{9.3}$$

and

$$T_{\mathrm{hom_R}} = (2g_{10_R}g_{11_R} + 2g_{01_R}g_{00_R})/(1 - 2\,\hat{p}\hat{q}) \tag{9.4}$$

Since there are equal numbers of homozygous and heterozygous runs in a population, the mean length of a run is simply the arithmetic average of the inverses of the two termination probabilities.

$$m_{\mathrm{C}} = (1/T_{\mathrm{het_C}} + 1/T_{\mathrm{hom_C}})/2 \tag{9.5}$$

and

$$m_{\mathrm{R}} = (1/T_{\mathrm{het_R}} + 1/T_{\mathrm{hom_R}})/2 \tag{9.6}$$

As in Chapter 8, these m values can be compared to the mean lengths of runs in the absence of disequilibrium. This in turn gives an indirect measure of the degree to which the variance of the character in question will be changed.

The effect of having all coupling and all repulsion disequilibrium in our two imaginary populations is quite different. Suppose that there are 100 loci segregating in each population, and both populations have an effective size N_e of 100. Recombination between adjacent loci is 0.01. At each locus, there is an advantageous allele in the process of replacing a less advantageous allele, and in order to make the calculations easier we will assume that each advantageous allele is at a frequency $p = 0.2$. Each advantageous allele is completely recessive.

Table 9.2 shows the effect of the two extreme cases of disequilibrium on the length of runs of homozygous and heterozygous loci. If all adjacent loci are in coupling disequilibrium, the length of a run is almost twice the length if all adjacent loci are in repulsion disequilibrium. Even under these latter extreme conditions, which would apply if the Muller ratchet and the Hill–Robertson effect were exerting their greatest influence on the populations, the reduction in the variance of homozygosity in the population should not be marked.

Of course, the assumptions in the above calculations are unrealistic. In any population there is likely to be a mixture of coupling and repulsion disequilibria between adjacent loci. It is possible to show, assuming the population is at steady state for sampling disequilibrium, that the Muller ratchet and the Hill–Robertson effect can rapidly be swamped by a small proportion of loci in coupling disequilibrium.

Consider such a population, in which the proportion C of the adjacent locus pairs are in coupling disequilibrium and the proportion $1 - C$ are in repulsion disequilibrium. Then, to see the effect on the variance of the selected character, we must determine the lengths of runs of heterozygous and homozygous loci

Table 9.2. The effect of sampling disequilibrium on the length of runs in a population in which $p = 0.2$ at all loci, $L = 100$, $N_e = 100$ and $r = 0.01$.

	m_{C}	m_{R}	m
a) All adjacent locus pairs are in coupling disequilibrium	2.504	5.322	3.913
b) All adjacent locus pairs are in repulsion disequilibrium	1.355	2.879	2.117

in such mixed populations. This can be done by taking the weighted mean of the lengths if all the loci are in coupling and if they are all in repulsion, which will be the weighted geometric mean:

$$m = m_C^C \cdot m_R^{(1-C)} . \tag{9.7}$$

I have shown in Fig. 9.2 the effect of different mixes of disequilibria on the average length of a run in populations like those examined in Table 9.2. Fig. 9.2 shows the range from complete repulsion to complete coupling disequilibrium. It will be seen that only a very small proportion of the adjacent locus pairs need be in coupling—about 15% in this example—to restore the variance of homozygosity to its binomial size and to completely wipe out the Hill–Robertson effect. It is not surprising, therefore, that Felsenstein and Yokoyama had to subject their populations to such high mutation rates and selection pressures to see any advantage for genes permitting recombination. Had their populations been allowed to accumulate appreciable amounts of coupling disequilibrium, the result would undoubtedly have been the loss of the recombination genes rather than their retention. It would be most informative to monitor the strength and sign of disequilibrium and the size of the variance of the selected character in simulations of this sort carried out in the future.

The Muller ratchet, enhanced by the Hill–Robertson effect, has been shown to operate under extreme conditions. I suspect, on the basis of the calculations given above, that it is unlikely to be very important for weakly selected genes that have been in the population for long periods of time and have therefore

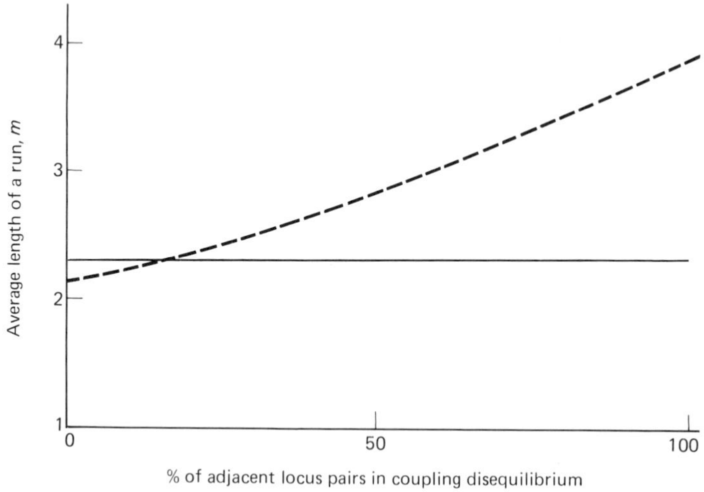

Fig. 9.2. The effect on the average length of a run of an increasing proportion of loci in coupling disequilibrium in the population of Table 9.2. The solid line marks the length of a run in the absence of disequilibrium.

accumulated coupling as well as repulsion disequilibrium. Once again we face Fisher's dilemma. A plentiful stream of strongly selected advantageous mutations may provide an advantage to recombination. But suppose the environment were stable. Under these conditions, favorable mutations are unlikely to be as plentiful or as strongly selected as they would be under secularly changing conditions. We are still forced to seek a *continuously-operating* advantage for recombination.

How truncation selection provides an advantage to recombination

Felsenstein (1965) was the first to examine the effect of truncation on the fixation of pairs of linked loci subject to directional selection. The truncation used was a sharp one, but with the addition of a normally-distributed environmental uncertainty factor. Felsenstein was able to show that this type of selection in a two-locus system would tend to generate repulsion disequilibrium. As a consequence, he found that in this system tight linkage reduced the probability of fixation at the linked loci.

Felsenstein's situation was a rather specialized one. If A_1 and B_1 are the rarer alleles at each locus, and are being selected against, then the majority of the population will consist of A_1B_1/A_2B_2, which will tend to have low fitness, A_1B_2/A_2B_2 and A_2B_1/A_2B_2 which will each tend to have intermediate fitness, and A_2B_2/A_2B_2 which will tend to have the highest fitness. The occasional homozygotes for the rare alleles will count for very little. Because of the environmental uncertainty factor, all the genotypes will tend to be misclassified and saved when they should be truncated or vice versa. But A_1B_2/A_2B_2 and A_2B_1/A_2B_2 will on average lie much closer to the truncation line, and will therefore survive proportionately more often by chance, than will A_1B_1/A_2B_2. There will thus be a tendency for repulsion gametes to increase in the population at the expense of coupling gametes.

This intriguing observation led to results quite opposite from those of the other types of selection employed by Felsenstein. When fitness was a direct function of the genotype, coupling disequilibrium was generated and the probability of fixation was increased. This suggests that the mode of selection may be centrally important to determining the fate of genes that regulate recombination.

The rest of this chapter will be devoted to observations and theory suggesting that the results of balancing truncation selection may also be very different from selection in which the genotype determines the fitness directly. It is shown that linkage disequilibrium actually hinders the movement of alleles to their selective equilibrium if the population is subjected to balancing truncation selection, and that this occurs even if the variance of the selected character is greater than the binomial variance. This phenomenon gives an advantage to high levels of recombination in a steady-state population when the Hill–Robertson effect may not be operating.

The results of computer simulations

The first indication we had that balancing truncation selection could lead to this effect came in some studies reported by Wills et al. (1970). All the simulated populations in those studies were subject to balancing truncation and replacement selection of the type we have already dealt with. Most of them involved genes on a linkage map 100 cM long. One run, however, was made with random assortment between loci and it was found, rather to our surprise, that the standard deviation of the loci around their selective equilibrium of 0.5 was smaller in this run (Fig. 9.3). For some reason, not fully understood at the time, balancing truncation selection was more effective at maintaining polymorphisms in the absence of linkage (and therefore in the absence of disequilibrium) than in its presence.

Wills and Miller (1976) wrestled with this problem further. A number of simulations were carried out, shown in Figs. 9.4 and 9.5, in which populations were started either with a uniform initial distribution of allele frequencies (Fig. 9.4) or with all the frequencies near 0.5 (Fig. 9.5). Selection was once again balancing truncation-and-replacement, and selection was symmetrical so that the equilibrium frequency at all loci was 0.5. The fate of each population, which consisted of 100 organisms and 100 loci, was followed for 200 generations. In the figures, X refers to populations in which the genes were linked on to a map 100 cM long, and R refers to populations in which the genes were free to assort at random.

It will be seen that in each case the number of loci fixed over time was smaller, and the standard deviation of the frequencies of the loci was smaller, in the populations with random assortment.

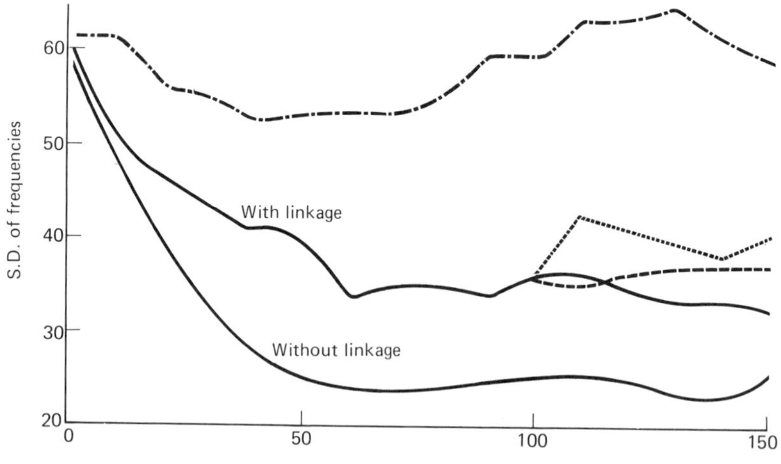

Fig. 9.3. The simulated populations were subject to balancing rank-order truncation selection with $\hat{p}$ at each locus equal to 0.5. The graph shows the standard deviation of the unfixed loci around this equilibrium point for runs with linkage and with random assortment. The top line shows a run without selection. Data from Wills *et al.* (1970).

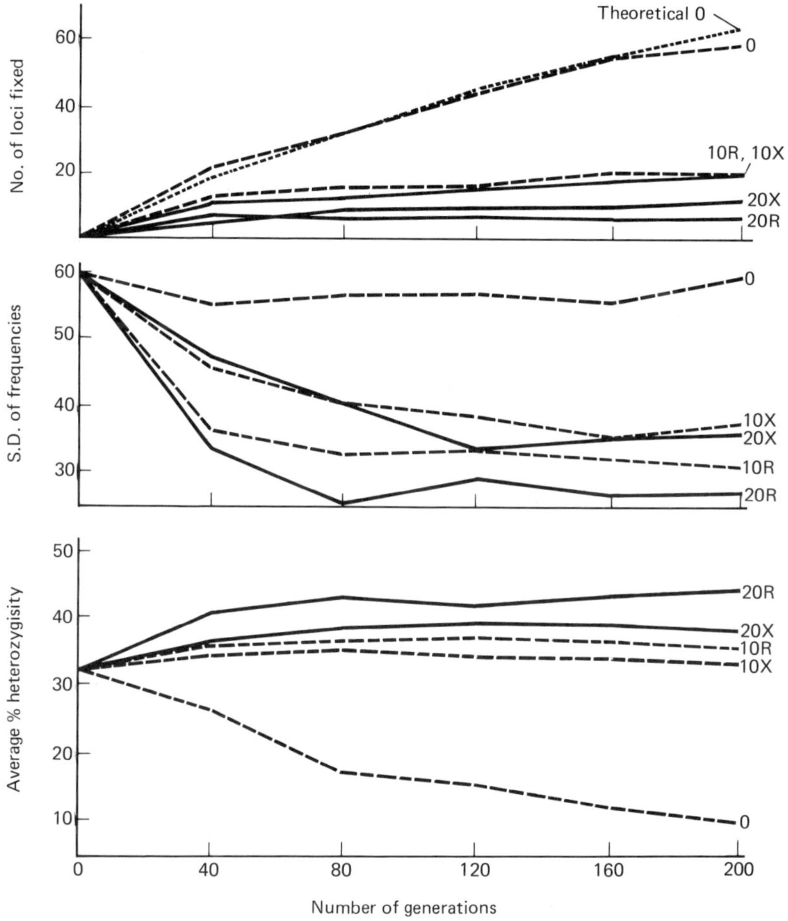

Fig. 9.4. Populations with N = 100 and 100 loci per chromosome, started with a uniform distribution of allele frequencies. Selection toward an equilibrium frequency of 0.5. The numerals refer to the intensity of selection in per cent; R = random assortment; X = linkage. The first graph shows the rate of loss of polymorphism with time. The second shows the standard deviation of the unfixed alleles around their selective-equilibrium frequency. The third shows the average heterozygosity per organism in each of the populations. The zero selection line is an average of the two runs made with and without linkage. Data from Wills and Miller (1976).

At the time, it was suggested that this was due to a modification of the Muller ratchet. While loci in coupling disequilibrium on the same side of their selective equilibrium will be moved rapidly towards that selective equilibrium, if they are in repulsion disequilibrium and on opposite sides of their selective equilibrium, then as one is moved towards its equilibrium the other is perforce moved away from it. In fact, however, the effect is seen with loci in coupling

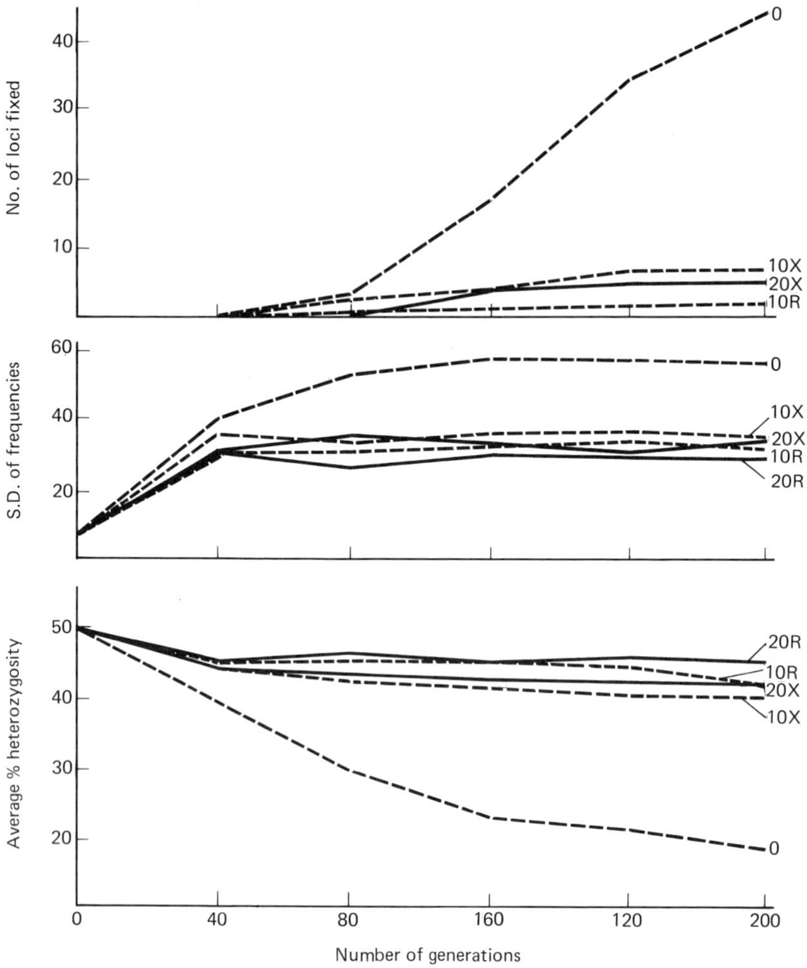

Fig. 9.5. Same conditions as described in legend to Fig. 9.4, but with each locus in the starting population near its equilibrium frequency of 0.5. Note that no loci were fixed in the run made with 20% selection and random assortment. Data from Wills and Miller (1976).

as well as in repulsion disequilibrium, so it is not confined to cases in which the variance of heterozygosity is reduced below the binomial level. What was not fully realized at the time was that this effect is confined to balancing truncation selection. Had simulations of the balancing multiplicative type been carried out, this effect would not have been seen. In fact, the opposite would have happened. Multiplicative selection is *more* efficient when the variance of heterozygosity is increased by disequilibrium, not less.

The effect of truncation selection on R

It is possible to quantify what happens to truncation selection in the presence of disequilibrium by using the approach taken in Chapter 6. It will be recalled (eq 6.11) that balancing truncation selection is simply a replacement of the rarer and more common homozygotes by heterozygotes. The proportions of these two types of homozygote are p_H^2 and q_H^2 respectively, where p_H and q_H are the frequencies of the rarer and commoner alleles averaged over all polymorphic loci. If $\hat{p} = 0.5$, then the ratio of these two values, R, is the ratio of the two types of homozygotes removed.

This replacement process is *unique* to balancing truncation selection. Further, it is not dependent on the precise mode of truncation. The change in heterozygosity due to a particular type of truncation and replacement selection is given by eqn (6.10), but any other mode of balancing truncation selection could have been substituted at that point without any alteration in the subsequent theory. All that is required is that at least some of the most homozygous organisms be removed from the population and be replaced by more heterozygous organisms. It is important to note that only polymorphic loci count towards fitness potential in truncation selection; at a monomorphic locus there is no way in which homozygotes can be replaced by heterozygotes.

It will be seen that if p_H^2 is less than q_H^2, there will be a net movement of loci towards their selective equilibrium since more common than rare alleles are being removed. If $p_H^2 = q_H^2$, however, there will be no net movement, and R will be one.

I wish to show that there are certain circumstances in which the ratio R is in fact one, leading to no net change in allele frequencies as a result of balancing truncation selection. Let me turn first to two extreme cases, in which crystallization of chromosomes has been allowed to occur.

Imagine two synchronously-reproducing multilocus populations in which many two-allele balanced polymorphisms are segregating, and in which there is random assortment between loci. In the first population, all the loci are being driven by balancing truncation selection towards an internal equilibrium of 0.5. In the second, they are all being driven by multiplicative selection towards the same internal equilibrium. These populations will be characterized by variances of loci around this selective equilibrium, σ_p^2. These values will be a function of the population size, which governs the increase in this variance due to drift, and the per-locus selection, which governs the decrease in the variance. Suppose further that the populations have reached steady state between selection and drift, and that C is small in each case so that the rate of loss of polymorphisms is negligible.

Now suppose that the rate of recombination is suddenly reduced to zero. What will be the effect on the populations? The most heterozygous zygotes will be at a selective advantage, and the most homozygous ones at a selective disadvantage. Gametes which in combination yield a high proportion of heterozygous loci will be selected for and will begin to increase in frequency in the population. They will not be subject to recombination, so that they will remain unaltered from one generation to the next. Eventually, two of these gametes will take over in each population, so that before a given round of selection half the organisms will be heterozygous for a large number of loci and the other half will be homozygous at all loci.

However, just as in the Franklin and Lewontin crystallizations discussed in the last chapter, some loci will have been forced to fixation. Consider a small segment of these two fittest gametes. When heterozygous, they might look like this:

Locus	1	2	3	4	5	6	7	8	9	10
Gamete 1	1	0	$\underline{0}$	1	$\underline{1}$	0	0	0	0	$\underline{1}$
Gamete 2	0	1	$\underline{0}$	0	$\underline{1}$	1	1	1	0	$\underline{1}$

Locus 3 will, when the population has crystallized, be fixed for the rarer allele, while loci 5 and 10 are fixed for the more common allele. All the rest of the loci will be at 0.5 for both alleles. During the crystallization some loci will be forced to fixation (and σ_{p}^2 will be increased accordingly), but at the end of the crystallization process as many rare as common alleles will be removed by selection each generation. No subsequent movement of alleles will occur.

The selection now acting on the two populations, however, is somewhat different. In both, a greater number of common alleles than of rare alleles will have been fixed. In the population subject to balancing truncation selection, those fixed loci will not contribute to fitness potential. All the unfixed loci will be at 0.5 for rare and common alleles, so that $p_H^2 = q_H^2$ and $R = 1$. In the population subject to multiplicative selection, the fixed loci still contribute to the determination of fitness. Homozygotes for the common alleles contribute more to the decrease in fitness than do homozygotes for the rare alleles. But because there are only two types of gamete in the population, the unequal contributions in fitness due to homozygosity for commoner and rarer alleles are powerless to alter allele frequencies at the unfixed loci.

For a multilocus population with a long genetic map, the disequilibria seen between adjacent loci will be the result of the sampling process. Such a population can be thought of as partway between the random assortment case and the completely crystallized case given above. The total proportion of crystallized loci in the population will be D/D_{max}, where D is the average of all the coupling and repulsion disequilibria between adjacent loci.

If we examine a small segment of the chromosome in such a population, a few loci long, we will find a situation similar to, but less extreme than, the completely crystallized case. Within that segment, two approximately complementing chromosomes will each be represented a number of times as a result of sampling, while the remainder of the chromosomes in the population will exhibit a variety of allelic combinations. Within the subgroup of complementing chromosome types, just as within the entire chromosomes of the previous example, loci will be either at 0.5 or carried to 'pseudofixation'. It should be emphasized that a pseudofixed locus which is represented by only a 0 or a 1 allele within the subgroup may not be fixed in the rest of the population. The process of pseudofixation can be defined as the change in allele frequencies, brought about by partial crystallization of the genome, at a locus which happens to have the same allele in both the chromosomes involved in the crystallization.

Now the difference between multiplicative and truncation selection becomes important. Let us deal with multiplicative selection first. If the population is being subject to multiplicative selection, all the loci within the subgroup are counted towards fitness. Because the 1 alleles are by definition the commonest at each locus, there will tend to be more 1 than 0 alleles driven to 'pseudofixation' within the subgroups.

Since both the pseudofixed 1 alleles and the pseudofixed 0 alleles contribute to the fitness of an organism carrying them, the variances of fitnesses in the population will be in part a result of the variance of these two classes of allele. Let the numbers of pseudofixed loci possessed by the average organism be μ_0 and μ_1, with $\mu_0 < \mu_1$. Then the variances can be defined as V_0 and V_1.

Both μ_0 and μ_1 are likely to be much less than 0.5 (since otherwise the population will be fixed at most loci), so that V_0 will be less than V_1. Thus, the effectiveness of selection against pseudofixed 1 alleles will be greater than that against pseudofixed 0 alleles, and the net effect will be a movement of the pseudofixed loci towards their selective equilibrium. This movement will be a result of selection acting on the crystallized part of the genome.

The basis of truncation selection is very different. The measure on which selection is based is the number of heterozygous loci possessed by each organism. If the mean number of heterozygous loci in the crystallized subgroups is $\mu_{H'}$, then the variance $V_{H'}$ on which truncation acts will be a function of this number. Since the pseudofixed 0 and 1 loci do not contribute to fitness potential their frequencies will be unchanged from one generation to the next by truncation selection. These alleles are *sequestered* from the effects of truncation selection. Because the frequencies of all the unfixed loci in the crystallized portion of the genome are at 0.5, truncation selection will have no effect on allele frequencies *within the crystallized portion*, though of course it will have an effect on alleles of chromosomes that have not yet crystallized.

The effect of disequilibrium on balancing truncation selection can now be quantified. If $\hat{p} \neq 0.5$, then R is adjusted by multiplying by the ratio $\hat{p}^2/\hat{q}^2$, as was pointed out earlier. In these cases, the effect of disequilibrium is still to alter R towards unity. The degree of alteration is a direct function of the proportion of gametes exhibiting disequilibrium in the population, which is D/D_{max}. Since R is a ratio, then the modified value of R due to disequilibrium,

R_D, is simply the weighted geometric mean of R for the loci not exhibiting disequilibrium and unity:

$$R_D = r^{(1 - D/D_{max})} \cdot (1)^{D/D_{max}} \tag{9.8}$$
$$= r^{(1 - D/D_{max})}.$$

When $\hat{p} \neq 0.5$, as we have seen, there will be different average strengths of disequilibrium between loci in coupling and loci in repulsion. Since we are interested in the total proportion of gametes involved in disequilibrium, we simply take the arithmetic average of the two D/D_{max} values—on the assumption that equal numbers of adjacent-locus pairs are in coupling and in repulsion disequilibrium.

Comparison of theory with computer simulations

This modified value of R can now be used in conjunction with the methods introduced in Chapter 6 to see whether the magnitude of the effect predicted by theory is comparable to that seen in computer simulations. Disequilibrium will have two effects on populations subjected to balancing rank-order truncation selection:

(a) Since equal numbers of locus pairs are in coupling and in repulsion disequilibrium, the variance of heterozygosity will be increased by sampling-generated disequilibrium. This will increase the effectiveness of truncation selection over a population without disequilibrium, since the variance of fitness potential will be increased.

(b) Working against this is the effect pointed out above, that increased disequilibrium will increase the subset of the population in which equal numbers of common and rare alleles will be removed by truncation selection.

The question is: what are the relative magnitudes of these effects? The simulations of Wills and Miller (1976) form an excellent body of data to test whether influence (b) will outweigh influence (a).

In Table 9.3, I have detailed the steps of the calculations necessary to check the theory against the eight simulations of Figs. 9.4 and 9.5. In four of the simulations the initial variance of loci around the selective equilibrium was specified to be small; in the other four it was specified to be large. Thus, the runs can be grouped into pairs in which the steady state value for σ_p is being approached from both sides. The final observed σ_p, averaged over the last 80 generations of the simulations, was used to calculate p_H. The intermediate values were calculated from the appropriate equations in Chapters 6 and 8, with the calculation of R modified by equation (9.8.) The final column of Table 9.3 shows the calculated difference between the increase in $\overline{H}$ due to selection and its decrease due to drift. Since the populations have had 200 generations to come to steady state, the difference between the expected increase and decrease of $\overline{H}$ should be small.

The correspondence to theory is excellent. In Case 1, subject to 20% trun-

Table 9.3. Calculation of the expected increase in heterozygosity due to selection compared with the expected decrease due to drift for simulations carried out by Wills and Miller (1976), Figs. 9.4 and 9.5 (pp. 265–66)

Case	Initial σ_p	Selection (%)	N_s	Linkage	L	Final σ_p (observed)	p_H	$\overline{H}$	r	D	$\Delta\overline{H}_s$
1	Small	20	80	None	100	0.12	0.38	0.4712	—	—	0.02795
2	Small	10	90	None	98	0.13	0.37	0.4662	—	—	0.01794
3	Small	20	80	100	95	0.135	0.365	0.4636	0.01053	±0.1196	0.03263
4	Small	10	90	100	93	0.14	0.36	0.4608	0.01075	±0.1133	0.02064
5	Large	20	80	None	95	0.14	0.36	0.4608	—	—	0.02864
6	Large	10	90	None	80	0.155	0.345	0.4520	—	—	0.01981
7	Large	20	80	100	90	0.175	0.325	0.4388	0.0111	±0.1172	0.03317
8	Large	10	90	100	80	0.19	0.31	0.4278	0.0125	±0.1066	0.02177

Case	$R/1 + R$	p'_H	$\overline{H}'$	$\Delta\overline{H}'$ ($\times 10^{-3}$)	$\overline{H}/2N_s = \Delta\overline{H}_D$ ($\times 10^{-3}$)	$\Delta\overline{H}' - \Delta\overline{H}_D$
1	0.2731	0.3863	0.4742	2.96	−2.95	10^{-5}
2	0.2565	0.3744	0.4684	2.23	−2.59	-3.6×10^{-4}
3	0.3595	0.3696	0.4660	2.4	−2.9	-5×10^{-4}
4	0.3477	0.3631	0.4625	1.74	−2.56	-8.2×10^{-4}
5	0.2404	0.3674	0.4649	4.06	−2.88	$+1.18 \times 10^{-3}$
6	0.2121	0.3507	0.4554	3.47	−2.51	$+9.6 \times 10^{-4}$
7	0.3151	0.3312	0.4430	4.18	−2.74	$+1.44 \times 10^{-3}$
8	0.2854	0.3147	0.4312	3.51	−2.38	$+1.13 \times 10^{-3}$

cation and replacement selection without linkage and with all loci initially near 0.5, the increase and decrease in heterozygosity are almost identical, as would be expected if the population were at steady state. Case 5, subject to the same amount of selection but started with loci uniformly distributed between 0 and 1, shows a slight excess of the effect of selection over that of drift. The observed final σ_p value in this run is 0.14, suggesting that the variance of loci around their equilibrium has not yet been reduced by selection to the balance point between selection and drift.

The same pattern is apparent in all the other runs. In cases in which the initial σ_p was small, drift is still slightly outweighing selection after 200 generations. In cases in which the initial σ_p was large, selection outweighs drift. The balance point between selection and drift will therefore lie between the observed values of σ_p for each pair of computer runs, at or very near the values predicted by theory.

In each case, it will be noted that the runs started with small initial σ_p are closer to the predicted steady-state for selection and drift than are those started with large initial σ_p. This would be expected, in view of the fact that the allele frequencies in the latter runs have further to shift on average and therefore will not be as close to steady state after 200 generations.

It is instructive to go a step further and compare truncation with multiplicative selection leading to the same steady-state between selection and drift. In Table 9.4, I have explored the consequences of the two types of selection, using the same numbers as in Case 1 of Table 9.3. As we saw, in Case 1 20% truncation and replacement selection is enough to keep a population of size 100 with 100 loci stable at $\sigma_p = 0.12$, in the absence of linkage. This forms Case A of the new table. In Case B, the population measures are kept the same but the loci are linked on to a map 100 cM long. It can immediately be seen that conditions are no longer stable when $\sigma_p = 0.12$. Now, the increase in $\overline{H}$ due to drift is greater than the decrease due to selection each generation. The result will be that over time σ_p will increase until these two factors just cancel each other out.

With a multiplicative model, the first thing that is seen is that much stronger per-locus selection is required to keep the population at $\sigma_p = 0.12$ in the absence of linkage. This can be calculated by rearranging equation (6.15) and solving. The per-locus selection coefficients turn out to be 0.1059, which will lead to an average fitness of the population of 0.0043. By contrast, truncation and replacement selection leading to the same steady state between selection and drift can be accomplished by reducing the average fitness of the population to only 0.8.

Because of steady-state disequilibrium, the average length of a run is increased from 2.0 to 2.63 in this population. This increases the per locus selection coefficients to 0.139 in the presence of disequilibrium, so that with multiplicative selection exactly the opposite effect occurs as with truncation selection. Now, in the presence of disequilibrium, the increase in the heterozygosity

Table 9.4. Calculation of the effects of steady-state disequilibrium for various populations with the same σ_p as case 1 of Table 9.3.

Truncation and replacement selection

Case	Selection	N_s	Linkage	L	σ_p	p_H	$\overline{H}$	r	D	$\Delta\overline{H}_s$	$R/1+R$
A	20%	80	None	100	0.12	0.38	0.4712	—	—	0.0279	0.2731
B	20%	80	100 cM	100	0.12	0.38	0.4712	0.01	±0.122	0.0358	0.3828

Case	p'_H	$\overline{H}'$	$\Delta\overline{H}'$	$\Delta\overline{H}_D$	$\Delta\overline{H}' - \Delta\overline{H}_D$
A	0.3863	0.4742	2.96×10^{-3}	2.96×10^{-3}	10^{-5}
B	0.3842	0.4732	1.98×10^{-3}	2.95×10^{-3}	-9.7×10^{-4}

Multiplicative selection

Case	Selection	N_e	Linkage	L	σ_p	p_H	$\overline{H}$	r	D	p'_H	$\overline{H}$
C	$s_1 = s_2 = 0.1059$	80	None	100	0.12	0.38	0.4712	—	—	0.3863	0.4742
D	$s_1 = s_2 = 0.1059$	80	20 cM	100	0.12	0.38	0.4712	0.01	±0.172	0.3885	0.4751

Case	$\Delta\overline{H}'$	$\Delta\overline{H}_D$	$\Delta\overline{H}' - \Delta\overline{H}_D$
C	2.96×10^{-3}	2.95×10^{-3}	10^{-5}
D	3.93×10^{-3}	2.95×10^{-3}	$+9.8 \times 10^{-4}$

due to selection is greater than the decrease due to drift. The result will be that over time the value for σ_p will decrease rather than increase.

The effect of disequilibrium on the retention of polymorphisms

We can now use the methods given in Chapters 6 and 8, in conjunction with the above arguments, to investigate how many polymorphisms can be retained by a population in the presence of steady-state disequilibrium. As in Chapter 6, we seek the number of loci L at which, for a given amount of truncation and replacement, selection will retain polymorphisms to a selective criterion C. As in those examples, we will set $C = \hat{p}/5$.

In Figs. 9.6, 9.7 and 9.8 the calculations have been carried out for populations in which all loci have a $\hat{p}$ of 0.5, 0.1 or 0.05. It can be seen that L is

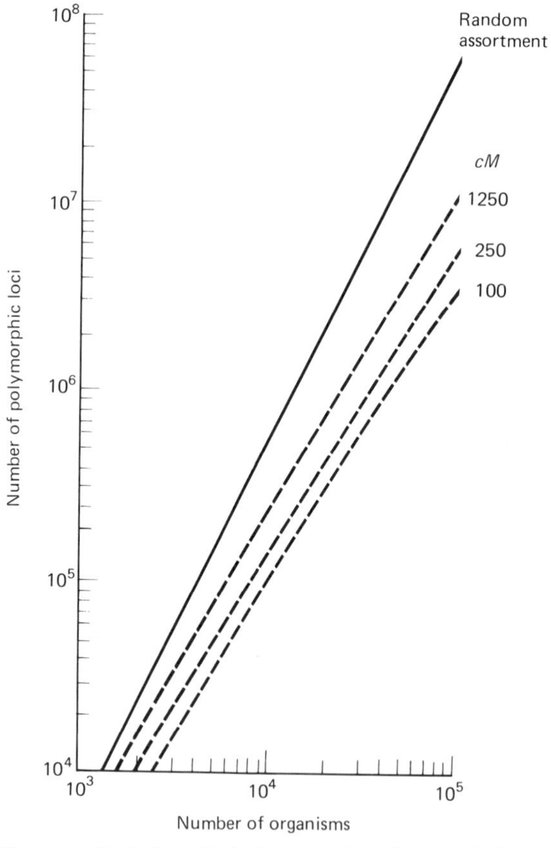

Fig. 9.6. The magnitude by which the capacity of a population to maintain polymorphisms to a selective criterion $C = \hat{p}/5$ is diminished by steady-state sampling disequilibrium. All the polymorphisms are being driven to $\hat{p} = 0.5$. The cM values refer to the total length of the genetic map. With random assortment, this length is infinite.

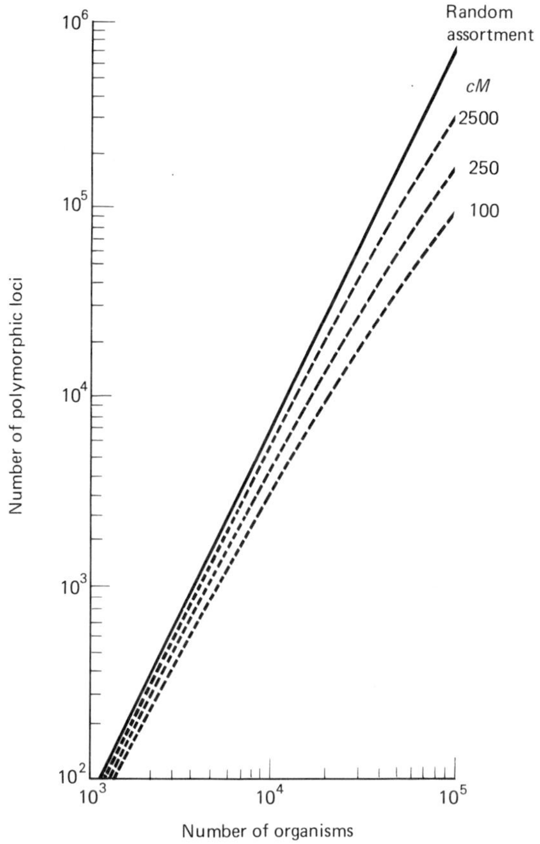

Fig. 9.7. As in Fig. 9.6, but with all loci being driven towards $\hat{p} = 0.1$.

reduced even if the genes are linked loosely, for example if the total length of the map is 2500 cM. The reduction becomes progessively more marked as the total length of the map is reduced. Suppose $\hat{p} = 0.5$ at all loci, $N_e = 10^5$, and all the genes are linked on a chromosome 100 cM long. Then the number of polymorphisms that can be retained to the criterion C is only $\frac{1}{50}$ that of the number that can be retained if all the loci are free to assort at random. This is a very large effect on the population. If such selection is operating, it seems likely to influence the fate of genes that govern the level of recombination.

How recombination might be selected for

We can now see how a continuously operating mechanism selecting for recombination may be present in outbreeding populations even when the environment is constant and new advantageous mutations are not arising at a high rate. The requirement for the mechanism to operate is simply that at least some loci in

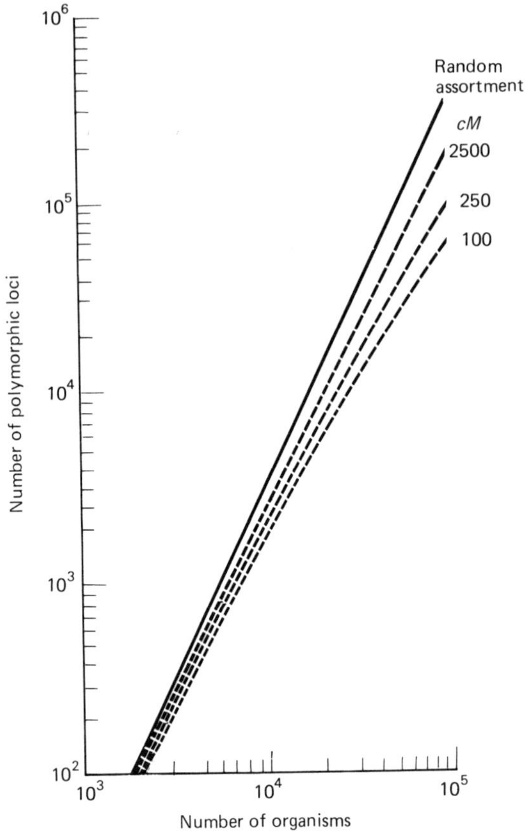

Fig. 9.8. As in Fig. 9.6, but with all loci being driven towards $\hat{p} = 0.05$.

the population are subject to balancing rank-order truncation selection. One particularly pleasing aspect of the calculations in the preceding section is that only small reductions in fitness are needed to give an advantage to recombination.

The fate of a gene reducing recombination may be imagined as follows. After it arises in a population it may, either by chance or because it is linked to an advantageous allele at a nearby locus, rise in frequency. If it does so, then the subgroup of the population carrying this allele (or the greatest number of whose ancestors carried this allele) will, as a result of the reduced rate of recombination, exhibit stronger disequilibrium between adjacent loci than will the remainder of the population not carrying the allele. Balancing truncation selection will be less effective in this subsample of the population, since truncation selection will result in a smaller increase in $\overline{H}$ than will the same amount of truncation acting on the remainder of the population. This subgroup of the population will diminish in numbers, taking the gene that reduces recombination with it.

The precise conditions under which this will happen may have to be explored—at least initially—through computer simulations. It seems certain, however, that the advantage to recombination given by truncation selection will be expressed even if overall selection is rather weak, a vivid contrast to the heroic measures that had to be applied to their simulated populations by Felsenstein and Yokoyama (1976) in order to demonstrate the Muller ratchet and the Hill–Robertson effect.

Factors influencing the advantage of recombination

Of course, the situation is not as simple as these calculations might suggest. A number of influences will be operating on any population to govern the amount of recombination exhibited by it. Let me suggest a list, which is by no means exhaustive.

(a) Balancing multiplicative selection has quite the opposite effect on selection for recombination from balancing truncation selection, though the effect is much weaker for a comparable reduction in fitness. For it to play a role, balancing truncation selection must be an important contributor to the overall reduction in fitness of the population due to selection. It must outweigh multiplicative selection in its effects on recombination.

(b) The selective advantage due to the formation of supergenes may locally outweigh any advantage for reduced recombination. At least over short regions of the genome, supergenes may appear, but crystallization is unlikely in large outbreeding populations.

(c) Directional truncation and multiplicative selection, whether acting against harmful alleles or acting on unconditionally advantageous alleles, may be enhanced in their effectiveness by disequilibrium. This will occur if an appreciable fraction of the locus pairs in the population exhibit coupling disequilibrium, as was pointed out on p. 262. Thus, the reduction in fitness due to these four types of selection must also not be large compared with the reduction in fitness due to balancing truncation selection if recombination is to be selected for.

(d) As was shown in Chapter 6 (pp. 141–45), rank-order selection is unlikely to be pure. That is, the shape of the selective curve imposed on the population will probably be influenced by the underlying distribution of fitness potentials. Some examples of this were seen in Fig. 6.2. If the fitness curve remains unaltered as the variance of fitness potential shrinks, then a population might shift from truncation selection to a simple proportional selection.

Some simple numbers will help further to illustrate the point. If $L = 100$ and $\hat{p} = 0.5$, then (in the absence of disequilibrium) 95% of the population will be expected to lie between two standard deviations of the mean heterozygosity—in this case between 40% and 60% heterozygosity. If we equate fitness potential with heterozygosity in such a population, then the spread of fitness potentials in such a population will be large relative to the mean. Thus, if the fitness curve imposed on the population is in part dependent on the spread of fitness potentials, then the contribution of this spread may be large.

Now consider a similar population in which $L = 10\,000$. The spread of fitness potentials relative to the mean is now much smaller. Ninety-five percent of the organisms in the population will now lie between 49% and 51% heterozygosity. This will not affect pure rank-order selection. But if the shape of the fitness curve is in part dependent on the spread of fitness potentials, then the efficiency of selection may be reduced. It seems highly likely that truncation will not be as effective in this case as in the case when L is small.

All these factors together will undoubtedly be operating to different degrees in different populations, along with other factors that have not yet been thought about. In addition, while we can now see how recombination may have the advantage in outbreeding organisms even when the population is at steady state, there remains the problem of how it might have been retained in habitually inbreeding or selfing organisms. Yeast, for example, which rarely undergoes sexual reproduction, has a very high rate of recombination. There is no evidence that the rate of recombination in habitually selfing plants has been reduced. It seems possible that sib selection after the occasional round of outbreeding (Williams 1975) may provide the answer here. Those organisms having the highest rate of recombination will be able to produce the most genetically diverse sibships in the next generation.

Despite these complexities, I think that the evolutionary advantage of recombination becomes far more understandable when it is realized that balancing truncation selection may confer a powerful and continuously acting advantage on organisms exhibiting random assortment.

10. Conclusions

I have often thought in my long practical career and at my age, how much pains and how many researches are necessary to probe to the depths a rather complicated phenomenon. The greatest difficulty comes from the fact that we are too much accustomed to attribute to a single cause that which is the product of several, and the majority of our controversies come from that.

Letter from Justus von Liebig to Emile Duclaux, 1872.
Quoted by Dubos 1950.

The reader may have gained the impression from this book that I am fixated on truncation selection and its consequences. Let me hasten to state that I do not for one moment discount the importance of drift, multiplicative and additive selection, and epistatic interactions between loci in the shaping of the genome and the process of evolution. I have concentrated on the ramifications of truncation selection for two reasons.

First, while truncation selection has been intensively investigated by quantitative geneticists, very little has been done to investigate its effects at the level of gene frequencies. This book provides some preliminary explorations of what will, I predict, become an important aspect of population-genetics theory. Adequate theory already exists for other types of selection, but some of the effects explored here are sufficiently *outré* to suggest that even more surprises may be in store for the theoretical investigator of truncation selection.

For example, it was shown in Chapter 6 that multiplicative selection can be more efficient than truncation selection at removing rare harmful variants from a population, while truncation selection will be more efficient than multiplicative selection if the variants are *conditionally* harmful so that they are more numerous and at higher frequency. Each type of selection acts optimally on a particular class of alleles.

The theory of Chapter 8 showed that there are many types of truncation selection that lead to *no change in disequilibrium* in linked systems of heterotic genes. This effect is often seen when large numbers of loci are segregating simultaneously, and is in marked contrast to conventional population genetics theory.

In Chapter 9, a comparison of balancing multiplicative and truncation selection showed that the former gives an advantage over the long term to increased linkage, while the latter gives an advantage to increased recombination. How universal are these effects, and how important might they be in shaping the genome? Only extensive further exploration of the theory of truncation will tell, but at the moment this type of selection can be seen to provide an advantage to recombination even if populations are in a constant environment.

There is a second important reason for concentrating on truncation selection. This is that such selection in one or another form can lead to the maintenance of very large amounts of genetic variability in natural populations without a severe reduction in fitness. As was set forth in Chapter 1, the rate, and indeed the very mechanisms of evolution will be different if populations are genetically diverse or genetically uniform. There may be a much larger pool of selected variability in the population, potentially capable of acting as modifiers of saltational changes, than has been permitted by most theory.

The effects of truncation selection will be difficult to look for. Does nature actually truncate? There is very little evidence on the subject, aside from the commonplace observation that organisms in nature can have fitnesses ranging from zero to much greater than one, something which other types of selection cannot provide if many loci must be selected simultaneously. In Chapter 8 I noted that truncation selection is unlikely to be acting on lethal or severe-subvital genes, so that experiments in the laboratory involving such markers may not detect truncation.

It is more probable that truncation acts primarily on selected functional polymorphisms, since, as we have seen, the selective pressures acting on these alleles are more dependent on the environment and on the remainder of the genotype. This increases the difficulty of detecting truncation selection.

There are, as was shown in Chapter 8, circumstances attainable with relative ease in laboratory populations in which the gene is the unit of selection. If conditional heterosis is widespread, then selection of such populations under severe conditions should reveal heterozygote excess at many loci. Other possible outcomes would make the interpretation of results much more difficult, but I would like to make a plea here for doing such experiments. Let me emphasize that while there is, as yet, no evidence *for* the operation of truncation selection on selected functional polymorphisms, there is no evidence *against* it either. I hope that more elegant ways to approach the problem will occur to the reader. Unlike the problem of the existence of neutral genes, if truncation selection is occurring in populations it should be possible to demonstrate it.

As genes continue to be sequenced, it is becoming more apparent daily what a powerful force purifying selection has been in the shaping of the genome. I have given arguments in Chapters 4 and 6 for why much of this purifying selection must be of the truncation type. In Chapters 6 and 7 I showed why there should be no impediment to the accumulation and maintenance of balanced polymorphisms by truncation selection, provided that the polymorphisms are not unconditionally heterotic. It will be fun to look for such polymorphisms.

It seems unlikely that we will ever be able to determine with certainty what proportion of alleles in a population is neutral and what proportion is subject to selection. But I have emphasized in this book that there are many exciting properties of the selected fraction of alleles that remain to be explored. It is to be hoped that we can say of population genetics and the study of genetic variation (Marquis 1927):

> toujours gai archy
> that is my motto
> there s more than one dance
> in the old dame yet

Bibliography

Abramowitz, M. and I. A. Stegun, (1965). *Handbook of Mathematical Functions.* Dover, New York.

Aird, I.; H. H. Bentall; and J. A. Fraser Roberts (1953). A relationship between cancer of the stomach and the ABO blood groups. *Br. med. J.* **i**:799–801.

Alikhanian, S. J. (1937). A study of the lethal mutations in the left end of the sex chromosome in *Drosophila melanogaster. Zool. Zh.* **16**:247–79.

Allard, R. W. (1975). The mating system and microevolution. *Genetics* **79** (suppl.): 115–26.

Allen, D. W. and J. H. Jandle (1961). Oxidative hemolysis and precipitation of hemoglobin. II. Role of thiols in oxidant drug action. *J. clin. Invest.* **40**:454–75.

Allison, A. C. (1954*a*). The distribution of the sickle cell trait in East Africa and elsewhere, and its apparent relationship to subtertian malaria. *Trans. R. Soc. trop. Med. Hyg.* **48**:312–18.

——— (1954*b*). Protection afforded by sickle cell trait against subtertian malarial infection. *Br. med. J.* **ii**:290–94.

——— (1960). Glucose-6-phosphate dehydrogenase deficiency in red blood cells of East Africans. *Nature* **186**:531–3.

——— (1964). Polymorphism and natural selection in human populations. *Cold Spring Harb. Symp. quant Biol.* **29**:137–49.

Alper, C. A. (1976). Inherited structural polymorphism in human C2: evidence for genetic linkage between C2 and Bf. *J. exp. Med.* **144**:1111–15.

Anderson, M. E. (1971). Sedimentation equilibrium experiments on self-association of hemoglobin from the lamprey *Petromyzon marinus. J. biol. chem.* **246**:4800–6.

——— and Q. H. Gibson (1971). A kinetic analysis of the binding of oxygen and carbon monoxide to lamprey hemoglobin. *J. biol. Chem.* **246**:4790–9.

Anderson, W. W. and T. K. Watanabe (1974). Selection by fertility in *Drosophila pseudoobscura. Genetics* **77**:559–64.

Arfors, K. E.; L. Beckman; and L. G. Lundin (1963). Genetic variations of human serum phosphatase. *Acta genet. Statist. med.* **13**:89-94.

Arniaz-Villena, A. and H. Festenstein (1975). Anti-H2 antisera raised across D-end incompatibilities inhibiting lymphoid complement receptors. *Folia Biol. Praha.* **21**:432–6.

Arnold, R. W. (1969). The effects of selection by climate on the land snail, *Cepaea nemoralis* (L.). *Evolution* **23**:370–78.

Ayala, F. J. (1975). Genetic differentiation during the speciation process. *Evol. Biol.* **8**:1-78.

——— and Th. Dobzhansky (1974). A new subspecies of *Drosophila pseudoobscura* (Diptera:Drosophilidae). *Pan Pacif. Ent.* **50**:211–19.

Bank, A. and P. A. Marks (1966). Excess α chain synthesis relative to β chain synthesis in thalassemia major and minor. *Nature* **212**:1198–200.

Barnes, B. W. (1968). Maternal control of heterosis for yield in *Drosophila melanogaster. Heredity* **23**:563–72.

Beardmore, J. A.; F. Lints; and A. L. al-Baldawi (1975). Parental age and heritability of sternopleural chaeta number in *Drosophila melanogaster. Heredity* **34**:71-82.

Beckenbach, A. T. and S. Prakash (1977). Examination of allelic variation at the hexokinase loci of *Drosophila pseudoobscura* and *D. persimilis* by different methods. *Genetics* **87**:743–61.

Beckman, L. (1968). Blood groups and serum alkaline phosphatase. *Ser. Hematol.* **i**:137–52.

——— and M. Grivea (1964). Haptoglobin variations in newborn children. *Acta genet. Statist. med.* **14**:159–67.

Beet, E. A. (1949). The genetics of the sickle-cell trait in a Bantu tribe. *Ann. Eugen.* **14**:279–84.

Benesch, R. and R. E. Benesch (1967). The effect of organic phosphates from the human erythrocyte on the allosteric properties of hemoglobin. *Biochem. biophys. Res. Commun.* **26**:162–7.

Berger, E. M. (1977). Are synonymous mutations adaptively neutral? *Am. Nat.* **111**:606–7.

——— (1978). Pattern and chance in the use of the genetic code. *J. Molec. Evol.* **10**:319–23.

Bernard, E. A.; M. S. Cohen; M. H. Gold; and J. K. Kim (1972). Evolution of ribonuclease in relation to polypeptide folding mechanisms. *Nature* **240**:395–8.

Bernstein, F. (1924). Ergenbnisse einer biostatischen zusammenfassenden Betrachtung über die erblichen Blutstrukturen des Menschen. *Klin. Wschr.* **3**:1495–7.

Beutler, E. (1978). Glucose-6-phosphate dehydrogenase deficiency. In *The metabolic basis of inherited disease* (eds. J. B. Stanbury, J. B. Wyngaarden, and D. S. Fredrickson) 4th edn, pp. 1430–51, McGraw-Hill, New York.

———; R. J. Dern; and C. L. Flanagan (1955). Effect of sickle cell trait on resistance to malaria. *Br. med. J.* **i**:1189–91.

Bienzle, U.; A. O. Lucas; O. Ayeni; and L. Luzzatto (1972). GGPD and malaria: greater resistance of females heterozygous for enzyme deficiency and of males with non-deficient variant. *Lancet* **i**:107–10.

Bijlsma-Meeles, E. and W. van Delden (1974). Intrapopulation and interpopulation selection concerning the alcohol dehydrogenase locus in *Drosophila melanogaster*. *Nature* **247**:369–71.

Birnboim, H. C. and R. N. Sederoff (1975). Polypyrimidine sequences in *Drosophila melanogaster* DNA. I. Detection of a cryptic satellite containing polypyrimidine-polypurine DNA. *Cell* **5**:173–81.

———; N. A. Straus; and R. R. Sederoff (1975). Characterization of polypyrimidines in *Drosophila* and L-cell DNA. *Biochemistry* **14**:1643–7.

Bishop, J. P.; J. G. Morton; M. Rosbash; and M. Richardson (1974). Three abundance classes in HeLa cell messenger RNA. *Nature* **250**:199-204.

Björkhem, I. (1972). On the role of alcohol dehydrogenase in omega-oxidation of fatty acids. *Eur. J. Biochem.* **30**:441–51.

Blatherwick, C. and C. Wills (1971). The accumulation of genetic variability in yeast. *Genetics* **68**:547–57.

Blundell, T. L. and S. P. Wood (1975). Is the evolution of insulin Darwinian or due to selectively neutral mutation? *Nature* **257**:197-203.

Boag, P. T. and P. R. Grant (1978). Heritability of external morphology in Darwin finches. *Nature* **274**:793–4.

Bodmer, W. F.; H. Cann; and A. Piazza (1972). Differential genetic variability among polymorphisms as an indicator of natural selection. In *Histocompatibility testing 1972* (Eds. J. Dausset and J. Colombani, pp. 753–67. Munksgaard, Copenhagen.

Boettcher, B. (1964). The Rh "deletion" phenotypes and the information they provide about the Rh genes. *Vox Sang.* **9**:641–52.

——— (1965). The Rh antigens of anthropoid apes in relation to Rh "deletion" phenotypes. *Am. J. hum. Genet.* **17**:308–10.

Bottini, E.; P. Lucarelli; and P. Pigram (1972). Interaction between placental alkaline phosphatase and ABO system polymorphisms during intrauterine life. *Am. J. hum. Genet.* **24**:495-504.

———; ———; and F. Gloria (1972). Placental alkaline phosphatase and ABO system polymorphisms: analysis of relationships among gene frequencies in human populations. Further evidence at the population level for an interaction between the two polymorphisms. *Am. J. hum. Genet.* **24**:505–13.

Boyer, S. H.; E. F. Crosby; A. N. Noyes; G. F. Fuller; S. E. Leslie; L. J. Donaldson; G. R. Vrablik: E. W. Schaefer; and T. F. Thurman (1970). Primate hemoglobins: some sequences and some proposals concerning the character of evolution and mutation. *Biochem. Genet.* **5**:405–48.

———; D. L. Rucknagel; D. J. Weatherall; and E. J. Watson-Williams (1963). Further evidence for linkage between the β and δ loci governing human hemoglobin and the population dynamics of linked genes. *Am. J. hum. Genet.* **15**:538–48.

Bridgen, J.; D. Snary; M. J. Crumpton; C. Barnstable; P. Goodfellow; and W. F. Bodmer (1976). Isolation and N. terminal amino acid sequence of membrane-bound human HL-A and HL-B antigens. *Nature* **261**:200–5.

Britten, R. J.; D. E. Graham; and B. R. Neufeld (1974). Analysis of repeating DNA sequences by reassociation. *Meth. Enzymol.* **29**:363-418.

Bruce Chwatt, L. J. (1948). Infection of reticulocytes by *Plasmodium falciparum* and *Plasmodium malariae* in hyperendemic indigenous malaria. *Ann. trop. Med. Parasit.* **42**:101–12.

Brues, A. M. (1954). Selection and polymorphism in the ABO blood groups. *Am. J. phys. Anthropol. (Ser.2)* **12**:559–97.

Bumpus, H. C. (1899). The elimination of the unfit as illustrated by the introduced sparrow, *Passer domesticus*. Eleventh lecture before the Marine Biological Laboratory, Woods Hole, 1898. Ginn & Co., Boston.

Bunn, H. F. and B. G. Forget (1975). Information on an untranslated portion of beta-chain gene provided by a frameshift hemoglobin mutant and sequence studies on normal beta-message-RNA. *Am. J. hum. Genet.* **27**:A23.

Butcher, A. C.; J. Croft; and M. Grindle (1972). Use of genotype-environmental interaction analysis in the study of natural populations of *Aspergillus nidulans*. *Heredity* **29**:263–83.

Butcher, G. A.; G. H. Mitchell; and S. Cohen (1973). Mechanism of host specificity in malarial infection. *Nature* **244**:40–2.

Cain, A. J. (1971). Color and banding morphs in subfossil samples of the snail *Cepaea*. In *Ecological genetics and evolution* (ed. R. Creed) pp. 65-92. Blackwell, Oxford.

——— and J. D. Currey (1963*a*). Area effects in Cepaea. *Phil. Trans. R. Soc. Ser. B* **246**:1-81.

——— ——— (1963*b*). The causes of area effects. *Heredity* **18**:467–71.

———; J. M. B. King; and P. M. Sheppard (1960). New data on the genetics of polymorphism in the snail *Cepaea nemoralis* L. *Genetics* **45**:393-411.

Carcassi, U.; R. Ceppellini; and F. Pitzers (1957). Frequenza della talassemia in quattro popolazioni sarde e suoi rapporti con la distribuzione dei gruppi sanguini e della malaria. *Boll. Ist. sieroter. milan.* **36**:206–18.

Carpenter, G. D. H. (1920). *Pseudacraea eurytus hobleyi*, its forms and its models on the islands of L. Victoria, and the bearing of the facts on the explanation of mimicry by Natural Selection. *Trans. R. Ent. Soc.* **67**:84-98.

——— (1949). *Pseudacraea eurytus* (L.) (Lep., Nymphalidae): a study of a polymorphic mimic in various degrees of speciation. *Trans. R. Ent. Soc.* **100**:71-133.

Carson, H. L. (1955). The genetic characteristics of marginal populations of *Drosophila*. *Cold Spring Harb. Symp. quant. Biol.* **20**:276–87.

Carter, M. A. (1968*a*). Thrush predation of an experimental population of the snail *Cepaea nemoralis* (L.). *Proc. Linn. Soc.* **179**:241–9.

——— (1968*b*). Studies on Cepaea II. Area effects and visual selection in *Cepaea nemoralis* (L.) and *Cepaea hortensis*. *Phil. Trans. R. Soc. Ser. B* **253**:397-446.

Cavalli-Sforza, L. L.; I. Barrai; and A. W. F. Edwards (1964). Analysis of human evolution under random genetic drift. *Cold Spring Harb. Symp. quant. Biol.* **29**:9-20.

——— and W. F. Bodmer (1971). *The genetics of human populations*. Freeman, San Francisco.

Cawley, R. M.; T. McKeown; and R. G. Record (1954). Parental stature and birth weight. *Am. J. hum. Genet.* **6**:448–56.

Chagnon, N. (1968). *Yanomamo: the fierce people*. Holt, Rinehart and Winston, New York.

Chalmers, J. N. M.; E. W. Ikin; and A. E. Mourant (1949). The ABO, MN and Rh blood groups of the Basque people. *Am. J. phys. Anthropol.* **7**:529–44.

Chanutin, A. and R. R. Curnish (1967). Effect of organic and inorganic phosphates on the oxygen equilibrium of human erythrocytes. *Archs Biochem.* **121**:96-102.

Chrispeels, M. and D. Sadava (1977). *Plants, Food and People.* Freeman, San Francisco.

Chung, C. S.; E. Matsunaga; and N. E. Morton (1960). The ABO polymorphism in Japan *Jap. J. Hum. Genet.* **5**:124–34.

Clarke, C. A. and P.M. Sheppard (1960*a*). The evolution of dominance under disruptive selection. *Heredity* **14**:73-87.

—— —— (1960*b*). The evolution of mimicry in the butterfly *Papilio dardanus. Heredity* **14**:163–73.

—— —— (1960*c*). Supergenes and mimicry. *Heredity* **14**:175–85.

Clegg, J. B. and D. J. Weatherall (1976). Molecular basis of thalassemia. *Br. med. Bull.* **32**:262–9.

Cobbs, G. and S. Prakash (1977). A comparative study of the esterase-5 locus in *Drosophila pseudoobscura, D. persimilis* and *D. miranda. Genetics* **85**:697-711.

Colman, R. W.; A. J. Osbahr; and R. E. Morris (1967). New vasoconstrictor, bovine peptide B, released during blood coagulation. *Nature* **215**:292–3.

Comstock, R. E. and R. H. Moll (1963). Genotype environment interactions. In *Statistical genetics and plant breeding* (eds. W. D. Hanson and H. F. Robinson) NAS-NRC Publication No. 982, pp. 164-196.

Conconi, F.; P. T. Rowley; L. del Senno; S. Pontremoli; and S. Volpato (1972). Induction of β-globin synthesis in the β-thalassemia of Ferrara. *Nature New Biol.* **238**:83–7.

Conterio, F. and L. L. Cavalli-Sforza (1960). Selezione per caratteri quantitavi nell'uomo. *Atti. Ass. Genet. Ital.* **5**:295-304.

Coyne, J. A. (1976). Lack of genic similarity between two sibling species of Drosophila as revealed by varied techniques. *Genetics* **84**:593-607.

Crosby, W. H. (1956). Newsletter—favism in Sardinia. *Blood* **11**:91–2.

Crow, J. F. (1972*a*). Darwinian and non-Darwinian evolution. In *Proceedings of 6th Berkeley symp. on mathematical statistics and probability* (eds. J. Neyman, E. L. Scott, and L. Le Cam) pp.1-22, University of California Press.

—— (1972*b*). The dilemma of nearly neutral mutations: how important are they for evolution and human welfare? *J. Hered.* **63**:306–16.

—— (1979). Minor viability mutants in Drosophila. *Genetics* **92**:5165–72.

—— and M. Kimura (1970). *An introduction to population genetics theory.* Harper & Row, New York.

—— —— (1979). Efficiency of truncation selection. *Proc. natn. Acad. Sci., U.S.A.* **76**:396–9.

Crozier, R. H.; L. A. Briese; M. A. Guerin; T. R. Harris; J. L. McMichael; C. H. Moore; P. R. Ramsey; and S. R. Wheeler (1972). Population genetics of hemoglobins S, C and A in Africa: equilibrium or replacement? *Am. J. hum. Genet.* **24**:156–67.

Crumpton, M. J.; D. Snary; F. S. Walsh; C. J. Barnstable; P. N. Goodfellow; E. A. Jones; and W. F. Bodmer (1978). Molecular structure of the gene products of the human HLA system: isolation and characterization of HLA-A, -B, -C and Ia antigens. *Proc. R. Soc. Ser. B* **202**:159–75.

Currey, J. D. and A. J. Cain (1968). Studies on *Cepaea.* IV. Climate and selection of banding morphs in *Cepaea* from the climate optimum to the present day. *Phil. Trans. R. Soc. Ser. B* **253**:483–98.

Dahms, T.; S. M. Horvath; M. Luzzana; L. Rossi-Bernardi; F. J. W. Roughton; and

G. Stella (1972). The regulation of oxygen affinity of human hemoglobin. *J. Physiol.* **223**:29P–31P.

Darwin, C. (1868). *The variation of animals and plants under domestication.* John Murray, London.

——— (1871). *The descent of man and selection in relation to sex.* John Murray, London.

Das, B. M.; D. Ranjan; and G. Flatz (1975). Predominance of the HbE gene in a Mongoloid population in Assam (India). *Humangenetik* **30**:187–91.

Datta, P. M.; P. Yadagini; and B. R. J. Rao (1978). Discovery of early Jurassic micro-mammals from upper Gondwana sequence of Prambita-Godavari Valley, India. *J. Geol. Soc. India* **19**:64.

Dausset, A. and A. Svejgaard (1977). *HLA and disease: predisposition to disease and clinical implications.* INSERM, Paris.

Dayhoff, M. O. (1972). *Atlas of protein sequence and structure.* National Biomedical Research Foundation, Washington.

de Beer, G. (1962). The origin of Darwin's ideas on evolution and natural selection. *Proc. R. Soc. Ser. B* **155**:321–38.

Deininger, P.; A. Esty; P. La Porte; and T. Friedmann (1979). Nucleotide sequence and genetic organization of the Polyoma late region: features common to the Polyoma early region and to SV40. *Cell* **18**:771–9.

de Jong, G. and W. Scharloo (1976). Environmental determination of selective significance or neutrality of amylase variants in *Drosophila melanogaster.* *Genetics* **84**:77–94.

Dempster, E. R. (1955). Maintenance of genetic heterogeneity. *Cold Spring Harb. Symp. Quant. Biol.* **20**:25–32.

Dienst, A. (1905). Das Eklampsiegift. *Zentbl. Gynäk.* **29**:353–64.

Diver, C. (1929). Fossil records on Medelian mutants. *Nature* **124**:183.

——— (1940). The problem of closely related species living in the same area. In *The new systematics* (ed. Julian Huxley) pp. 308–28, Clarendon Press, Oxford.

Dobzhansky, Th. (1951). *Genetics and the origin of species,* 3rd edn. rev. Columbia, New York.

——— (1959*a*). Evolution of genes and genes in evolution. *Cold Spring Harb. Symp. Quant. Biol.* **24**:15–30.

——— (1959*b*). Variation and evolution. *Proc. Am. Phil. Soc.* **103**:252–63.

——— (1970). *Genetics of the evolutionary process.* Columbia, New York.

——— and H. Levene (1955). Genetics of natural populations. XXIV. Developmental homeostasis in natural populations of *Drosophila pseudoobscura. Genetics* **40**:797–808.

——— and M. L. Queal (1938). Genetics of natural populations. I. Chromosomal variation in populations of *Drosophila pseudoobscura* inhabiting isolated mountain ranges. *Genetics* **23**:239–51.

——— and B. Spassky (1944). Genetics of natural populations. XI. Manifestation of genetic variants in *Drosophila pseudoobscura* in different environments. *Genetics* **29**:270–90.

——— ——— (1963). Genetics of natural populations. XXXIV. Adaptive norm, genetic load and genetic elite in *Drosophila pseudoobscura. Genetics* **48**:1467–86.

——— and A. H. Sturtevant (1938). Inversions in the chromosomes of *Drosophila pseudoobscura. Genetics* **23**:28–64.

Doherty, P. C.; D. Götze; G. Trinchieri; and R. M. Zinkernagel (1976). Models for recognition of virally modified cells by immune thymus-derived lymphocytes. *Immunogenetics* **3**:517–24.

——— and R. M. Zinkernagel (1975). Enhanced immunological surveillance in mice heterozygous at the H-2 gene complex. *Nature* **256**:50–2.

Doolittle, R. F.; R. Chen; C. Glasgow; G. Mross; and M. Weinstein (1970). The molecular constancy of fibrinopeptides A and B from 125 individual humans. *Hum. Genet.* **10**:15–29.

Downie, A. W.; G. Meiklejohn; L. S. Vincent; A. R. Rao; B. V. S. Babu; and C. H. Kempe (1965). Smallpox frequency and severity in relation to A, B and O blood groups. *Bull. Wld. Hlth. Org.* **33**:623–5.

Dozy, A. M.; Y. W. Kam; S. H. Embury; W. C. Mentzer; W. C. Wang; B. Lubin; J. R. Davis, Jr.; and H. M. Koenig (1979). α-Globin gene organization in blacks precludes the severe form of α-thalassemia. *Nature* **280**:605–7.

Drake, J. W. (1970). *The molecular basis of mutation.* Holden-Day, San Francisco.

Dubos, R. J. (1950). *Louis Pasteur, free lance of science.* Little, Brown, Boston.

Dungern, E. von and L. Hirszfeld (1910). Ueber Vererbung gruppenspezifischer Strukturen des Blutes. *Z. Immunforsch.* **6**:284–92.

Dunn, L. C. (1965). *A short history of genetics.* McGraw-Hill, New York.

Dupont, B.; J. A. Hansen; and C. Whitsett (1976). Association between HLA and diseases. In *Clinical evaluation of immune function in man.* (Eds. S. D. Litwin, C. L. Christian and G. W. Suskind) pp. 97–132. Grune & Stratton, New York.

East, E. M. (1910). A Mendelian interpretation of variation that is apparently continuous. *Am. Nat.* **44**:65–82.

Ednington, G. M. and W. M. Laing (1957). Relationship between hemoglobins C and S and malaria in Ghana. *Br. med. J.* **2**:143–5.

Efstratiadis, A.; F. D. Kafatos; and T. Maniatis (1977). The primary structure of rabbit beta-globin mRNA as determined from cloned DNA. *Cell* **10**:571–85.

Eisely, L. (1961). *Darwin's century: evolution and the men who discovered it.* Anchor, New York.

Ewens, W. E. (1972). The sampling theory of selectively neutral alleles. *Theor. Pop. Biol.* **3**:87–112.

Felsenstein, J. (1965). The effect of linkage on directional selection. *Genetics* **52**:349–63.

——— (1974). The evolutionary advantage of recombination. *Genetics* **78**:737–56.

——— (1976). The theoretical population genetics of variable selection and migration. A. Rev. Genet. **10**:253–80.

——— and S. Yokoyama (1976). Evolutionary advantage of recombination. II. Individual selection for recombination. *Genetics* **83**:845–59.

Fiers, W.; R. Contreras; F. Duerinck; G. Haegeman; D. Iserantant; J. Merregaert; W. Min Jou; F. Molemans; A. Raeymaekers; A. Van den Berghe; G. Volckaert; and M. Ysebaert (1976). Complete nucleotide sequence of bacteriophage MS2 RNA: primary and secondary structure of the replicase gene. *Nature* **260**:500–7.

——— ———; G. Haegeman; R. Rogiers; A. Van de Voorde; H. Van Herreweghe; G. Volckaert; and M. Ysebaert (1978). Complete nucleotide sequence of SV40 DNA. *Nature* **273**:113–20.

Fisher, R. A. (1918). The correlation between relatives on the supposition of Mendelian inheritance. *Trans. R. Soc. Edinb.* **52**:399–433.

——— (1922). On the dominance ratio. *Proc. R. Soc. Edinb.* **42**:321–41.

——— (1927). On some objections to mimicry theory: statistical and genetic. *Trans. Ent. Soc. Lond.* **75**:269–78.

——— (1928). The possible modification of the response of the wild-type to recurrent mutations. *Am. Nat.* **62**:115–26.

——— (1930). *The genetical theory of natural selection.* Oxford University Press, London.

——— (1954). Retrospect of the criticisms of the theory of natural selection. In *Evolution as a process* (eds. J. Huxley, A. C. Hardy, and E. B. Ford) pp. 84–98. Allen & Unwin, London.

——— (1947). The Rhesus factor. *Am. Sci.* **35**:95–102, 113.

——— and E. B. Ford (1947). The spread of a gene in natural conditions in a colony of the moth *Panaxia dominula* L. *Heredity* **1**:143–74.

Fitch, W. M. (1971). Toward defining the course of evolution—minimum changes for a specific tree topology. *Syst. Zool.* **20**:406–16.

——— (1976). Molecular evolutionary clocks. In *Molecular evolution* (ed. F. J. Ayala) pp. 160–78. Sinauer, Sunderland, Ma.

——— and E. Margoliash (1967). Construction of phylogenetic trees. *Science* **155**:279–84.

——— and E. Markowitz (1970). An improved method for determining codon variability in a gene and its application to the rate of fixation of mutations in evolution. *Biochem. Genet.* **4**:579–93.

——— and J. V. Neel (1969). The phylogenetic relationships of some Indian tribes of Central and South America. *Am. J. hum. Genet.* **21**:384–97.

Flatz, G.; C. Pik; and S. Sringam (1965*a*). Hemoglobinopathies in Thailand. I. A study of patients with the thalassemia syndrome and their families. *Br. J. Haematol.* **11**:216–26.

———; C. Pik; and S. Sringam (1965*b*). Hemoglobin E and β-thalassemia: their distribution in Thailand. *Ann. hum. Genet.* **29**:151–70.

Ford, E. B. (1937). Problems of heredity in the Lepidoptera. *Biol. Rev.* **12**:461–503.

——— (1940). Genetic research in the Lepidoptera. *Ann. Eugen.* **10**:227–52.

——— (1959). Evolution in progress. In *Evolution after Darwin* (ed. Sol Tax), Vol. I, pp. 181–96, University of Chicago Press.

——— (1975). *Ecological genetics,* 4th edn. Chapman and Hall, London.

——— and P. M. Sheppard (1969). The *medionigra* polymorphism of *Panaxia dominula. Heredity* **24**:561–69.

Franklin, I. R. and R. C. Lewontin (1970). Is the gene the unit of selection? *Genetics* **65**:707–34.

Friedman, M. J. (1978). Erythrocytic mechanism of sickle cell resistance to malaria. *Proc. natn. Acad. Sci. U.S.A.* **75**:1994–7.

Friedmann, T.; A. Esty; P. La Porte; and P. Deininger (1979). The nucleotide sequence and genome organization of the Polyoma early region: extensive nucleotide and amino acid homology with SV40. *Cell* **17**:715–24.

Frydenberg, O. (1963). Population studies of the lethal mutant in *Drosophila melanogaster.* I. Behavior in populations with discrete generations. *Hereditas* **50**:89–116.

Fujiki, K. (1975). Estimation of gene frequencies from segregation data and a study of selection in ABO blood groups. *Jap. J. hum. Genet.* **19**:291–316.

Furlar, M.; T. Seelick; and E. A. Beck (1976). Clottability and cross linking reactivity of fibrin(ogen) following differential release of fibrinopeptides A and B. *Thromb. Haemostos.* **36**:582–92.

Gall, J. G.; E. H. Cohen; and D. D. Atherton (1973). The satellite DNAs of *Drosophila virilis. Cold Spring Harb. Symp. quant. Biol.* **38**:417–22.

Gedde-Dahl, T.; A. L. Grimstad; S. Gunderson; and E. Vogt (1967). A possible crossing over or mutation in the MNSs blood group system. *Acta Genet. Statist. Med.* **17**:193–210.

Geiringer, H. (1944). On the probability theory of linkage in Mendelian heredity. *Ann. Math. Stat.* **15**:25–57.

Gibson, J. B. and B. P. Bradley (1974). Stabilizing selection in constant and fluctuating environments. *Heredity* **33**:293–302.

Gilles, H. M.; K. A. Fletcher; R. G. Hendrick; R. Linder; S. Reddy; and N. Allan (1967). Glucose-6-phosphate dehydrogenase deficiency, sickling, and malaria in African children in South Western Nigeria. *Lancet* **i**:138–40.

Gillespie, J. H. (1976). A general model to account for enzyme variation in natural populations. II. Characterization of the fitness functions. *Am. Nat.* **110**:809–21.

——— (1977*a*). Sampling theory for alleles in a random environment. *Nature* **266**:443–45.

——— (1977*b*). A general model to account for enzyme variation in natural populations. III. Multiple alleles. *Evolution* **31**:85–90.

——— (1978*a*). A general model to account for enzyme variation in natural populations. IV. The quantitative genetics of viability mutants. In *Measuring selection in natural populations* (ed. F. B. Christiansen, and T. M. Fenchel) pp. 301–14. Springer-Verlag, New York.

——— (1978*b*). A general model to account for enzyme variation in natural populations. V. The SAS–CFF model. *Theoret. Pop. Biol.* **14**:1–45.

——— and C. Langley (1974). Multilocus behavior in random environments. I. Random Levene models. *Genetics* **82**:123–37.

Godson, G. N.; B. G. Barrell; R. Staden; and J. C. Fiddes (1978). Nucleotide sequence of bacteriophage G4 DNA. *Nature* **276**:236–47.

Goodman, M.; G. W. Moore; and G. Matsuda (1975). Darwinian evolution in the genealogy of haemoglobin. *Nature* **253**:603–7.

Gould, S. J. and D. S. Woodruff (1978). Natural history of *Cerion*. VIII. Little Bahama Bank—a revision based on genetics, morphometrics, and geographic distribution. *Bull. Mus. Comp. Zool. Harv.* **148**:371–415.

Grant, P. R. (1972). Centripetal selection and the house sparrow. *Syst. Zool.* **21**:23–30.

Grunstein, M.; P. Schedl; and L. Kedes (1976). Sequence analysis and evolution of sea urchin (*Lytechinus pictus* and *Strongylocentrotus purpuratus*) histone H4 messenger RNA's. *J. molec. Biol.* **104**:351–70.

Haldane, J. B. S. (1927). A mathematical theory of natural and artificial selection: V. Selection and mutation. *Proc. Camb. Phil. Soc.* **23**:838–44.

——— (1932). *The causes of evolution.* Harper, London.

——— (1939*a*). The theory of the evolution of dominance. *J. Genet.* **37**:365–74.

—— (1939*b*). The equilibrium between mutation and random extinction. *Ann. Eugen.* **9**:400–5.

—— (1949). Disease and evolution. *Ricerca scient.* **19**: Suppl., 1:3–10.

—— (1954). The measurement of natural selection. *Proc. 9th Int. Cong. Genet.*, pp. 480–7.

—— (1957). The cost of natural selection. *J. Genet.* **55**:511–24.

—— (1962). The selection of double heterozygotes. *J. Genet.* **58**:125–8.

—— and S. D. Jayakar (1963). Polymorphism due to selection of varying direction. *J. Genet.* **58**:237–42.

Harlan, H. V. and M. L. Martini (1938). The effect of natural selection on a mixture of barley varieties. *J. agric. Res.* **57**:189–99.

Harris, H. (1966). Enzyme polymorphisms in man. *Proc. R. Soc. Ser. B* **164**:298–310.

—— D. A. Hopkinson and E. B. Robson (1974). The incidence of rare alleles determining electrophoretic variants: data on 43 enzyme loci in man. *Ann. hum. Genet.* **37**:237–53.

Harrison, G. A.; A. J. Boyce; R. W. Hornabrook; and W. J. Craig (1976). Associations between polymorphic variety and disease susceptibility in two New Guinea populations. *Ann. hum. Biol.* **3**:253–67.

Hartl, D. L. (1977). Mutation selection balance with stochastic selection. *Genetics* **86**:687–96.

Haverkorn, M. J. and W. R. O. Goslings (1969). Streptococci, ABO blood groups and secretor status. *Am. J. hum. Genet.* **21**:360–75.

Hedrick, P. W. (1974). Genetic variation in a heterogeneous environment. I. Temporal heterogeneity and the absolute dominance model. *Genetics* **78**:757–70.

—— M. E. Ginevan and E. P. Ewing (1976). Genetic polymorphism in heterogeneous environments. *Ann. Rev. Ecol.* **7**:1–32.

Herrick, J. B. (1910). Peculiar, elongated and sickle shaped red corpuscles in a case of severe anemia. *Archs. int. Med.* **6**:517–21.

Heywood, J. D.; M. Karon and S. Weissman (1964). Amino acids: incorporation into α and β chains of hemoglobin by normal and thalassemic reticulocytes. *Science* **146**:530–31.

Hill, A. V. (1910). The possible effects of the aggregation of the molecules of hemoglobin on its dissociation curves. *J. Physiol. Lond.* **40**:iv–vii.

Hill, W. G. (1974*a*). Disequilibrium among several linked neutral genes in finite population. I. Mean changes in disequilibrium. *Theoret. pop. Biol.* **5**:366–92.

—— (1974*b*). Disequilibrium among several linked neutral genes in finite population. II. Variances and covariances of disequilibria. *Theoret. pop. Biol.* **6**:184–98.

—— and A. Robertson (1966). The effect of linkage on limits to artificial selection. *Genet. Res.* **8**:269–94.

—— ——(1968). Linkage disequilibrium in finite populations. *Theoret. appl. Genet.* **38**:226–31.

Hiorns, R. W. and G. A. Harrison (1977). The combined effects of selection and migration in human evolution. *Man* **12**:438–45.

Hiraizumi, Y. (1964). Prezygotic selection as a factor in the maintenance of variability. *Cold Spring Harb. Symp. Quant. Biol.* **29**:51–60.

Hirszfeld, L. (1928). Konstitutionsserologie und Blutgruppenforschung. Translation In *U. S. Army Medical Reasearch Laboratory*. Selected contributions to the literature of blood groups and immunology. Vol. 3, p. 1. Fort Knox, Kentucky, 1969.

——— and H. Hirszfeld (1919). Serological differences between the blood of different races. The result of researches on the Macedonian front. *Lancet* **ii**:675–9.

Holmquist, R.; T. H. Jukes; H. Moise; M. Goodman; and G. W. Moore (1976). The evolution of the globin family genes: concordance of stochastic and augmented maximum parsimony genetic distances for α hemoglobin, β hemoglobin and myoglobin phylogenies. *J. molec. Biol.* **105**:39–74.

Housman, D.; B. G. Forget; A. Skoultchi; and E. J. Benz, Jr. (1973). Quantitative deficiency of chain specific globin messenger ribonucleic acids in the thalassemia syndromes. *Proc. natn. Acad. Sci., U.S.A.* **70**:1809–13.

Huehns, E. R. and E. M. Shooter (1965). Human hemoglobins. *J. Med. Genet.* **2**:48–90.

Hunt, W. A. and R. K. Selander (1973). Biochemical genetics of hybridization in European house mice. *Heredity* **31**:11–33.

Ingram, V. M. (1957). Gene mutations in human hemoglobin: the chemical difference between normal and sickle cell hemoglobin. *Nature* (Lond.) **180**:326–8.

Jenkin, F. (1867). The origin of species. *N. Br. Rev.* **46**:277–318.

Jenkins, M. E.; R. B. Scott; and R. L. Baird (1960). Studies in sickle cell anemia. XVI. Sudden death during sickle cell anemia crises in young children. *J. Pediat.* **56**:30–8.

Jick, H.; D. Slone; B. Westerholm; W. H. W. Inman; M. P. Vessey; S. Shapiro; G. P. Lewis; and J. Worcester (1969). Venous thromboembolic disease and ABO blood type. *Lancet* **i**:539–42.

Johannsen, W. (1907). Does hybridization increase fluctuating variability? In *Report of the Third International Conference on Genetics,* pp. 98–103. Spottiswood, London.

Johnson, G. B. (1976). Hidden alleles at the α glycerophosphate dehydrogenase locus in *Colias* butterflies. *Genetics* **83**:149–67.

——— (1977). Characterization of electrophoretically cryptic variation in the alpine butterfly *Colias meadii. Biochem. Genet.* **15**:665–93.

Johnson, R. F. and R. K. Selander (1964). House sparrows: rapid evolution of races in North America. *Science* **144**:548–50.

Jones, J. S. and A. J. Irving (1975). Gene frequencies, genetic background and environment in Pyrenean populations of *Cepaea nemoralis* (L). *Biol. J. Linn. Soc.* **7**:249–59.

———; B. H. Leith; and P. Rawlings (1977). Polymorphism in *Cepaea:* a problem with too many solutions? *Am. Rev. Ecol. Syst.* **8**:109–43.

Judd, B. H.; M. W. Shen; and T. C. Kaufman (1972). The anatomy and function of a segment of the X chromosome of *Drosophila melanogaster. Genetics* **71**:139–56.

——— and M. W. Young (1973). An examination of the one cistron: one chromomere concept. *Cold Spring Harb. Symp. Quant. Biol.* **38**:573–79.

Jukes, T. H. and R. Holmquist (1972). Estimation of evolutionary changes in certain homologous peptide chains. *J. molec. Biol.* **64**:163–79.

Kacian, D. L.; R. Gambino; L. W. Dow; E. Grossbarn; C. Natta; F. Ramirez; S. Spiegelman; P. A. Marks; and A. Bank (1973). Decreased globin messenger

RNA in thalassemia detected by molecular hybridization. *Proc. natn. Acad. Sci. U.S.A.* **70**:1886–90.

Kafatos, F. C.; A. Efstratiadis; B. G. Forget; and S. M. Weissmann (1977). Molecular evolution of human and rabbit hemoglobin mRNA's. *Proc. natn. Acad. Sci. U.S.A.* **74**:5618–22.

Kan, Y. W.; A. M. Dozy; H. E. Varmus; J. M. Taylor; J. P. Holland; L. E. Lie-Injo; J. Ganesan; and D. Todd (1975). Deletion of α-globin structural loci. *Nature* **255**:255–6.

Karn, M. N. and L. S. Penrose (1951). Birth weight and gestation time in relation to maternal age, parity and infant survival. *Ann. Eugen.* **15**:147–64.

Kettlewell, H. B. D. (1961). The phenomenon of industrial melanism in the Lepidoptera. *A. Rev. Ent.* **6**:245–62.

Kilmartin, J. V. (1976). Interaction of hemoglobin with protons, CO_2 and 2, 3-diphosphoglycerate. *Br. med. Bull.* **32**:209–12.

Kimura, M. (1956). Rules for testing stability of a selective polymorphism. *Proc. natn. Acad. Sci., U.S.A.* **42**:336–40.

——— (1968*a*). Evolutionary rate at the molecular level. *Nature* **217**:624–6.

——— (1968*b*). Genetic variability maintained in a finite population due to mutational production of neutral and nearly neutral isoalleles. *Genet. Res.* **11**:247–69.

——— (1971). Theoretical foundations of population genetics at the molecular level. *Theoret. pop. Biol.* **2**:174–208.

——— (1977). Preponderance of synonymous changes as evidence for the neutral theory of molecular evolution. *Nature* **267**:275–6.

——— and J. F. Crow (1964). The number of alleles that can be maintained in a finite population. *Genetics* **49**:725–38.

——— ——— (1978). The effect of overall phenotypic selection on the genetic change at individual loci. *Proc. natn. Acad. Sci. U.S.A.* **75**:6168–71.

——— and T. Ohta (1971). On the rate of molecular evolution. *J. molec. Evol.* **1**:1–17.

King, J. L. (1967). Continuously distributed factors affecting fitness. *Genetics* **55**:483–92.

——— (1972). Genetic polymorphisms and environment. *Science* **176**:545.

——— and T. H. Jukes (1969). Non-Darwinian evolution. *Science* **164**:788–98.

Kirk, R. L. (1971). A haptoglobin mating frequency, segregation and ABO blood group interaction analysis for additional series of families. *Ann. hum. Genet.* **34**:329–37.

———; H. Kinnus; and N. E. Morton (1970). Interaction between the ABO blood groups and haptoglobin systems. *Am. J. hum. Genet.* **22**:384–9.

Kirkman, H. N. (1971). Glucose-6-phosphate dehydrogenase deficiency. *Adv. hum. Genet.* **2**:1–60.

Kleihauer, E. (1970). The hemoglobins. In *Physiology of the perinatal period; functional and biochemical development in mammals.* Vol. 1 (ed. U. Stave), pp. 255–97. Appleton-Century-Crofts, New York.

Klein, J. (1974). Genetic polymorphism of the histocompatibility-2-loci of the mouse. *A. Rev. Genet.* **8**:63–77.

Knoll, A. H. and E. S. Barghoorn (1975). Pre-cambrian eukaryotic organisms—reassessment of the evidence. *Science* **190**:52–4.

Kochwa, S.; R. E. Rosenfield; L. Tallal; and L. R. Wasserman (1961). Isoagglutinins associated with ABO erythroblastosis. *J. clin. Invest.* **40**:874–83.

Koehn, R. K.; R. Milkman; and J. B. Mitton (1976). Population genetics of marine pelecypods. IV. Selection, migration and genetic differentiation in the blue mussel *Mytilus edulis. Evolution* **30**:2–32.

———; F. J. Turano; and J. B. Mitton (1973). Population genetics of marine pelecypods. II. Genetic differences in microhabitats of *Modiolus demissus. Evolution* **27**:100–5.

Kojima, K. (1971). The distribution and comparison of "genetic loads" under heterotic selection and simple frequency dependent selection in finite populations. *Theoret. pop. Biol.* **2**:159–73.

———; J. Gillespie; and Y. N. Tobari (1970). A profile of *Drosophila* species' enzymes assayed by electrophoresis. I. Number of alleles, heterozygosities, and linkage disequilibrium in glucose metabolizing systems and some other enzymes. *Biochem. Genet.* **4**:627–37.

Laird, C. D. and B. J. McCarthy (1968). Magnitude of interspecific nucleotide sequence variability in Drosophila. *Genetics* **60**:303–22.

Lamotte, M. (1952). Le rôle des fluctuations fortuites dans la diversité des populations naturelles de *Cepaea nemoralis* (L.). *Heredity* **6**:333–43.

——— (1959). Polymorphism of natural populations of *Cepaea nemoralis. Cold Spring Harb. Symp. quant. Biol.* **24**:65–86.

Landsteiner, K. (1900). Zur Kenntnis der antifermentativen lytischen und agglutinierenden Wirkingen des Blutserums und der Lymphe. *Zentralb. Bakteriol.* **27**:357–62.

——— and A. S. Wiener (1940). An agglutinable factor in human blood recognized by immune sera for rhesus blood. *Proc. Soc. exp. Biol. N.Y.* **43**:233.

——— ———(1941). Studies on an agglutinogen (Rh) in human blood reacting with anti-Rhesus sera and with human isoantibodies. *J. exp. Med.* **74**:309–20.

Langley, C. H. and J. F. Crow (1974). Direction of linkage disequilibrium. *Genetics* **78**:937–41.

——— and W. M. Fitch (1974). An examination of the constancy of the rate of molecular evolution. *J. molec. Evol.* **3**:161–77.

———; Y. N. Tobari; and K. Kojima (1974). Linkage disequilibrium in natural populations of *Drosophila melanogaster. Genetics* **78**:921–36.

Langman, M. J. S. and R. Doll (1965). ABO blood group and secretor status in relation to clinical characteristics of peptic ulcers. *Gut* **6**:270–3.

Laurent, T. C. and B. Blombäck (1958). On the significance of the release of two different peptides from fibrinogen during clotting. *Acta chem. scand.* **12**:1875–7.

Lauritsen, J. G.; N. Grunnet; and O. M. Jensen (1975). Materno-fetal ABO incompatibility as a cause of spontaneous abortion. *Clin. Genet.* **7**:308–16.

Lehmann, H. and R. G. Huntsman (1974). *Man's haemoglobins.* North Holland, Oxford.

Leigh, E. G., Jr. (1973). The evolution of mutation rates. *Genetics Suppl.* **73**:1–18.

Lerner, I. M. (1958). *The genetic basis of selection.* John Wiley, New York.

Levene, H. (1953). Genetic equilibrium when more than one ecological niche is available. *Am. Nat.* **87**:331–3.

Lewontin, R. C. (1964). The interaction of selection and linkage. I. General considerations. Heterotic models. *Genetics* **49**:49–67.

——— (1974*a*). *The genetic basis of evolutionary change.* Columbia University Press, New York.

——— (1974*b*). The analysis of variance and the analysis of causes. *Am. J. hum. Genet.* **26**:400–11.

——— and J. L. Hubby (1966). A molecular approach to the study of genic heterogeneity in natural populations. II. Amount of variation and degree of heterozygosity in natural populations of *Drosophila pseudoobscura. Genetics* **54**:595–609.

——— and K. Kojima (1960). The evolutionary dynamics of complex polymorphisms. *Evolution* **14**:458–72.

Livingstone, F. B. (1967). *Abnormal hemoglobins in human populations.* Aldine, Chicago.

Luisada, A. (1941). Favism. A singular disease chiefly affecting the red blood cells. *Medicine* **20**:229–50.

Luzzatto, L. (1975). Inherited hemolytic states: glucose-6-phosphate dehydrogenase deficiency. *Clin. Hematol.* **4**:83–108.

——— and A. Afolayan (1971). Genetic variants of human-erythrocyte glucose-6-phosphate deficiency. II. *In vitro* and *in vivo* function of the A^- variant. *Biochemistry* **10:**415–20.

———; E. S. Nwachuku-Jarrett; and S. Reddy (1970). Increased sickling of parasitized erythrocytes as mechanism of resistance against malaria in the sickle cell trait. *Lancet* **i**:319–21.

———; E. A. Usanga; and S. Reddy (1969). Glucose-6-phosphate dehydrogenase deficient red cells: resistance to infection by malarial parasites. *Science* **164**:839–42.

McDonald, J. F. and F. J. Ayala (1974). Genetic response to environmental heterogeneity. *Nature* **250**:572–4.

McReynolds, L.; B. W. O'Malley; A. D. Nisbet; J. E. Fothergill; D. Girol; S. Fields; M. Robertson; and G. G. Brownlee (1978). Sequence of chicken ovalbumin mRNA. *Nature* **273**:723–8.

Marquis, D. (1927). *archy and mehitabel.* Doubleday, New York.

Margoliash, E. and E. L. Smith (1965). Structural and functional aspects of cytochrome *c* in relation to evolution. In *Evolving genes and proteins* (eds. V. Bryson and H. J. Vogel) pp. 221–42. Academic Press, New York.

Mather, K. (1953). The genetical structure of populations. *Symp. Soc. exp. Biol.* **7**:66–95.

——— and L. G. Wigan (1942). The selection of invisible mutations. *Proc. Roy. Soc. Ser. B.* **131**:50–64.

Matsunaga, E. and S. Itoh (1958). Blood groups and fertility in a Japanese population, with special reference to intra-uterine selection due to maternal–fetal incompatibility. *Ann. hum. Genet.* **22**:111–31.

Maxam, A. M. and W. Gilbert (1977). A new method for sequencing DNA. *Proc. natn. Acad. Sci., U.S.A.* **74**:560–4.

Maynard Smith, J. (1966). Sympatric speciation. *Am. Nat.* **100**:637–50.

——— (1968). 'Haldane's dilemma' and the rate of evolution. *Nature* **219**:1114–16.

——— (1978). *The evolution of sex.* Cambridge University Press, Cambridge.

Mayr, E. (1942). *Systematics and the origin of species.* Columbia University Press, New York.

——— (1959*a*). The emergence of evolutionary novelties. In *Evolution after Darwin* (ed. Sol Tax), Vol. 1, pp. 349–80. University of Chicago Press, Chicago.

——— (1959*b*). Where are we? *Cold Spring Harb. Symp. Quant. Biol.* **24**:1–14.

Milkman, R. D. (1967). Heterosis as a major cause of heterozygosity in nature. *Genetics* **55**:493–5.

——— (1973). A competitive selection model. *Genetics* **74**:727–32.

——— (1978). Selection differentials and selection coefficients. *Genetics* **88**:391–403.

Miller, L. H.; J. A. Dvorak; T. Shirioshi; and J. R. Durocher (1973). Influence of erythrocyte membrane components on malaria merozoite invasion. *J. exp. Med.* **138**:1597–1601.

———; S. J. Mason; J. A. Dvorak; M. H. McGinniss; and I. K. Rothman (1975). Erythrocyte receptors for *Plasmodium knowlesii* malaria: Duffy blood group determinants. *Science* **189**:561–3.

Miller, M. J.; J. V. Neel; and F. B. Livingstone (1956). Distribution of parasites in the red cells of sickle cell trait carriers infected with *Plasmodium falciparum. Trans. R. Soc. Trop. Med. Hyg.* **50**:294–6.

Minawa, A. and A. J. Birley (1975). Genetical and environmental diversity in *Drosophila melanogaster. Nature* **255**:702–4.

Min Jou, W. and W. Fiers (1976). Studies on the bacteriophage MS2. XXXIII. Comparison of the nucleotide sequences in related bacteriophage RNAs. *J. molec. Biol.* **106**:1047–60.

Moll, R. H.; M. F. Lindsey; and H. F. Robinson (1963). Estimates of genetic variances and level of dominance in maize. *Genetics* **49**:411–23.

Moore, G. W.; M. Goodman; C. Callahan; R. Holmquist; and H. Moise (1976). Stochastic *versus* augmented maximum parsimony method for estimating superimposed mutations in the divergent evolution of protein sequences. Methods tested on cytochrome *c* amino acid sequences. *J. molec. Biol.* **105**:15–37.

Morton, N. E. (1964). Genetic studies of Northwestern Brazil. *Cold Spring Harb. Symp. quant. Biol.* **29**:69–79.

——— and C. S. Chung (1959). Are the MN blood groups maintained by selection? *Am. J. hum. Genet.* **11**:237–51.

———; J. F. Crow; and H. J. Muller (1956). An estimate of the mutational damage in man from data on consanguineous marriages. *Proc. natn. Acad. Sci., U.S.A.* **42**:855–63.

———; H. Krieger; and M. P. Mi (1966). Natural selection on polymorphisms in Northeastern Brazil. *Am. J. Hum. Genet.* **18**:153–71.

Motulsky, A. G. (1960). Metabolic polymorphisms and the role of infectious diseases in human evolution. *Human Biol.* **32**:28–62.

Mourant, A. E.; A. C. Kopec; and K. Domaniewska-Sobczak (1976). *The distribution of the human blood groups and other polymorphisms.* 2nd edn., Oxford University Press, London.

——— ——— ——— (1978). Blood groups and diseases. Oxford University Press, London.

Mross, G. A. and R. F. Doolittle (1967). Amino acid sequence studies on artiodactyl fibrinopeptides. II. Vicuna, elk, muntjak, pronghorn antelope and water buffalo. *Arch. Biochem. Biophys.* **122**:674–84.

——— ———; and B. F. Roberts (1970). Gibbon fibrinopeptides: identification of a glycine-serine allelism at position B-3. *Science* **170**:468–70.

Mukai, T. (1964). The genetic structure of natural populations of *Drosophila melanogaster*. I. Spontaneous mutation rate of polygenes controlling viability. *Genetics* **50**:1–19.

——— (1968). Genetic variance of the degree of dominance of mutant polygenes controlling viability in Drosophila. *Genetics* **60**:206.

——— (1969). The genetic structure of natural populations of *Drosophila melanogaster*. VII. Synergistic interaction of spontaneous mutant polygenes controlling viability. *Genetics* **61**:749–61.

———; L. E. Mettler; and S. Chigusa (1971). Linkage disequilibrium in a local population of *Drosophila melanogaster*. *Proc. Natn. Acad. Sci., U.S.A.* **68**:1065–9.

——— and O. Yamaguchi (1974). Genetic structure of natural populations of *Drosophila melanogaster*. XI. Genetic variability in a local population. *Genetics* **76**:339–66.

Muller, H. J. (1950). Our load of mutations. *Am. J. hum. Genet.* **2**:111–76.

——— (1964). The relation of recombination to mutational advance. *Mut. Res.* **1**:2–9.

Murayama, M. (1966). Molecular mechanism of red cell sickling. *Science* **153**:145–9.

——— and R. M. Nilbandian (1973). *Sickle cell hemoglobin: molecule to man.* Little, Brown, Boston.

Nance, W. E. (1963). Genetic control of hemoglobin synthesis. *Science* **141**:123–30.

Neel, J. V.; M. Layrisse; and F. Salzano (1977). Man in the tropics: the Yanomama Indians. In *Population structure and human variation* (ed. G. A. Harrison) International Biological Program, Vol. II, pp. 109–42. Cambridge University Press, Cambridge.

——— and E. A. Thompson (1978). Founder effect and number of private polymorphisms observed in Amerindian tribes. *Proc. natn. Acad. Sci., U.S.A.* **75**:1904–8.

——— and R. H. Ward (1970). Village and tribal genetic distances among American Indians, and the possible implication for human evolution. *Proc. natn. Acad. Sci., U.S.A.* **65**:323–30.

——— ——— (1972). The genetic structure of a tribal population, the Yanomama Indians. VI. Analysis by F statistics (including a comparison with the Makiritare Xavante). *Genetics* **72**:639–66.

Nei, M. (1975). Molecular population genetics and evolution. North Holland, New York.

———; T. Maruyama; and R. Chakraborty (1975). The bottleneck effect and genetic variability in populations. *Evolution* **29**:1–10.

Nienhuis, A. W.; P. Turner; and E. J. Benz, Jr. (1977). Relative stability of α and β globin messenger RNA's in homozygous β^+ thalessemia. *Proc. natn. Acad. Sci. U.S.A.* **74**:3960–4.

Ohta, T. (1973). Slightly deleterious mutant substitutions in evolution. *Nature* **246**:96–7.

——— and M. Kimura (1969). Linkage disequilibrium at steady state determined by random genetic drift and recurrent mutation. *Genetics* **63**:229–38.

——— ——— (1970). Development of associative overdominance through linkage disequilibrium in finite populations. *Genet. Res.* **16**:165–77.

——— ——— (1971). Behavior of neutral mutants influenced by associated overdominant loci in finite populations. *Genetics* **69**:247–60.

Old, J.; J. Longley; W. G. Wood; J. B. Clegg; and D. J. Weatherall (1977). The molecular basis for acquired hemoglobin H disease. *Nature* **269**:524–5.

Osborne, R. H. and F. V. de George (1959). *Genetic basis of morphological variation.* Harvard University Press, Cambridge.

Ottolenghi, S.; P. Comi; B. Gaglioni; P. Tolstoshev; W. G. Lanyon; G. J. Mitchell; R. Williamson; G. Russo; S. Musumeci; G. Schiliro; G. A. Tsistrakis; S. Charache; W. G. Wood; J. B. Clegg; and D. J. Weatherall (1976). $\delta\beta$-thalassemia is due to a gene deletion. *Cell* **9**:71–80.

———; W. G. Lanyon, J. Paul; R. Williamson; D. S. Weatherall; J. B. Clegg; J. Pritchard; S. Pootrakul; and H. B. Wong (1974). The severe form of α-thalassemia is caused by a hemoglobin gene deletion. *Nature* **251**:381–2.

Paciorkiewicz, M. (1970). The correlation between blood groups and some infectious diseases in children. [In Polish] *Pediatria Pol.* **45**:943–50.

Pardue, M. L. and J. G. Gall (1970). Chromosomal localization of mouse satellite DNA *Science* **168**:1356–8.

Pasvol, G.; D. J. Weatherall; and R. J. M. Wilson (1978). Cellular mechanism for the protective effect of hemoglobin S against P. *falciparum* malaria. *Nature* **274**:701–3.

Pauling, L.; H. A. Itano; S. J. Singer; and I. C. Wells (1949). Sickle cell anemia, a molecular disease. *Science* **110**:543–8.

Peacock, W. J.; R. Appels; P. Dunsmuir; A. R. Lohe; and W. L. Gerlach (1977). Highly repeated DNA sequences: chromosomal localization and evolutionary conservatism. In *International Cell Biology* 1976-77 (eds. B. R. Brinkley and K. R. Porter) pp. 494–506. Rockefeller, New York.

Peckham, M. (1959). *The origin of species, by Charles Darwin: a variorum text.* University of Pennsylvania Press, Philadelphia.

Perutz, M. F. (1970). Stereochemistry of cooperative effects in haemoglobin. *Nature* **228**:726–39.

———; H. Muirhead; L. Mazzarella; R. A. Crowther; J. Greer; and J. V. Kilmartin (1969). Identification of residues responsible for the alkaline Bohr effect in haemoglobin. *Nature* **222**:1240–3.

Pettenkofer, H. J. and R. Bickerrich (1960). Über Antigen Gemeinschaften zwischen menschlichen Blutgruppen ABO and den Erregern gemeingefährlicher Krankheiten. *Zbl. Bakt. Parasitkde* **179**:433–6.

———; B. Stöss; W. Helmbold; and F. Vogel (1962). Alleged causes of the present day world distribution of the human ABO blood groups. *Nature* **193**:445–6.

Powell, J. R. (1971). Genetic polymorphisms in varied environments. *Science* **174**:1035–6.

——— (1972). Genetic polymorphism and environment (reply). *Science* **176**:545.

Prakash, S. (1973). Patterns of gene variation in central and marginal populations of *Drosophila robusta. Genetics* **75**:347–69.

——— and M. Levitan (1973). Associations of alleles of the esterase-1 locus with gene arrangements of the left arm of the second chromosome in *Drosophila robusta. Genetics* **75**:371–9.

———; R. C. Lewontin; and J. L. Hubby (1969). A molecular approach to the study of genic heterozygosity in natural populations. IV. Patterns of genic variation in central, marginal and isolated populations of *Drosophila pseudoobscura. Genetics* **61**:841–58.

Propping, P. (1972). Comparison of point mutation rates in different species with human mutation rates. *Humangenetik* **16**:43–8.

Provine, W. B. (1971). *The origins of theoretical population genetics.* University of Chicago Press.

Punnett, R. C. (1915). *Mimicry in butterflies.* Cambridge University Press.

Race, R. R. and R. Sanger (1975). *Blood groups in man,* 5th edn., Davis, Philadelphia.

Ramirez, F.; J. V. O'Donnell; P. A. Marks; A. Bank; S. Musumeci; G. Schiliro; G. Pizzarelli; G. Russo; B. Lippiom; and R. Gambino (1976). Abnormal or absent β mRNA in β° Ferrara and gene deletion in $\delta\beta$ thalassemia. *Nature* **263**:471–5.

Raper, A. B. (1959). Further observations on sickling and malaria. *Trans. R. Soc. trop. Med. Hyg.* **53**:110–17.

Reeder, R. H.; D. D. Brown; P. K. Wellauer; and I. B. Dawid (1976). Patterns of ribosomal DNA spacer length are inherited. *J. molec. Biol.* **105**:507–16.

Rendel, J. and C. Stormont (1964). Variants of ovine alkaline serum phosphatases and their association with the R-O blood groups. *Proc. Soc. exp. Biol. Med.* **115**:853–6.

Rensch, B. (1959). *Evolution above the species level.* Methuen, London.

Renz, M. and L. A. Day (1976). Transition from non-cooperative to cooperative binding of histone H1 to DNA. *Biochemistry* **15**:3220–8.

Repetski, J. A. (1978). A fish from the upper Cambrian of North America. *Science* **200**:529–31.

Richmond, R. and J. R. Powell (1970). Evidence of heterosis associated with an enzyme locus in a natural population of Drosophila. *Proc. natn. Acad. Sci., U.S.A.* **67**:1264–7.

Rigas, D. A. and R. D. Koler (1961). Decreased erythocyte survival in hemoglobin H disease as a result of the abnormal properties of hemoglobin H: the benefit of splenectomy. *Blood* **18**:1–17.

Ritter, H. and K. Hinkelmann (1966). Zur Balance des Polymorphismus der Haptoglobine. *Hum. genet.* **2**:21–4.

Roberts, D. F.; W. F. Billewicz; and I. A. McGregor (1978). Heritability of stature in a West African Population. *Ann. hum. Genet.* **42**:15–24.

Robertson, F. W. (1957). Studies in quantitative inheritance. X. Genetic variation of ovary size in *Drosophila. J. Genet.* **55**:410–27.

——— and J. H. Sang (1944). The ecological determinants of population growth in a *Drosophila* culture. I. Fecundity of adult flies. *Proc. R. Soc. Ser. B* **132**:258–89.

Robinson, H. F. and R. E. Comstock (1955). Analysis of genetic variability in corn with reference to probable effects of selection. *Cold Spring Harb. Symp. quant. Biol.* **20**:127–36.

Romero-Herrera, A. E.; H. Lehmann; K. A. Joysey; and A. E. Friday (1973). Molecular evolution of myoglobin and the fossil record: a phylogenetic synthesis. *Nature* **246**:389–95.

Rossmann, M. G.; D. Moras; and K. W. Olsen (1974). Chemical and biological evolution of nucleotide-binding proteins. *Nature* **250**:194–99.

Roughton, F. J. W. and J. C. Kendrew (1949). *Haemoglobin (Bancroft Memorial Conference),* Butterworth, London.

Rucknagel, D. L. and J. V. Neel (1961). The hemoglobinopathies. *Prog. med. Genet.* **1**:158–260.

Russo, G.; F. Mollica; L. Pavone; and G. Schiliro (1972). Hemolytic crisis of favism in Sicilian females heterozygous for G6PD deficiency. *Pediatrics* **49**:854–9.

Ryder, L. P. and A. Svejgaard (1976). Associations between HLA and disease. Reprinted in Mourant et al. (1978), pp. 275–98.

Sederoff, R.; L. Lowenstein; and H. C. Birnbaum (1975). Polypyrimidine segments in *Drosophila melanogaster* DNA. II. Chromosome location and nucleotide sequence. *Cell* **5**:183–94.

Saugstad, L. F. (1977). Heterozygote advantage for the phenylketonuria allele. *J. med. Genet* **14**:2–24.

Seeberg, P. H.; J. Shine; J. A. Martial; J. D. Baxter; and H. M. Goodman (1977). Nucleotide sequence and amplification in bacteria of structural gene for rat growth hormone. *Nature* **270**:486–94.

Sheppard, P. M. (1951*a*). Fluctuations in the selective value of certain phenotypes in the polymorphic land snail *Cepaea nemoralis* (L). *Heredity* **5**:125–34.

——— (1951*b*). A quantitative study of two populations of the moth *Panaxia dominula* (L). *Heredity* **5**:349–78.

——— (1952). Natural selection in two colonies of the polymorphic land snail, *Cepaea nemoralis. Heredity* **6**:233–38.

——— (1959). The evolution of mimicry. *Cold Spring Harb. Symp. Quant. Biol.* **24**:131–40.

——— (1967). *Natural selection and heredity.* Hutchinson University Library, London.

Shine, J.; P. H. Seeburg; J. A. Martial; J. D. Baxter; and H. M. Goodman (1977). Construction and analysis of recombinant DNA from human chorionic somatomammotropin. *Nature* **270**:494–9.

Simmons, M. J.; C. R. Preston; and W. R. Engels (1981). Pleiotropic effects on fitness of mutations affecting viability in *Drosophila melanogaster. Genetics* (in press).

Simpson, G. G. (1944). *Tempo and mode in evolution.* Columbia University Press, New York.

——— (1953). *The major features of evolution.* Columbia University Press, New York.

Singh, R. S.; R. C. Lewontin; and A. A. Felton (1976). Genetic heterogeneity within electrophoretic "alleles" of xanthine dehydrogenase in *Drosophila pseudoobscura. Genetics* **84**:609–29.

Singh, S. M. and E. Zouros (1978). Genetic variation associated with growth rate in the American oyster *(Crassostrea virginica)*. *Evolution* **32**:342–53.

Siniscalco, M.; L. Bernini; B. Latte; and A. G. Motulsky (1961). Favism and thalassemia in Sardinia and their relationship to malaria. *Nature* **190**:1179–80.

Slatkin, M. (1972). On treating the chromosome as the unit of selection. *Genetics* **72**:157–68.

Smith, G. P. (1973). Unequal crossover and the evolution of multigene families. *Cold Spring Harb. Symp. quant. Biol.* **38**:507–14.

Smithes, O. (1964). Chromosomal rearrangements and protein structure. *Cold Spring Harb. Symp. quant. Biol.* **29**:309–19.

Smole, W. J. (1976). *The Yanomama Indians.* University of Texas Press, Austin.

Snell, G. D. (1968). The H-2 locus of the mouse—observations and speculations concerning its comparative genetics and its polymorphism. *Folia Biol. Praha* **14**:335–58.

Soulé, M. (1973). The epistasis cycle: a theory of marginal populations. *A. Rev. Ecol. Syst.* **4**:165–87.

Southern, E. M. (1975). Long range periodicities in mouse satellite DNA. *J. molec. Biol.* **94**:51–91.

Spector, W. S., ed. (1956). *Handbook of biological data.* Saunders, Philadelphia.

Spielman, R. S.; E. C. Migliazza; and J. V. Neel (1974). Regional linguistic and genetic differences among Yanomama Indians. *Science* **184**:637–44.

Springer, G. F.; R. E. Horton; and M. Forbes (1959). Origin of anti-human blood group β agglutinins in White Leghorn chicks. *Proc. 7th Cong. Int. Soc. Blood Transf.* 529–35.

———; B. Williamson; and W. C. Brandes (1961). Blood group activity of gram negative bacteria. *J. exp. Med.* **113**:1077–93.

Stebbins, G. L. and R. C. Lewontin (1972). Comparative evolution at the levels of molecules, organisms and populations. In *Proceedings of 6th Berkeley symp. on mathematical statistics and probability* (eds. J. Neyman, E. L. Scott and L. LeCam) pp. 23–42, University of California Press.

Strobeck, C. and K. Morgan (1978). The effect of intragenic recombination on the number of alleles in a finite population. *Genetics* **88**:829–44.

Sturtevant, A. H. (1965). *A history of genetics.* Harper & Row, New York.

Sukamaran, P. K.; H. R. Master; J. V. Andevia; V. Balakrishnan; and L. D. Sanghvi (1966). ABO blood groups in active cases of smallpox. *Indian J. med. Sci.* **20**:119–22.

Sved, J. A. (1968*a*). Possible rates of gene substitution in evolution. *Am. Nat.* **102**:283–93.

——— (1968*b*). The stability of linked systems of loci with small population size. *Genetics* **59**:543–63.

——— (1971). Estimate of heterosis in *Drosophila melanogaster*. *Genet. Res.* **18**:97–105.

———; T. E. Reed; and W. F. Bodmer (1967). The number of balanced polymorphisms that can be maintained in a natural population. *Genetics* **55**:469–81.

Tanner, M. J. and D. J. Anstee (1976). The membrane change in En(a$^-$) human erythrocytes. Absence of the major erythrocyte sialoglycoprotein. *Biochem. J.* **153**:271–7.

Teisberg, P.; I. Akesson; B. Olafsen; T. Gedde-Dahl; and E. Thorsby (1976). Genetic polymorphism of C4 in man and localization of a structure C4 locus to the HLA gene complex of chromosome 6. *Nature* **264**:253–4.

Thoday, J. M. (1958). Homeostasis in a selection experiment. *Heredity* **12**:401–15.

Thorsby, E. (1974). The human major histocompatibility system. Transplant. Rev. **18**:51–129.

Tovey, G. H. (1945). A study of the protective factors in heterospecific blood group pregnancy and their role in the prevention of hemolytic disease of the newborn. *J. Pathol. Bacteriol* **57**:295–305.

Tracey, M. L. and F. J. Ayala (1974). Genetic load in natural populations—is it compatible with the hypothesis that many polymorphisms are maintained by natural selection? *Genetics* **77**:569–89.

———; N. F. Bellet; and C. D. Graven (1975). Excess allozyme homozygosity and breeding population structure in the mussel *Mytilus californianus. Mar. Biol.* **32**:303–11.

Trager, W. and J. B. Jensen (1978). Cultivation of malarial parasites. *Nature* **273**:621–2.

Turner, J. R. G. (1967). Why does the genome not congeal? *Evolution* **21**:645–56.

Ullrich, A.; J. Shine; J. Chirgwin; R. Pictet; E. Tischer; W. J. Rutter; and H. M. Goodman (1977). Rat insulin genes: construction of plasmids containing the coding sequences. *Science* **196**:1313–18.

Vana, L. R. and A. G. Steinberg (1975). Haptoglobin–ABO interaction: a possible explanation for the excess of Hp 1 among offspring of ABO incompatible matings. *Am. J. hum. Genet.* **27**:224–32.

Van Valen, L. (1974). Molecular evolution as predicted by natural selection. *J. molec. Evol.* **3**:89–101.

Voak, D. (1969). The pathogenesis of ABO hemolytic disease of the newborn. *Vox Sang.* **17**:481–513.

Vogel, F. and M. R. Chakravartti (1966). ABO blood groups and smallpox in a rural population of West Bengal and Bihar (India). *Humangenetik* **3**:166–80.

———; H. J. Pettenkoffer; and W. Helmbold (1960). Uber die Populationsgenetik der ABO-Blut-gruppen. 2. Mitterlunz, Genhäufizkeit und Epidemische Erkrankungen. *Acta genet. Statist. med.* **10**:267–94.

——— and R. Rathenberg (1974). Spontaneous mutation in man. In *Advances in human genetics* (eds. H. Harris and K. Hirschhorn) Vol. 5, 1975, pp. 223–318. Plenum Press, New York.

de Vries, H. (1901–3). *The mutation theory: experiments and observations on the origin of species in the vegetable kingdom* (Trans. J. B. Farmer and A. D. Darbishire) Open Court. Chicago, 1909–1910.

Walker, P. M. B. (1971). Repetitive DNA in higher organisms. In *Progress in biophysics and molecular biology* (eds. J. A. V. Butler and D. Noble) Vol. 23, pp. 145–90. Pergamon Press, New York.

Wallace, B. (1968). *Topics in population genetics.* Norton, New York.

——— (1975). Hard and soft selection revisited. *Evolution* **29**:465–73.

——— and C. Madden (1953). The frequencies of sub- and supervitals in experimental populations of *Drosophila melanogaster. Genetics* **35**:456–70.

Ward, R. H. (1972). The genetic structure of a tribal population, the Yanomama Indians. V. Comparison of a series of genetic networks. *Ann. hum. Genet.* **36**:21–43.

Wasi, P.; S. Na-Nakorn; P. Pootrakul; and V. Panich (1972). Incidence of hemoglobin Thai: a re-examination of the genetics of α-thalassemic disease. *Ann. hum. Genet.* **35**:467–70.

Watkins, W. M. (1966). Blood group substances. *Science* **152**:172–81.

Watterson, G. A. (1978). An analysis of multiallelic data. *Genetics* **88**:171–9.

Weatherall, D. J. and J. B. Clegg (1972). *The thalassemia syndromes.* Blackwell, Oxford.

——— ——— (1976). Molecular genetics of human hemoglobin. *A. Rev. Genet.* **10**:157–78.

Weismann, A. (1893). *The germ plasm, a theory of heredity* (Transl. W. N. Ponker and H. Roñnefeldt) Walter Scott, London.

Whipple, G. H. and W. L. Bradford (1936). Mediterranean disease–thalassemia (erythroblastic anemia of Cooley): associated pigment abnormalities simulating hemochromatosis. *J. Pediat.* **9**:279–311.

White, J. M. and J. V. Dacie (1971). The unstable hemoglobins. Molecular and clinical features. *Progr. Hematol.* **7**:69–109.

Wickler, W. (1968). *Mimicry in plants and animals.* Weidenfeld and Nicolson, London.

Wiener, A. S. (1943). Distribution and heredity of variants of the Rh type. *Science* **98**:182–4.

——— (1960). Modern blood group mythology. *J. forens. Med.* **7**:166–76.

———; J. Moor-Jankowski; and E. B. Gordon (1964). Blood groups of apes and monkeys. IV. The Rh–Hr blood types of anthropoid apes. *Am. J. hum. Genet.* **16**:246–53.

———; ———; F. C. Cadigan, Jr.; and E. B. Gordon (1968). Comparison of the A-B-H blood group specificities and M-N blood types in man, gibbons *(Hylobates)* and siamangs *(Symphalangus)*. *Transfusion* **8**:235–43.

———; ———; and W. W. Socha (1972). Principles of blood grouping in apes and monkeys—human, simian and cross-immune types. *Transplant. Proc.* **4**:101–5.

——— and L. G. Unger (1955). Excess of group O mothers in ABO hemolytic disease with special reference to the role of the blood factor C. *Exp. Med. Surg.* **13**:204–24.

Williams, G. C. (1975). *Sex and evolution. Princeton Monographs in Population Biology 8* (ed. R. M. May), Princeton University Press

Wills, C. (1966). The mutational load in two natural populations of *Drosophila pseudoobscura*. *Genetics* **53**:281–94.

——— (1973). In defense of naive pan-selectionism. *Am. Nat.* **107**:23–34.

——— (1975*a*). Marginal overdominance in Drosophila. *Genetics* **81**:177–89.

——— (1975*b*). Selective pressures on isozymes in Drosophila. In *Isozymes* (ed. C. L. Markert) Vol. IV, pp. 517–31. Academic Press, New York.

——— (1978). Rank-order selection is capable of maintaining all genetic polymorphisms. *Genetics* **89**:403–17.

——— and C. Miller (1976). A computer model allowing maintenance of large amounts of genetic variability in Mendelian populations. II. The balance of forces between linkage and random assortment. *Genetics* **82**:377–99.

——— and L. Nichols (1971). Single gene heterosis in Drosophila revealed by inbreeding. *Nature* **233**:123–5.

——— ——— (1972). How genetic background masks single-gene heterosis in Drosophila. *Proc. natn. Acad. Sci., U.S.A.* **69**:323–5.

———; J. Crenshaw; and J. Vitale (1970). A computer model allowing maintenance of large amounts of genetic variability in Mendelian populations. I. Assumptions and results for large populations. *Genetics* **64**:107–23.

———; J. Phelps; and R. Ferguson (1975). Further evidence for selective differences between isoalleles in Drosophila. *Genetics* **79**:127–41.

Wilson, A. C.; S. S. Carlson; and T. J. White (1977). Biochemical evolution. *A. Rev. Biochem.* **46**:573–639.

Wood, W. G. (1976). Hemoglobin synthesis during human fetal development. *Br. med. Bull.* **32**:282–7.

Wright, S. (1929*a*). Fisher's theory of dominance. *Am. Nat.* **63**:274–9.

——— (1929*b*). The evolution of dominance. *Am. Nat.* **63**:556–61.

——— (1929*c*). Evolution in a Mendelian population. *Anat. Rec.* **44**:287.

——— (1931). Evolution in Mendelian populations. *Genetics* **16**:97–159.

——— (1939). The distribution of self-sterility alleles in populations. *Genetics* **24**:538–52.

——— (1940). The statistical consequences of Mendelian heredity in relation to speciation. In *The new systematics* (ed. Julian Huxley) pp. 161–83. Clarendon Press, Oxford.

——— (1943). Isolation by distance. *Genetics* **28**:114–38.

——— (1948). On the role of directed and random changes in gene frequency in the genetics of populations. *Evolution* **2**:279–94.

——— (1968). *Evolution and the genetics of populations,* Vol. I., *Genetic and biometric foundations.* University of Chicago Press.

——— (1969). *Evolution and the genetics of populations,* Vol. II, *The Theory of gene frequencies.* University of Chicago Press.

——— (1970). Random drift and the shifting balance theory of evolution. In *Mathematical Topics in Population Genetics* (ed. K. Kojima) pp. 1–31. Springer-Verlag, Berlin.

——— (1975). *Evolution and the genetics of populations*, Vol. III, *Experimental Results and Evolutionary Deductions.* University of Chicago Press.

Yamazaki, T. (1972). Detection of single gene effect by inbreeding. *Nature New Biol.* **240**:53–4.

——— (1977). The effects of overdominance in linkage in a multilocus system. *Genetics* **86**:227–36.

——— and T. Maruyama (1972). Evidence for the neutral hypothesis of protein polymorphism. *Science* **178**:56–8.

——— ——— (1974). Evidence that enzyme polymorphisms are selectively neutral, but blood group polymorphisms are not. *Science* **183**:1091–2.

Yunis, J. J. and W. G. Yasmineh (1971). Heterochromatin, satellite DNA, and cell function. *Science* **174**:1200–9.

Zuckerkandl, E. (1976). Evolutionary processes and evolutionary noise at the molecular level. II. A selectionist model for random fixations in proteins. *J. molec. Evol.* **7**:269–312.

——— and L. Pauling (1962). Molecular disease, evolution and genic heterogeneity. In *Horizons in biochemistry* (eds. M. Kosha and B. Pullman) pp. 189–225. Academic Press, New York.

Index